NATO ASI Series

Advanced Science Institutes Series

A series presenting the results of activities sponsored by the NATO Science Committee, which aims at the dissemination of advanced scientific and technological knowledge, with a view to strengthening links between scientific communities.

The Series is published by an international board of publishers in conjunction with the NATO Scientific Affairs Division

A Life Sciences B Physics	Plenum Publishing Corporation London and New York
C Mathematical and Physical Sciences D Behavioural and Social Sciences E Applied Sciences	Kluwer Academic Publishers Dordrecht, Boston and London
F Computer and Systems Sciences G Ecological Sciences H Cell Biology I Global Environmental Change	Springer-Verlag Berlin Heidelberg New York Barcelona Budapest Hong Kong London Milan Paris Santa Clara Singapore Tokyo

PARTNERSHIP SUB-SERIES

1. Disarmament Technologies	Kluwer Academic Publishers
2. Environment	Springer-Verlag / Kluwer Academic Publishers
3. High Technology	Kluwer Academic Publishers
4. Science and Technology Policy	Kluwer Academic Publishers
5. Computer Networking	Kluwer Academic Publishers

The Partnership Sub-Series incorporates activities undertaken in collaboration with NATO's Cooperation Partners, the countries of the CIS and Central and Eastern Europe, in Priority Areas of concern to those countries.

NATO-PCO DATABASE

The electronic index to the NATO ASI Series provides full bibliographical references (with keywords and/or abstracts) to about 50000 contributions from international scientists published in all sections of the NATO ASI Series. Access to the NATO-PCO DATABASE compiled by the NATO Publication Coordination Office is possible in two ways:

- via online FILE 128 (NATO-PCO DATABASE) hosted by ESRIN, Via Galileo Galilei, I-00044 Frascati, Italy.
- via CD-ROM "NATO Science & Technology Disk" with user-friendly retrieval software in English, French and German (© WTV GmbH and DATAWARE Technologies Inc. 1992).

The CD-ROM can be ordered through any member of the Board of Publishers or through NATO-PCO, Overijse, Belgium.

Series F: Computer and Systems Sciences, Vol. 149

Springer
Berlin
Heidelberg
New York
Barcelona
Budapest
Hong Kong
London
Milan
Paris
Santa Clara
Singapore
Tokyo

Emergent Computing Methods in Engineering Design

Applications of Genetic Algorithms and Neural Networks

Edited by

Donald E. Grierson

Department of Civil Engineering, University of Waterloo
Waterloo, Ontario N2L 3G1, Canada

Prabhat Hajela

Department of Mechanical Engineering, Aeronautical Engineering and Mechanics, Rensselaer Polytechnic Institute
Troy, New York 12180-3590, USA

Springer
Published in cooperation with NATO Scientific Affairs Division

Proceedings of the NATO Advanced Research Workshop on Emergent Computing Methods in Engineering Design, held in Nafplio, Greece, August 25–27, 1994

Library of Congress Cataloging-in Publication Data

Emergent computing methods in engineering design: applications of genetic algorithms and neural networks / edited by Donald E. Grierson, Prabhat Hajela.

p. cm. — (NATO ASI series. Series F, Computer and systems sciences ; vol. 149)

"Published in cooperation with NATO Scientific Affairs Division."

"Proceedings of the NATO Advanced Research Workshop on Emergent Computing Methods in Engineering Design, held in Nafplio, Greece,August 25–27, 1994"—T.P. verso.

Includes bibliographical references.

1. Engineering design—Congresses. 2. Computer-aided design—Congresses. 3. Genetic algorithms—Congresses. 4. Neural networks (Computer science) —Congresses. I. Grierson, D. E. II. Hajela, Prabhat, 1956– . III. North Atlantic Treaty Organization. Scientific Affairs Division. IV. NATO Advanced Research Workshop on Emergent Methods in Engineering Design (1994 : Nauplion, Greece) V. Series: NATO ASI series. Series F, Computer and systems sciences ; no. 149.

TA174.E44 1996

620'.0042—dc20 96-6924

CIP

ISBN 978-3-642-08239-9

CR Subject Classification (1991): I.0, I.2, J.2, J.6, D.2, G.1

Printed in Germany

Foreword

This volume contains the text of papers presented at the NATO Advanced Research Workshop on Emergent Computing Methods in Engineering Design, held in Nafplio, Greece, August 25-27, 1994. The workshop convened together some thirty or so researchers from Canada, France, Germany, Greece, Israel, Taiwan, The Netherlands, United Kingdom and the United States of America, to address issues related to the application of such emergent computing methods as genetic algorithms, neural networks and simulated annealing in problems of engineering design. The volume is essentially organized into three parts, with each part having some theoretical papers and other papers of a more practical nature. The first part, which comprises the largest number of papers, deals with genetic algorithms and evolutionary computing and presents subject matter ranging from proposed improvements to the computing methodology to specific applications in engineering design. The second part deals with neural networks and considers such topics as their application as approximation tools in design, their adaptation in control system design and theoretical issues of interpretation. The third part of the volume presents a collection of papers that examine such diverse topics as the combined use of genetic algorithms and neural networks, the application of simulated annealing techniques, problem decomposition techniques and the computer recognition and interpretation of emerging objects in engineering design.

The location and format of the workshop provided for a stimulating exchange of ideas among the participants that culminated in a final roundtable discussion that examined emergent computing methods in engineering design from a variety of viewpoints. The record of the roundtable discussion is presented as the final paper in this volume.

The editors hope that this volume will provide readers with an appreciation for the tremendous potential of emergent computing methods in engineering design, perhaps even stimulating some to join the ever-expanding research effort in this exciting field of computer application.

January 1996 D.E. Grierson , P. Hajela

Participants at the NATO ARW, Nafplio, Greece, 1994

Table of Contents

Part I

Part II

Part III

Local Ecological-Like Evolutionary Computing Framework For Global Robustness

Yuval Davidor and Oren Ben-Kiki
Schema - Evolutionary Algorithms Ltd.
61A Hadar Street, Herzelia 46326, Israel.

1 Introduction

In the quest for robustness for Evolutionary Computation (EC), many models and operators were explored. Most of these models introduced some kind of an active operator(s) designed to improve the balance between the drive for convergence and the efficient use of the information contained within the population. This quest proved over and over again to be an elusive enterprise. More specifically, it was realized that a successful application of an EC model to optimization problems usually requires considerable 'hacking' to avoid premature convergence when moderate computing resources are available or utilized.

In this paper, a different approach to the achievement of robustness in EC models, and in specific Genetic Algorithms (GAs), is presented. This approach, called the ECO(logical) framework, is motivated by the observation that the genetic/evolutionary mechanisms do not operate in a vacuum, nor have evolved independently from the limitations and governing rules of ecology. Furthermore, since some special goodness, if not optimality, in the combined effect of the genetic operators is assumed, it is only natural to expect some inherent importance in the combined effect of the infrastructure the genetic rules operate. Accordingly, the ECO framework constitutes a general framework for EC simulation which is based on simplified ecological concepts of evolution.

This ecologically motivated EC framework bare some resemblance to cellular automata models where the dynamics of the system is governed by local interaction rules. However, unlike cellular automata, the rules in the ECO environment are simplified evolutionary information processing rules.

2 Background

The ECO framework was originally proposed in order to improve the robustness of function optimization GAs by introducing emergent speciation [6]. Serendipitously, it turned out that the ECO paradigm exhibits fundamental robustness properties which makes it an attractive generic framework for many EC models.

Researchers in numerous models (and publications) have attempted to suggest that there are additional operators necessary to restrict mating and selection or alternatively, specific activation profiles for the genetic operators (such as varying the mutation rate), if controlled convergence is desired. However, while parallelism became an accessible technology, it was soon realized that the traditional set of genetic operators is unable to achieve robustness.

Realizing that other 'circumstantial' constraints considerably affect the performance of EC, the ECO framework was created to check the role of these additional aspects of biological evolution.

The paper is arranged in the following way: Section 3.0 briefly summarizes the operators suggested in the past to solve (or at least improve) the robustness and control over convergence in GAs. Section 4.0 discusses the ECO framework and present its skeleton structure. Finally, section 5.0 summarize the ECO framework and its message to EC models with reference to experimental results performed and published elsewhere.

3 Historical 'Genetic' Operators To Control Convergence

Since the early days of the '70's, two important aspects about GAs have been realized. One, that controlled convergence is an illusive goal, difficult to achieve with global operators. Two, that fine-tuning of the GA operators is not the key to solve this problem. Similar problems of course can be experienced with other EC models.

In other words, it is suggested that until EC models will inherently exhibit self adaptation through mutual interaction of the operators, it will be difficult to claim they possess the precious quality true adaptiveness. Mechanisms that provide some control over convergence have been suggested as early as the first simulations of GAs started to be experimented. The main idea was that there should be an active mechanism that will control and prevent an unbalanced proliferation of population members (or segments of genetic material) on the expense of global search.

It is clear that to some extent such control can be obtained through an operator which is based on an explicit testing of genotype diversity. However, it is equally clear that such a mechanism is computationally prohibitive. The 'battle' for controlled convergence focused on the attempt to obtain the desired controlled property indirectly in order to minimize computational overhead. The following is a brief history of the mechanisms that were suggested as an answer to the control of convergence issue.

As early as 1970, Cavicchio, suggested a pre-selection mechanism, a mechanism which replaces parent members in the population with their offspring [3]. The idea is that since offspring contain the same bit diversity as their parents, replacing parents with offspring reduce premature loss of bit diversity. A similar mechanism, the crowding scheme, was suggested by De Jong in which an offspring replaces the most similar string in bit terms (hamming distance) from a randomly drawn sub-population of size CF (crowding factor) from the main population [11].

Mauldin's uniqueness operator helped to maintain diversity by incorporating a 'censorship' operator with which the insertion of an offspring into the population is possible only if the offspring is genotypically different from all members of the population at a specified number of loci (hamming distance) [17]. Related ideas in a classifier system environment include Booker's sharing [2] and Schaffer's VEGA model [20].

Until the mid '80's globally monitoring genotypic diversity was regarded as the proper way to solve the instability exhibited by GAs, and the only concern was how to reduce the computation cost of such operations while increasing their effectiveness. Goldberg and Richardson incorporated some of the ideas mentioned above in a mechanism they called a Sharing Function [10,13]. This mechanism modifies the reproduction probability of a population member according to how many population members occupy a similar 'niche' of the solution space (either genotypic or phenotypic).

Eshelman and Schaffer introduced a more direct mating restriction with an operator they called incest prevention [12]. This operator prevented mating between two randomly paired population members if they are too similar to each other. However, this operator requires overhead and does not suggest how to promote fruitful mating, just how to prevent unfruitful ones.

Although the above mechanisms partly improve the convergence, they suffer from the fact that they employ additional operators sometimes at heavy additional computing costs. Also, they are not applicable to all problem domains and evolutionary models, representations and so forth. More importantly, most of the above mechanisms are only applicable to binary representations, and in the sense necessitate setting up additional magic number(s) such as niche count, CF, and so forth prior to implementation.

Muhlenbein and Gorges-Schleuter explored a particular hardware dependent simulation they called ASPARAGOS [18,21]. The underlying idea of ASPAROGOS is that multiple populations are held on a network of transputers, and act as interacting sub-populations. There are other relevant works involving parallel implementations of GAs, but mostly they just recapitulate previous approaches [4,15,19,25].

There are other parallel GAs which attempt to parallelize GAs and which attempted to parallelize GAs according to standard parallelization rules [5,22,1,7,16].

4 The Eco Framework Paradigm

An early observation in the field of EC is that EC is one of the few fields which is inherently parallel that was made serial due to hardware constraints. This observation goes further than stating that EC can be paralleled easily. It actually suggests much more. If the motivation to use EC models is driven by the believe that nature's 'blind' evolution is an adaptive process which promotes the persistence of combination of treats that work well together, then the same arguments can be directed towards the operators, environment and so forth in which it operates. It is clearly difficult to accept that what's applicable on the parameter level is not operative on the macro level. Therefore, it has to be seriously considered that if EC works, and especially works well as most of us believe it does, then the totalism of its operation has to be respected.

In other words, the combination of genetic operators cannot be separated from the specifics of the environment they operate in, and like wise for their effectiveness which is optimal in some sense in respect to the combination of genetic operators, structure of genetic material, ecological constraints, and so on.

It is difficult to accept that the micro evolutionary process is not affected by, and therefore have been shaped as a result of the macro evolutionary process. The fact that hens in Nafplio, Greece do not usually mate with hens in Australia, but do with hens in the heighbouring village, bare principal rule in the success of an evolutionary process. Thus, the 'goodness' of a given genetic operator cannot be measured in isolation from the rest of the genetic process and environment, and that further, the quality of a given genetic operator greatly depends, by and large, on the entire set of genetic process and framework. As a result of this observation, the ECO framework is suggested as a step towards this holistic relation in an evolutionary process and success.

The most fundamental difference between common EC models and the ECO framework is that all mechanisms in ECO that manipulate population members are based on grid position and grid proximity. By and large, in standard EC models population members are processed globally, such as global proportionate reproduction, global mate selection, global survival selection, and so forth. In the ECO framework, all genetic processes are turned local.

Turning the genetic processes local is achieved by introducing a geographical topology, and associating a position in that topology to each population member and admissibility to genetic interaction as a result thereof. This leads to genetic interactions which are a function of the relative position in that topology.

4.1 The Genetic Operators Turn Local

The ECO model requires local genetic interactions. Thus, reproduction, recombination, and selection have to be changed from global to local mechanisms. This section discusses the changes necessary to turn the genetic operators local.

4.1.1 Reproduction

The ECO reproduction is a type of steady state reproduction [23,24,26]. At each reproduction cycle, a grid element i,j is selected at random. The population member which occupies the i,j'th node is called the first parent, and together with the 8 population members adjacent to this i,j'th member forms the local and temporarily active sub-population.

The average fitness of this 9 member sub-population is calculated (first parent + 8 neighbours). The relative fitness of the first parent at the i,j'th node relative to the average fitness of the 9-member sub-population determines how many offspring are going to be produced in the current reproduction cycle. There are many mechanisms possible, one of which is: The number of offspring is the relative fitness of the first parent to its 9-member sub-population rounded probabilistically. For example, if the relative fitness of a given first parent is 1.4, then this first parent will reproduce at least one offspring, and another one with probability of 0.4, while a relative fitness of 0.5 means probability 0.5 for no offspring and 0.5 for 1 offspring.

4.1.2 Mating

After determining how many offspring the first parent will produce, a second parent is selected (with replacement) probabilistically relative to fitness from the sub-population excluding the first parent for each offspring produced.

Please note that the type of crossover mechanism, whether it be a two point, uniform, PMX, or other problem specific crossover is not mentioned nor the type of mutations because it is not relevant to the ECO interaction rules. By thus defining the choice of parents one is free to use whatever crossover, mutation, or other mechanisms one wishes. The mechanisms used in this work are those suitable for continuous representations as discussed in [14].

4.1.3 SELECTION

After producing the offspring and calculating their fitness, each offspring is introduced into the grid by selecting at random a grid element from the 9 grid elements of the i,jth environment. The population member in the selected element and the offspring are in conflict because of limited habitat resources (as only one population member is allowed at each grid element). The conflict is resolved according to the relative fitness of the two opponents.

This selection process is repeated for all offspring sequentially so there is the possibility that a second or latter offspring will replace an earlier invading offspring. This selection pressure bears some similarity to Cavicchio's pre-selection mechanism.

4.2 Eco Effects On Convergence

Resulting from the locality of operators, some strong granularity is expected in the dynamics of the ECO GA simulation. Since the 'effective' population size for each of the population members is small, much smaller than the total population size, it is reasonable to expect fast local (in grid terms) convergence.

Reproduction and sampling are proportionate to a very small number of population members. Members of superior fitness (superior, relative to local grid environment) have a much more pronounced sampling bias (a similar effect to that achieved by scaling operators used in generational GAs, but without the problems of premature global loss of diversity). The fast local convergence resulting from the small effective population size is amplified by the ECO's selection pressure; leaving offspring in the same local environment as their parents. Thus, depositing the biased genetic material in its origin environment, and enhancing local convergence.

So, if one would monitor the local environment of a given population member, one would observe the growth of genotypic isomorphism in its immediate vicinity. The local speedy convergence occurs in many parts of the grid towards different genetic makeup, and produces a speciation like phenomenon as the genetic diversity in many local neighbourhoods becomes diminished. The genotypic granularity of the grid becomes more pronounced, and small 'islands' of genetically isomorphic colonies begin to emerge.

As the degree of genotypic isomorphism increases inside an 'island', a process of specialization takes place. The sampling bias is reduced as the relative fitness approaches unity. Crossover becomes less disruptive as there are many shared schemata in the local neighbourhood. Mutations, if active, supply a greedy optimization. The overall effect is that of local optimization with in, or in close proximity to, the convex hall defined by the reducing diversity of the local environment.

On the perimeter of high genetic isomorphic regions a different process takes place. Since reproduction is biased towards highly fit members, if the average fitness of an isomorphic region is higher than its surrounding, it will expand. If the region's average fitness is lower than the fitness of its surrounding, it will shrink. The mating on the perimeter of a genetic highly isomorphic island is relatively dramatic compared to that inside islands. Since the genetic material that surrounds an island may be very different to that within it, mating on the borderline of an island results in offspring of diverse genetic makeup which explore new areas of the solution space.

So at initial stages one can observe the formation of small islands of similar genetic material. As the simulation progresses the islands of relative high fitness grow and take over grid space that previously was occupied by low fitness members. Also, within islands there will be specialization and the occasional discovery of locally (in phenotypic sense) optimized members, and on the perimeter of islands the creation and test of new genotypic combinations. Once such a new genotype of high relative fitness (to its local grid environment) is created, it will start to reproduce and expand.

To summarize the ECO paradigm, the ECO simulation is characterized by the formation of genetically isomorphic islands, the growth of islands of relative high fitness, the continued convergence and local optimization within islands, and discovery of new rewarding genetic formations on the borders of islands. In the next section a chronicle of an ECO simulation is presented with the intention to demonstrate the unique convergence dynamics described in this section.

5 Summary, Results And Future Work

The ECO framework discussed in this paper presents a new synthesis of the conventional genetic operators used in EC models. In this framework all operators are based on local Euclidean topology interaction. The choice of a 2D topology rather than a grid with higher dimensionally is not arbitrary and so is the range for the local genetic interaction. The number of dimensions in the topology and the range for the genetic interaction determines the size of the sub-population. The 2D configuration and first immediate neighbour interaction radius was chosen to achieve a balance between the overall number of sub-populations and their size. This new population interaction arrangement results in a rapid local convergence while maintaining both global diversity and speciation as an emergent property of the simulation. A too large local sub-population, whether the result of high dimensionally of the grid or a long genetic interaction range, and the ECO improved control over convergence property becomes diminished. Of course the other extreme of a very small sub-population and the genetic process becomes too weak.

The ECO framework as presented, does not claim to maintain niche & species ad infinitum. Further of course, it is not suggested the set of genetic interactions and topological constraints introduced here are the full or complete in any sense set of ecological interactions. Only the minimal, but fundamental subset of the entire repertoire to exist in nature was applied. Other topological constraints such as those known as fire wall constraints, island and migration models deserve careful consideration.

The ECO does allow, however, the emergence of sub-population niches in order to enhance the efficiency and robustness of an EC simulation. The effect of the ECO framework on convergence and stability has been tested in several domains and in most of them showed substantial improvements. In [7] the ECO was applied to Job Shop Scheduling with important improvements, and in [8] the ECO framework was applied to a dynamic optimization real world problem again with considerable improvements (also see [9] for additional results). The achievement of a complete niche & species equilibrium would advance the EC technology a giant step forward, and toward this goal the ECO paradigm is a step forward. Depending also on the solution's space topology and the core EC model used, the speed of local convergence is linked to the size of the grid. The ECO framework is not usually sensitive to the size of the grid over a large range of simulation conditions, but nevertheless, is affected by it.

References

1. Baluja, S. (1993) Structure and performance of fine-grain parallelism in genetic search, 5th International Conference on Genetic Algorithms, 155-162.

2. Booker, L. B. (1982) Intelligent behavior as an adaptation to the task environment, (Doctoral dissertation, University of Michigan) Dissertation Abstracts International, 43(2), 469B.
3. Cavicchio, D. J. (1970) Adaptive search using simulated evolution, (Doctoral dissertation, University of Michigan) Unpublished doctoral dissertation.
4. Cohoon, J. P., et al. (1987) Punctuated equilibria: A parallel genetic algorithm, 2nd International Conference on Genetic Algorithms, 148-154.
5. Collins, R. J. and Jefferson, D. R. (1991) Selection in massively parallel genetic algorithms, 4th International Conference on Genetic Algorithms, 249-256.
6. Davidor, Y. (1991) A naturally occurring niche & species phenomenon: The model and first results, 4th International Conference on Genetic Algorithms, 257-263.
7. Davidor, Y., Yamada, T., and Nakano, R. (1993) The ECOlogical framework II: Improving GA performance at virtually zero cost, 5th International Conference on Genetic Algorithms, 171-176.
8. Davidor, Y. (1993) An ecological model for evolutionary computing, Japanese Journal on Systems, Control, and Engineering, Vol. 37, No. 8, pp. 468-474.
9. Davidor, Y. (in press) Computing with an Ecological model of evolution, in Ray Paton (Ed.), Computing with biological metaphors, Chapman & Hall.
10. Deb, K. and Goldberg, D. E. (1989) An investigation of niche and species formation in genetic function optimization, 3rd International Conference on Genetic Algorithms, 42-50.
11. De Jong, K. (1975) An analysis of the behavior of a class of genetic adaptive systems, (Doctoral dissertation, University of Michigan) Dissertation Abstracts International 36(10), 5140B.
12. Eshelman, L. J. and Schaffer, J. D. (1991) Preventing premature convergence in genetic algorithms by preventing incest, 4th International Conference on Genetic Algorithms, 115-122.
13. Goldberg, D. E. and Richardson, J. (1987) Genetic algorithms with sharing for multimodal function optimization, 2nd International Conference on Genetic Algorithms, 41-49.
14. Grossman, T. and Davidor, Y. (1992) An investigation of a genetic algorithm in continuous parameter space (CS92-20), The Weizmann Institute.
15. Jog, P. and Van Gucht, D. (1987) Parallelisation of probabilistic sequential search algorithms, 2nd International Conference on Genetic Algorithms, 170-176.
16. Maruyama, T, Hirose, T., and Konagaya, A., (1993) A fine-grained parallel genetic algorithm for distributed parallel systems, 5th International Conference on Genetic Algorithms, 184-190.

17. Mauldin, M. L. (1984) Maintaining diversity in genetic search, National Conference on Artificial Intelligence, 247-250.

18. Muhlenbein, H. (1989) Parallel genetic algorithms, population genetics and combinatorial optimization, 3rd International Conference on Genetic Algorithms, 416-421.

19. Pettey, C. C. and Leuze, M. R. (1989) A theoretical investigation of a parallel genetic algorithm, 3rd International Conference on Genetic Algorithms, 398-405.

20. Schaffer, J. D. (1985) Multiple objective optimization with vector evaluated genetic algorithms, 1st Int. Conf. on Genetic Algorithms, 93-100.

21. Schleuter, M. G. (1989) ASPARAGOS, An asynchronous parallel genetic optimization strategy, 3rd International Conference on Genetic Algorithms, 422-427.

22. Spiessens, P. and Manderick, B. (1991) A massively parallel genetic algorithm: Implementation and first analysis, 4th International Conference on Genetic Algorithms, 279-286.

23. Syswerda, G. (1989) Uniform crossover in genetic algorithms, 3rd International Conference on Genetic Algorithms, 2-9.

24. Syswerda, G. (1991) A study of reproduction in generational and steady state genetic algorithms, In G. J. E. Rawlins (Ed.), International Workshop on the Foundations of Genetic Algorithms , Bloomington: Morgan Kaufmann.

25. Tanese, R. (1989) Distributed genetic algorithms, 3rd International Conference on Genetic Algorithms, 434-439.

26. Whitley, D. (1988) GENITOR: a different genetic algorithm, Rocky Mountain Conference on Artificial Intelligence,

Genetic Algorithms for the Design of Composite Panels

R.T. Haftka, R. Le Riche and P. Harrison

Department of Aerospace and Ocean Engineering
Virginia Polytechnic Institute and State University
Blacksburg, Virginia 24061, USA

Abstract. The design of composite laminates is a combinatorial optimization problem because ply thicknesses are typically discrete and ply orientations are often limited to a small set of angles. Additionally, laminate design problems often have a number of different local optima of similar performance. Genetic algorithms are well suited for this problem because they can handle the combinatorial nature of the problem and they permit the designer to obtain many of the local optima, and thus have a choice in selecting the most suitable design. However, the cost of genetic algorithms is usually high. To address this cost problem, it may be beneficial to tune the genetic algorithm on simple problems that can be analyze inexpensively. The paper here summarizes the results of several investigations that used this approach. Next, the paper summarizes applications to more complex problems where the lessons learned from the simple problems were used to advantage.

Keywords: Genetic Algorithms, Stochastic search, composites design

1. Introduction

Because of their high strength to weight ratio and the possibility of tailoring their behavior, notably by selecting fiber orientations, the optimization of composite laminated plates has received a growing attention in the last two decades. When designing composite laminated plates, typical design variables are the fiber orientations and the number of layers. Manufacturing constraints often limit the choice of fiber orientations to 0°, ±45° and 90°, and fix the basic ply thickness, making the minimum weight design of a laminate a discrete optimization problem.

Composite laminate performance is characterized by a number of parameters which is usually smaller than the number of design variables, and so different sets of design variables can produce similar results, i.e., there are many optimal or near-optimal designs. Also, for nonlinear problems, linearization may yield sub-optimal designs because of the existence of many different locally optimal designs (Nagendra et al. 1992, Shin et al. 1989). For this reason, genetic algorithms, which are more robust at finding global or near global optima, have been recently explored for designing composite laminated plates (Ball et al. 1993, Callahan and Weeks 1992, Kogiso et al. 1994a,b, Le Riche and Haftka 1993).

Genetic algorithms (GAs) are search techniques based on a simulation of natural reproduction genetics operating on a population of designs. As optimization algorithms, they are global in scope and can work on discrete as well as continuous design problems. They also have the advantage or generating a number of optimal and near-optimal designs during a single search.

The major drawback of genetic algorithms is the large number of analyses that they require. To alleviate this problem, it helps to tune the genetic algorithm to a particular problem by the selection of specialized operators, and by tuning the parameters of different operators. This tuning of the GA has to be performed on a simple version of the actual problem of interest. Then, it is hoped that the same GA will work efficiently on similar but more costly problems. Le Riche and Haftka (1993) undertook such tuning of a GA for an unstiffened laminate of given thickness designed for maximum failure load. By using simply supported boundary conditions, the analysis could be reduced to a simple algebraic formula, so that millions of analyses could be carried out. A tuned GA was found that could reach optimum designs with high reliability in about 1000 analyses. However, when this algorithm was applied to the design of stiffened composite panels, the optimum design could not be obtained even with 10,000 analyses.

The poor performance of the algorithm for the stiffened panel was traced to several causes. First, the stiffened panel design dealt with the dual problem of minimum weight design subject to strength and buckling constraints. Unlike the maximum failure load design for given thickness, this problem required variable string length. Second, constraints had to be represented by the use of a penalty function. Third, the stiffened panel had two plates (skin and blade) that had to be designed rather than one. The objective of the present work is to describe several recent modifications to the basic genetic algorithm developed by Le Riche and Haftka (1994a) and to show how these modifications improved the performance for stiffened panel design.

2. Simply Supported Single Plate

The simply supported laminate shown in Figure 1 is loaded in the x and y direction by N_x and N_y, respectively. The laminate is composed of N plies, each of thickness t, and is balanced and symmetric. Its longitudinal and lateral dimensions are a and b, respectively.

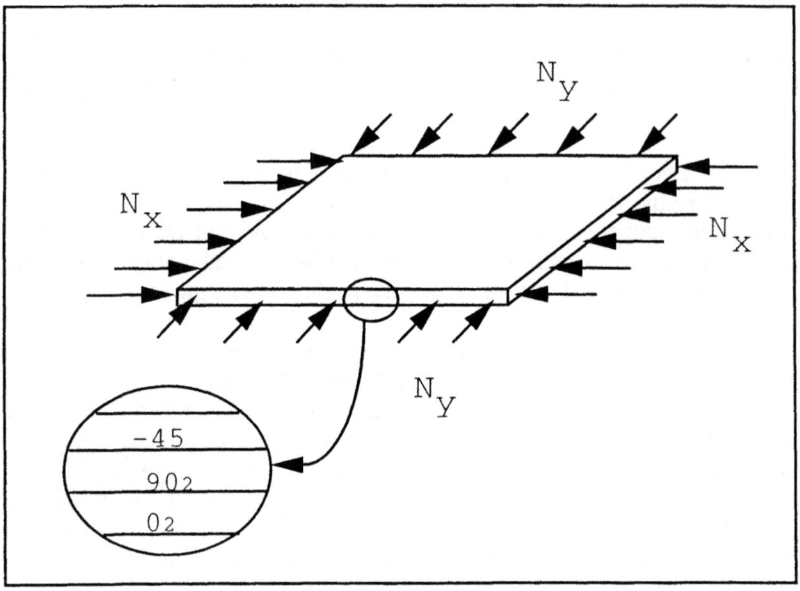

Figure 1: Simply supported laminated plate subjected to normal inplane loads.

The laminate buckles into m and n half-waves in the x and y directions, respectively, when the loads reach the values $\lambda_b(m, n)N_x$ and $\lambda_b(m, n)N_y$. The buckling load factor $\lambda_b(m, n)$ is given in terms of the flexural stiffness coefficients D_{ij} as

$$\lambda_b(m,n) = \pi^2 \frac{\left[D_{11}(m/a)^4 + 2(D_{12} + 2D_{66})(m/a)^2 + D_{22}(n/b)^4\right]}{(m/a)^2 N_x + (n/b)^2 N_y}.$$

The smallest value of λ_b over any m and n is the critical buckling load factor λ_{cb}. If λ_{cb} is larger than 1 the laminate can sustain the actual applied loads N_x and N_y without buckling.

The strength of the plate is predicted here by the maximum strain criterion. That is, the plate is assumed to fail if any of the ply strains exceeds its allowable value. The principal strains in the ith layer are calculated from classical lamination theory for symmetric and balanced laminates. The critical strength failure load factor λ_{cs} is defined as,

$$\lambda_{cs} = \min_{i} \left[\min\left(\frac{\varepsilon_1^{ua}}{f|\varepsilon_1^i|}, \frac{\varepsilon_2^{ua}}{f|\varepsilon_2^i|}, \frac{\gamma_{12}^{ua}}{f|\gamma_{12}^i|} \right) \right],$$

where ε_i^{ua} are the ultimate allowable strains, and a safety factor $f = 1.5$ is used. If λ_{cs} is smaller than 1, the plate is assumed to fail. The critical load factor λ_{cr} is,

$$\lambda_{cr} = \min(\lambda_{cb}, \lambda_{cs}).$$

To alleviate matrix cracking problems we also require that there are no more than four contiguous plies with the same fiber orientation.

The purpose of the optimization is to find the thinnest symmetric and balanced laminate satisfying the 4-ply contiguity constraint that will not fail because of buckling or excessive strains. The symmetry constraint is readily accommodated by optimizing only one half of the laminate, the other half being obtained by symmetry. The balance constraint is enforced by associating +45° and -45° plies into one ±45° stack. Stacks of 0_2°, ±45°, and 90_2° are our basic building blocks.

Genetic algorithms solve unconstrained optimization problems, so that constraints need to be either incorporated into the objective function via penalty functions. The following sections describe the genetic algorithm and then the penalty formulations.

2.1 Basic Genetic Algorithm

The basic genetic algorithm developed in [Kogiso et al. 1994b], [Le Riche and Haftka 1993], and [Nagendra et al. 1993] is summarized as the basic GA. A genetic algorithm works on a population of designs each coded in the form of a string. The lengths of all the design strings have to be the same. Because the designs can have different number of plies, empty plies are added as "padding" to make all designs have the same nominal number of plies. The coding used here is the standard stacking sequence notation with the additional symbol E for an empty ply. Because our basic unit is a stack of two plies, laminates can be composed only of 0_2, ±45, 90_2 and E_2 stacks.

The GA starts with the generation of an initial random population of designs. In a second step, the objective function values of the designs are evaluated. Each design is assigned a fitness value depending on the objective function values of all the designs in the population. It is then processed by genetic operators to create a new population. The selection operator selects parent strings based on their fitness and passes them on to the operators performing the genetic recombination. The probability of a design of being chosen as parent increases with its fitness. The old population is then replaced by the new one, with one exception. The best of the previous generation is cloned into the next generation, which corresponds to an "elitist" version of the genetic algorithm. This process is repeated until convergence.

2.1.1 Selection Scheme

The first step in creating a new generation is the selection of sets of parent designs for mating. Each individual has a fitness value that determines its probability of being chosen as a future parent. Here, the fitness was chosen to be the relative rank in terms of the objective function Φ in the population. The fitness assigned to the ith best individual of m designs is then equal to $[2(m + 1-i)]/(m^2 + m)$, so that the sum of all fitnesses is equal to one.

2.1.2 Crossover

Crossover is the operation used to trade characteristics of parent designs and create offsprings by exchanging parts of the parents' strings. The crossover used in the basic genetic algorithm is a two-point crossover, where two break points are chosen randomly. One offspring is created by splicing the parents' three substrings, i.e., the first substring comes from the first parent, the second from the second parent, and the third from the first parent (see Figure 2). Crossover is applied with a probability equal to one, based on the results presented in [Le Riche and Haftka 1993].

$$\textbf{Parent 1}: \quad [E_2/E_2/E_2/E_2/\pm 45/ \overset{\vee}{\underset{\wedge}{}} \pm 45/90_2/ \overset{\vee}{\underset{\wedge}{}} 90_2/0_2]_s$$

$$\textbf{Parent 2}: \quad [E_2/90_2/\pm 45/90_2/0_2/ \overset{\vee}{\underset{\wedge}{}} 0_2/\pm 45/ \overset{\vee}{\underset{\wedge}{}} \pm 45/\pm 45]_s$$

$$\textbf{Child}: \quad [E_2/E_2/E_2/E_2/\pm 45/ \overset{\vee}{\underset{\wedge}{}} 0_2/\pm 45/ \overset{\vee}{\underset{\wedge}{}} 90_2/0_2]_s$$

Figure 2: Traditional two-Point crossover

2.1.3 Mutation

The design generated by crossover is next subjected to the mutation operator. Mutation performs random changes in an individual's string with a low probability (0.01 in this paper). Its purpose is to bring in completely new bit combinations, thereby protecting the population against becoming uniform. Each bit in the string has "probability of mutation" chances of changing. Figure 3 gives an example where two bits are mutated.

Before Mutation : $[E_2/90_2/\underline{\pm\mathbf{45}}/90_2/0_2/0_2/\underline{\pm\mathbf{45}}/\pm 45/\pm 45]_s$

After Mutation : $[E_2/90_2/\underline{\mathbf{0_2}}/90_2/0_2/0_2/\underline{\mathbf{E_2}}/\pm 45/\pm 45]_s$

Figure 3: gives an example where two bits are mutated.

2.1.4 Permutation

Permutation is a new operator devised for laminate design in [Le Riche and Haftka 1993]. Permutation swaps the location of two pairs of stacks without changing the flexural properties of the laminate. First two points in the string are selected, creating three substrings in the chromosome. The two pairs of stacks at the two extremities of the central substring are then swapped (Figure 4). In accord with results presented in [Le Riche and Haftka 1993], permutation is applied with a probability of 1.

Before Permutation : $[E_2/\pm 45/\overset{\vee}{\underset{\wedge}{}}\pm 45/90_2/\pm 45/0_2/0_2/90_2/\overset{\vee}{\underset{\wedge}{}}\pm 45]_s$

After Permutation : $[E_2/\pm 45/\overset{\vee}{\underset{\wedge}{}}90_2/0_2/\pm 45/0_2/90_2/\pm 45/\overset{\vee}{\underset{\wedge}{}}\pm 45]_s$

Figure 4 Permutation operation

2.1.5 Objective Function Formulation: Handling of Constraints

For the present laminate design problem, the objective is to minimize the number of plies N. The three constraints of buckling, strength failure, and the 4-ply contiguity constraint are incorporated into the objective function through penalty functions. The penalty should be kept as low as possible, just above the limit below which infeasible designs become optimal (Le Riche and Haftka 1993, Richardson et al. 1989). The flaw of heavily penalizing infeasible designs is that it limits the exploration of the design space to feasible regions, precluding short cuts through infeasible regions.

The contiguity constraint is enforced by multiplying the objective by $P_C^{n_c}$ where P_C is the penalty parameter for the ply contiguity constraints, and n_C is the number of stacks of two plies in excess of the constraint value of two stacks (i.e., 4 plies) in one half of the laminate.

The treatment of the failure and stability constraints is more complicated for two reasons. First, because of the discrete nature of the problem, there may be several feasible designs of the same minimum thickness. Of these designs, we define the optimum to be the design with the largest failure load. Therefore, as in [Kogiso et al. 1994b], the objective function is linearly reduced in proportion to the failure margin for designs that satisfy or almost satisfy the stability and failure constraints. For such designs the objective function Φ_0 is defined as

$$if\, \lambda_{cr} \geq (1-\delta), \qquad \Phi_0 = P_c^{n_c}\left\{N + \varepsilon\left[(1\text{-}\delta) - \lambda_{cr}\right]\right\}$$

δ is a small tolerance parameter added for more flexibility in the definition of feasibility, and ε is a small parameter for rewarding constraint margin. When the stability or failure constraints are violated, it is possible to estimate the additional thickness needed to achieve feasibility from simple scaling rules (see Kogiso et al., 1994b). Based on the scaling rule, the objective function Φ_0 is defined as

$$if\, \lambda_{cr} < (1-\delta), \qquad \Phi_0 = P_c^{n_c}\left\{\frac{N}{\lambda_{cr}^{P_l}}\right\}$$

where P_l is an exponent used in the scaling law.

The main problem of the previous objective function formulation is that for some load cases, the critical load of infeasible designs that are one stack lighter than the optimum feasible designs is very close to 1. Then P_l needs to be very large for the objective function Φ_0 of feasible optimum designs to be smaller than the objective function of good infeasible designs. But a large P_l impairs the efficiency of the genetic search (Le Riche and Haftka 1993, Richardson et al. 1989).

The objective function and operators described so far are those used in [Kogiso et al. 1994a], [Kogiso et al. 1994b], [Le Riche and Haftka 1993], and [Nagendra et al. 1993]. We will now consider modifications for better efficiency in minimum weight laminate design.

2.2 Improvements to Basic GA

Le Riche and Haftka (1994a) studied exhaustively the variable-thickness single plate problem and proposed the following changes for improving the performance of the basic GA.

2.2.1 Modified Objective Function Formulation

The previously outlined problem of needing large values of P_l is avoided by adding a small constant penalty S for violating the strength or stability constraints. That is, the modified objective function Φ is defined as

$$if\, \lambda_{cr} \geq (1-\delta), \quad \Phi = P_c^{n_c}\{N + \varepsilon[(1-\delta) - \lambda_{cr}]\},$$

$$if\, \lambda_{cr} < (1-\delta), \quad \Phi_0 = P_c^{n_c}\left\{\frac{N}{\lambda_{cr}^{P_l}}\right\} + S.$$

S and $(1/\lambda_{cr}^{P_l})$ are both penalties, but their action differs in that $(1/\lambda_{cr}^{P_l})$ is function of the distance to feasibility while S is not. If $S > 0$, P_l can be chosen smaller than the minimum value of 1 derived in [Kogiso et al. 1994b]. Le Riche and Haftka found that a value of $S = 1$ and $P_l = 0.5$ were the best combination of those tried.

2.2.2 Modified Selection Procedure

In order to preserve population diversity, a test is added so as to never mate two identical designs, which would probably create a clone.

2.2.3 Crossover for Varying Thickness Laminates

The two-point crossover has a problem for the varying thickness case in that the cutting points can fall in the empty part of the laminate, reducing the potential for crossover to truly recombine the properties of the parents. After testing eight alternatives, Le Riche and Haftka found that a singlepoint crossover, where the cutting point is restricted to fall in the non-empty part of the thicker laminate was the best. This crossover was dubbed 'one-point thick crossover'.

2.2.4 Differentiated Mutation

The traditional mutation operator, previously described, has a biased and uneven effect depending on the string length (i.e. maximal laminate thickness) and number of digits denoting full stacks (i.e. actual laminate thickness). To correct this, the mutation operator is differentiated into thickness changing mutation and orientation changing mutation. Three separate mutation probabilities are used: a probability of adding plies per string, a probability of deleting plies per string, and a probability of changing the fiber orientation per digit. The thickness mutations add or delete two stacks somewhere in the laminate if the corresponding probabilistic test is passed. For change of fiber orientation, a test is implemented for each digit coding two full stacks, and if it is passed, the fiber orientation is changed.

2.2.5 Single-Stack-Swap Permutation

The original algorithm (Le Riche and Haftka 1993) has been developed for a class of relatively easy problems for which only one load case was considered in the optimization problem. However, for more difficult problems where many loadings are considered when designing a laminate, the basic version of permutation appears to shuffle the digits too much. Accordingly, Le Riche and Haftka (1994b) proposed to have a permutation operator that swaps a single pair of stacks instead of two pairs (cf. Figure 5).

Before Permutation : $[E_2/\pm 45/\pm 45/\underline{\mathbf{90_2}}/\pm 45/0_2/90_2/\underline{\mathbf{0_2}}/\pm 45]_s$

After Permutation : $[E_2/\pm 45/\pm 45/\underline{\mathbf{0_2}}/\pm 45/0_2/90_2/\underline{\mathbf{90_2}}/\pm 45]_s$

Figure 5: Single-stack swap permutation operator

2.3 Effect of Improvements

Genetic algorithm performance will be discussed in terms of "price of the search" and "reliability". The reliability of the algorithm is the probability it has of finding a practical optimum. A practical optimum is defined here as an optimal weight design with λ_{cr} within 0.1% of λ_{cr} of the global optimum. The price of the search is the number of analyses necessary to reach 80% reliability, i.e., to have 80% probability of finding a practical optimum.

We consider a graphite-epoxy plate with longitudinal and lateral dimensions $a = 20in.$ and $b = 5in.$ respectively. The material properties are: $E_1 = 18.5 \times 10^6\ psi$; $E_2 = 1.89 \times 10^6\ psi$; $G_{12} = 0.93 \times 10^6\ psi$; $\nu_{12} = 0.3$; t = 0.005*in.* (*t* is the basic ply thickness); $\varepsilon_1^{ua} = 0.008$; $\varepsilon_2^{ua} = 0.029$; $\gamma_{12}^{ua} = 0.015$. The maximum thickness for a laminate is assumed to be 64 plies (i.e. string length = 64/4 = 16).

Four different load cases are considered: load case 1, N_x = 13,000 *lb/in* and N_y = 1,625 *lb/in;* load case 2, N_x = 12,500 *lb/in* and N_y = 3,125 *lb/in*; load case 3, N_x = 9,800 *lb/in* and N_y = 4,900 *lb/in;* and a multiple load case where the loadings [N_x = 12,000 *lb/in*, N_y = 1,500 *lb/in*], [N_x = 10,800 *lb/in*, N_y = 2,700 *lb/in*] and [N_x = 9,000 *lb/in*, N_y = 4,500 *lb/in*] are considered simultaneously, i.e. the optimal design should be able to withstand each of these loadings. For each load case and each tuning of the genetic algorithm 200 independent searches are performed, each of them stopped after 6000 analyses.

The optimum designs for all load cases had 48 plies. Among the single load cases, load case 1 was found to be easy for the genetic algorithm because strain constraints are by far the most critical. There are dozens of practical optima that are different combinations of the same pool of fiber orientations. Load case 2 is more difficult since it has only 3 practical optima. Although load case 3 has 13 practical optima, it was found to be misleading for the genetic algorithm (Kogiso et al. 1994b): practical optimal designs are composed of +45, -45 and 90 degree plies exclusively. However, in the course of the search, the algorithm has difficulties removing the 0 degree plies because the strain constraint is critical in these plies. As the number of 0 degree plies in the laminate is reduced, the strains in the remaining 0 degree plies increase. However, once the last 0 degrees ply has been removed, the strain constraint is no longer critical. This kind of problem is known as a singular optimum and as being "GA-deceptive", leading to convergence to non-optimal designs. The multiple load problem is also a difficult test case because four failure modes (strain failure due to the first load set and buckling failure due to the three load sets) occur in the optimal region of the design space. Many active constraints around the optimum design means that many of the design points surrounding the optimum will be infeasible, therefore penalized, which lowers their probability of being selected for reproduction. However, some of these infeasible designs might have given a quick access to the optimum. The GA exhibits difficulties of a different nature for load case 3 and for the multiple load case. The GA needs to be able to make radical changes to the designs to allow removal of several 0° plies and solve load case 3. On the other hand, the large number of constraints in the multiple load case makes too much random exploration of the design space a rather inefficient strategy compared to a more careful juxtaposition of building blocks.

Le Riche and Haftka (1994a) describe the effect of each of the modifications on the genetic search. Here we will show only the combined effect of all the modifications averaged over the four load cases. The parameters common to the basic and modified GA are population size = 8, $P_c = \sqrt{10/9}$ and the probabilities of crossover and permutation equal to one. The probabilities of mutation is 0.01 for the basic GA, while the modified GA uses this probability for stack change, but has a probability of 0.05 for stack addition or deletion. For penalty parameters the basic GA has $P_l = 2$, $S = 0$, and $\delta = 0.0$, while the modified GA has $P_l = 0.5$, $S = 1.0$, $\delta = 0.005$. The two versions of the algorithm are compared in Figure 6 in terms of the reliability as a function of the number of analyses. The plot results from 200 independent runs for each one of the 4 test case described above.

The intersections of the curves with the 80% reliability line represent the price of the associated searches. The modified GA costs 1450 analyses, compared to 3300 for the basic GA, which represents a 56% decrease in the price of the search.

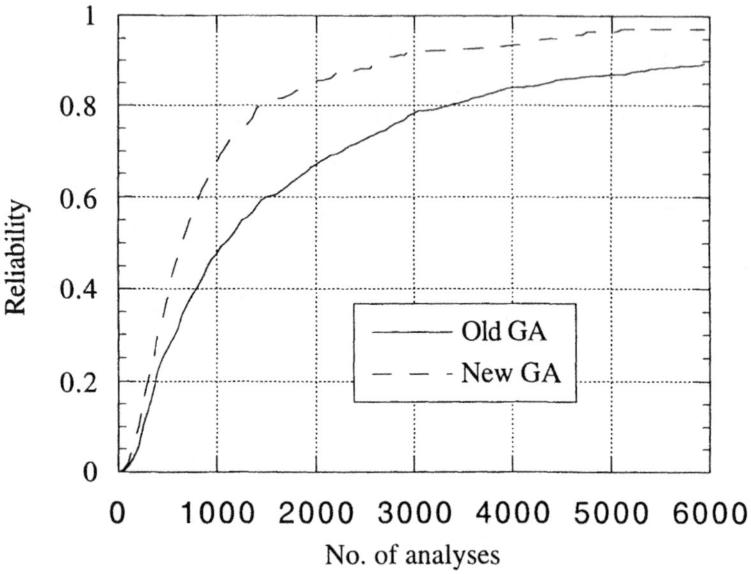

Figure 6: Comparison of the basic and modified versions of the GA, 200 runs.

3. Short Multiple Runs

We use $r_1(n)$ to denote the reliability of the algorithm after n analyses. With an elitist strategy, $r_1(n)$ is a monotonically increasing function. A certain percentage of genetic optimizations converge fast, and so for these cases the later generations are wasted. Other genetic optimization get trapped in bad local minima, and it takes an extremely large number of generations to extricate then from these traps. Because the later stage of a genetic search shows slow progress $r_1(n)$ tends to become flat as n increases. It is a case of diminishing returns. This occurs because some searches converge to the wrong region of the design space and take very long to extricate themselves out of that region. To reduce the chances of getting stuck in the wrong region, it may be more economical to make several short searches rather than have n large enough so that every single search would have time to reach a practical optimum. Le Riche and Haftka (1994b) explored the possibility of using this approach

We denote the reliability obtained by repeating the genetic optimization p times, each time with n searches as $r_p(n_t)$, where $n_t = (pn)$ is the total number of analyses, also called price of the search. Then, because the p searches are independent

$$r_p(nt) = 1 - (1 - r_l(n))^p$$

The problem of maximizing the reliability $r_p(n_t)$ for a given search price n_t can be formulated as

$$\underset{n,p}{\text{Maximize}} \qquad r_p(pn) = 1-(1-r_l(n))^p$$
$$\text{such that} \qquad np \le n_t .$$

For an elitist GA, $r_p(n_t)$ is an increasing function of n_t, and the inequality constraint in the dual problem becomes an equality because it will always be active. Consequently, $n = n_t/p$, and there is only one variable left, p. For the examples here we found the optimum number of searches $p_o$which maximizes $r_p(n_t)$) for a given $n_t = pn$ by enumeration, trying all values between 1 and 20. To simplify our notation we use $r_o(n_t)$ to mean the reliability that can be achieved with n_t analyses when the optimum number of searches is carried out.

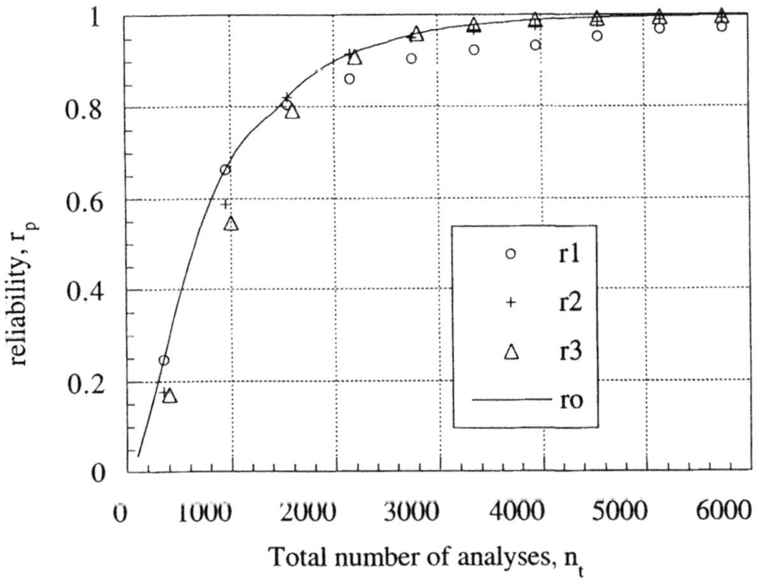

Figure 7: reliability versus total number of analysis (n_t), modified GA.

Figure 7 shows the reliability $r_p(n_t)$ versus the total number of analyses $n_t =$ (pn) for one, two, three and the optimum number p_o of searches, for the modified GA. Figure 7 shows that, for $R = 80\%$, one or two searches with the modified GA are optimal and require about 1450 analyses. However, we see that a reliability $R = 90\%$ may be achieved by two searches of 1050 analyses each, versus one search of 2650 analyses. That is, the number of analyses can be reduced by 20% by making two runs instead of one. The advantage of using several shorter searches increases when higher reliability is sought. R = 95% is yielded by three searches of 900 analyses each ($n_t = 2700$), versus one search of 4600 analyses, which represents 41% saving.

Another important conclusion appears to emerge from Figure 7: in order to obtain the complete $r_o(n_t)$ curve, it is sufficient to calculate $r_2(n_t)$ until the optimum number of searches p_o changes from 1 to 2. The optimal number of searches p_o changes from 1 to 2 at $n_t = 1450$ analyses, and from 2 to 3 at $n_t = 2550$ analyses. The portion of the curve $r_o(n_t) = r_2(n_t)$ ($p_o = 2$) needs only information from $r_1(n_t)$ up to the point $n_t = 2550/2 = 1275 < 1450$. Furthermore, if the target reliability is high enough, it is even possible to stop generating $r_1(n_t)$ earlier. For example, in order to achieve 80 percent reliability, it is necessary to run 1450 analyses. But if the target reliability is raised to 90 percent, 1050 analyses will be sufficient,t, and for 95 percent reliability only 900 analyses are needed. Practically, this means that GAs need not to be tested over a very large number of analyses when establishing $r_1(n_t)$. This conclusion, however, depends on the shape of the $r_1(n_t)$ curve, which for our two examples, has a negative second derivative d^2r_1/d^2n. If the derivative changes sign as in a sigmoid curve, or even if it becomes close to zero, p_o may not increase monotonically with n_t.

Le Riche and Haftka (1994b) also explored the possibility of specially tuning the GA in order to take advantage of repeated runs. The idea is to maximize the speed at which the GA makes progress early on. One way of forcing the GA to make fast progress at the beginning of the run is to increase the selection pressure. Although this strategy will not work on every problem, it is expected to apply to many cases. By increasing the selection pressure, we allow more runs to converge rapidly, but we also increase the chance that they will converge to the wrong answer. Which effect will dominate is problem dependent. However, when the number of analyses nt is small, we expect the first effect to be more important than when n_t is high.

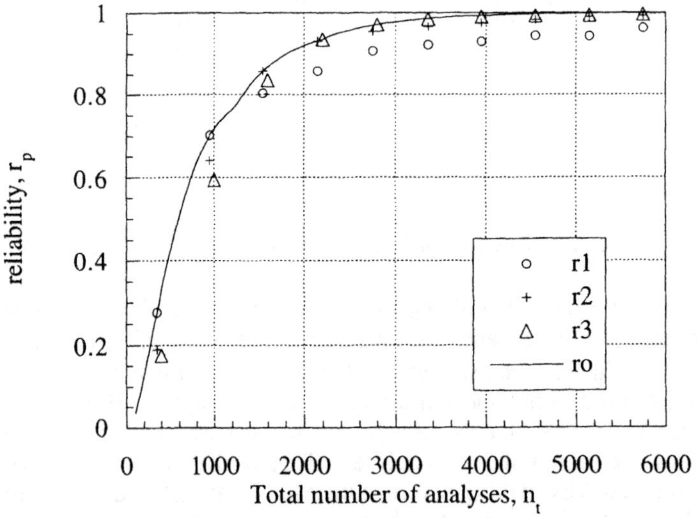

Figure 8: reliability versus total number of analyses (n_t),(modified) GAvar: fitness = square of the rank.

The GA used on Figure 8 is in every aspects similar to the modified GA, except for the selection designs which is based on the rank of the design squared, i.e. the probability of the i-th best individual in the population to be selected as parent is $6(m + 1 - i)^2/[m(m + 1)(2m + 1)]$. Comparison of Figures 7 and 8 indicates that the increased selection pressure improves the reliability for one search in the early stages of the optimization and lowers it in the later stages of the optimization. However, this reduced reliability is of no consequence because we can use two short runs instead of a single search. Thus, when the GA is run twice for 650 analyses it has a reliability of 0.8015%. For $R = 80\%$, the increased selection pressure with two runs reduces the cost of the search by about 10% (1300 against 1450 analyses). However, with a single run, the increased selection pressure increases the cost by about 10% (1590 against 1450 analyses to achieve 80% reliability).

4. Stiffened Panel Design

4.1 Blade Stiffened Plate

Nagendra et al. (1993) employed the basic GA to minimize the weight of a composite plate stiffened by 4 blade stiffeners (see Figure 9) subject to buckling and strength constraints. The design variables were the stacking sequences of the blade and skin laminates as well as the height of the blade. The two laminates were constrained to be balanced and symmetric. Additionally, for improved damage tolerance, the skin laminate did not include 0° plies, and both laminates were required to have ±45° as the outside stack. The flange laminate was the same as the blade laminate. The buckling and stress analyses were performed by the PASCO program (Stroud and Anderson, 1981). Two strings representing the laminates and one string representing the blade height as a 9-digit trinary number were concatenated together to form the chromosome. The population size was 20 and the other parameters were the same as for the single-plate problem.

Nagendra et al. (1993) performed several dozens runs of the basic GA for 3500 analyses each, a limit dictated by the desire not to run the program longer than 24 hours (on an IBM Risc 6000 Computer). The best design obtained had a weight of 25.19 lbs. This compared favorably with a 26.08 lb design obtained by continuous optimization and rounding of ply thicknesses.

Nagendra et al. (1994) revisited the problem with the modified GA. Besides the improvements discussed in the Single-plate Section, they implemented two changes to cater to the existence of several laminates. The single-stack-swap permutation was specialized into two: intralarninate swap with a probability of 0.75, and interlaminate swap with a probability of 0.55. Additionally, a twopoint crossover was implemented separately for each one of the two substrings.

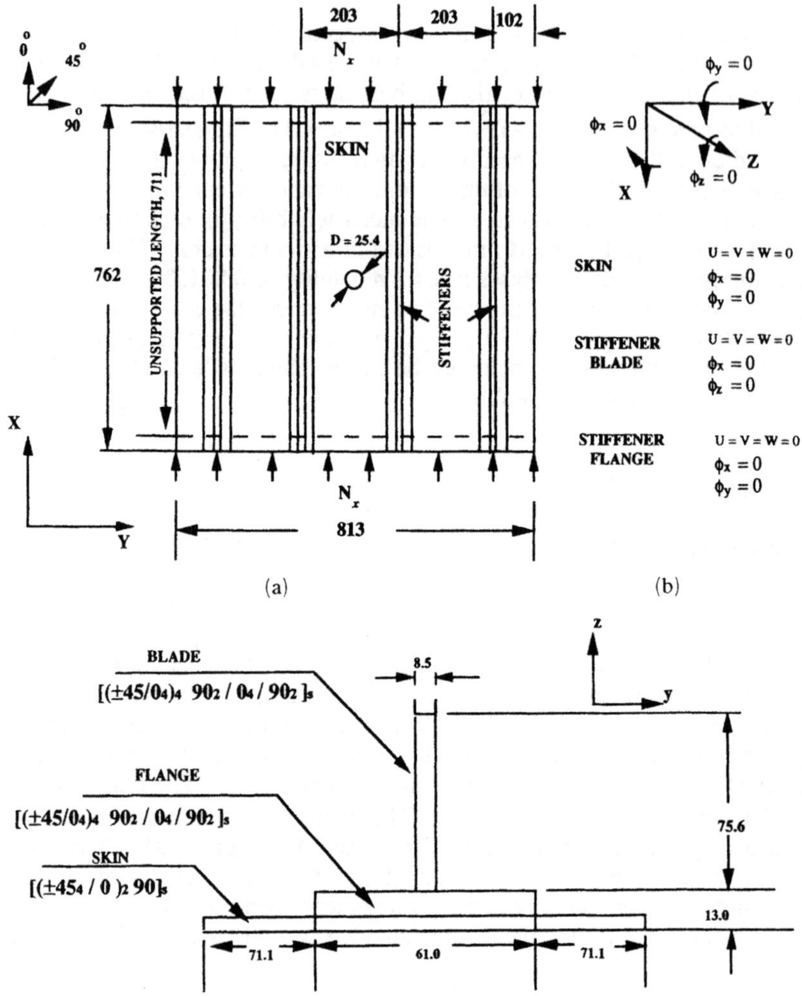

Figure 9: Geometry of blade-stiffened plate

The parameters of the genetic algorithm were optimized by solving first a set of simplified problems with the blade laminate and the blade height fixed. After some experimentation they settled on a probability of crossover of 0.95, and 0.01 for all the mutation probabilities (stack deletion, stack addition, and orientation change). With this parameters the best design found weighed 24.14 lbs, a substantial reduction over the best design obtained with the basic GA.

With a continuous blade height, the weight could vary continuously, so that a practical optimum was defined on the basis of the weight. Two definitions of the practical optimum were used, based on a 1 percent and 2 percent weight difference. With a 2 percent difference, a reliability of about 60 percent was achieved after 3000 analyses. For a 1 percent practical optimum, the same number of analyses produced only about 35 percent reliability.

The genetic optimizations produced 10 designs within 2 percent of the optimal weight with substantial variations in their characteristics. The blade height varied between 2.92 in. and 3.19 in., the skin laminate varied between 32 and 40 plies, and the blade laminate varied between 68 and 80 plies.

4.2 Hat Stiffened Curved Panel

Harrison et al. (1991) applied the GA to the design of a hat stiffened composite panel. The optimization problem addressed in the proposed paper is that of minimum weight design of a hat stiffened curved composite panel (see Figure 10). The panel is a 90° arc of a circular cylindrical panel of radius 120 in. with an axial length of 22 in. and hat stiffeners spaced at 11 in. intervals. As shown in Figure 10, the axial and tangential coordinates are x and y, respectively. The boundary conditions are simple support on all four edges with a combination of normal and shear loadings, N_x = -2400 lb/in and N_{xy} = 160 lb/in. The panel was analyzed with the PANDA2 program by Bushnell (1987). The design variables include the stacking sequences of three laminates and four geometric dimensions.

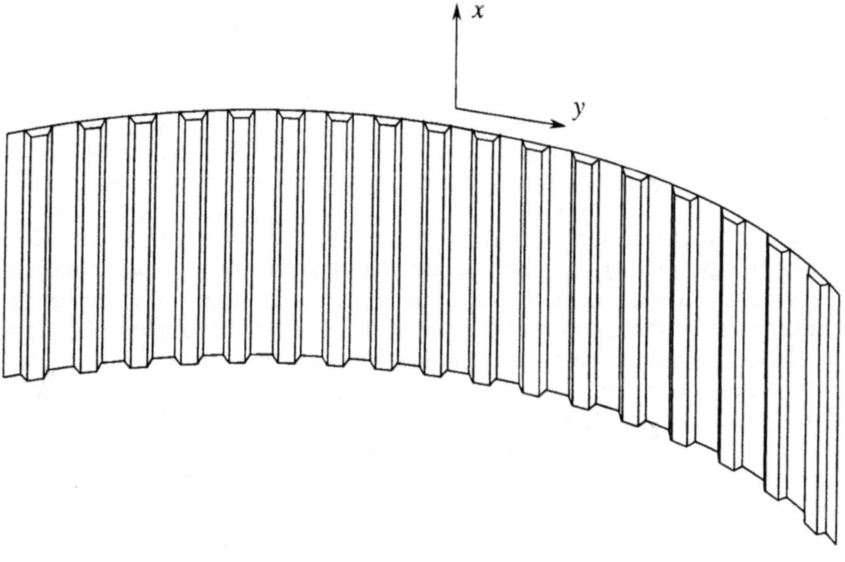

Figure 10: Geometry of hat-stiffened curved panel

The material description of the panel skin and stiffeners require the specification of the layups of three separate sublaminates. Because the laminates can be thinner than in the single-plate example, the plies were not grouped in stacks of two plies. Instead, each of the three sublaminates is encoded as a string with each element in the string being a ply orientation or the letter E for an empty location. To ensure balance without ±45 stacks, the value 45 is used to represent both of these fiber orientations through an alternation scheme. That is, the first occurrence of a 45 in an individual is a +45° ply, the second is a *P*45° ply, the third is a +45°, and so on. Coupled with a penalty for an odd number of occurrences of a 45, this approach leads to balanced laminates without each +45° ply having to be next to a *P*45° ply. As an example of this layup representation scheme, the laminate $[90/0/45/-45/0/45/90]_s$ is represented in a string of length ten as (E E E 90 0 45 45 0 45 90).

The geometry of the hat stiffeners are described by four dimensions. These four continuous variables were coded in a string with four floating point genes rather than as binary or trinary numbers.

Two-point crossover was used for each sublaminate, and an averaging single-point crossover for the substring representing the continuous variables. The averaging single-point crossover generates a random number *r* between 0 and 4. The first *r* genes are taken from the first parent, the *r*+1 gene is averaged as a random convex combination of the corresponding genes of the two parents, and the rest of the genes come from the second parent. For designs that do not violate the buckling and strength constraints, the objective function which is maximized is $(1-\mu)/W$. For design which violate a constraint, the objective function is multiplied by $(1 + m_c)^3$, where m_c is the most critical constraint margin calculated by PANDA2 (negative for constraint violation).

Although PANDA2 uses analysis methods that are closed form where possible, evaluation of a design still requires approximately 45 seconds of CPU on the CONVEX C240 computer used in this study. For a population of 30 individuals, each generation requires around 22 minutes of CPU time. Therefore, it is desirable to minimize the number of generations analyzed by PANDA2 in the search. Harrison et al. (1994) applied response surface methodology (RSM) to approximate the analysis (m_c in particular) of individuals in intermediate generations in an attempt to improve the convergence of this optimization problem in terms of the number of calls to PANDA2. RSM is used to determine relationships between independent variables and the response of a system governed by them. This is accomplished through constructing a quadratic response surface from the results of a series of analyses. The coefficients of the quadratic polynomial are determined through a least squares fit of PANDA2 results for individuals generated in the search. The independent variables used in the response surface are a combination of the four stiffener dimensions and five lamination parameters for each of the three laminates. Therefore, in total there are 19 independent variables, requiring the determination of 210 polynomial coefficients. Therefore, for a population of M

= 30 individuals, at least seven generations must be completed before a response surface can be fitted. However, since the individuals that are used in this fit are somewhat randomly located in the design space, fifteen generations (450 points) were completed prior to the first least squares fit.

Figure 11 shows a convergence comparison of the average of two runs with and without the RSM approximation. It is seen that substantial improvement in cost and weight is achieved.

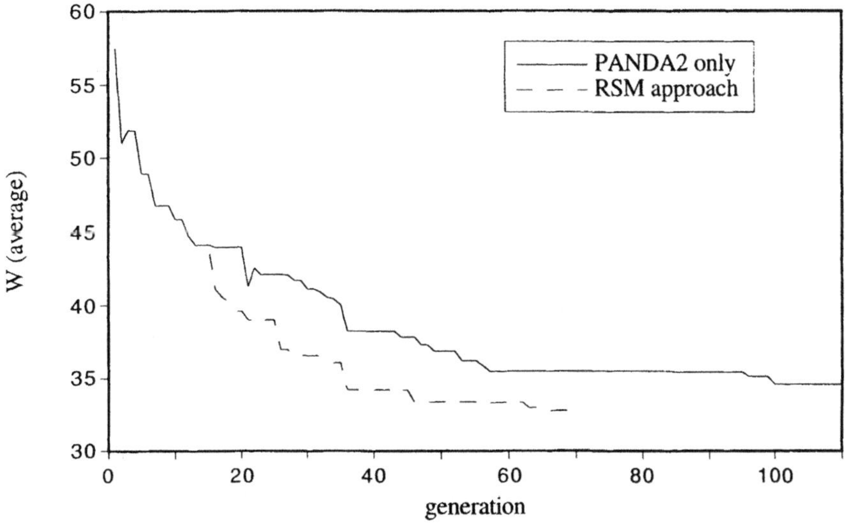

Figure 11: Weight convergence comparison with approximated generations omitted.

5. Concluding Remarks

Genetic algorithms are powerful tools for the design or composite lanlirlates, belt they require a large number of analyses. This paper described several approaches to reducing the cost annul increasing the reliability of the GAs. Specialized operators devised for composite laminates, as well as tuning of the parameters of standard operators can help. A combination of a fixed and variable penalty function formulation was helpful too.

The reliability of the genetic optimization was investigated by large scale testing on an unstiffened single laminate. Improvement in the penalty formulation and the genetic operators were shown to reduce computational cost by more then a factor of two. Additionally, it was shown that to obtain high reliability, multiple short runs may be more effective than a single long run.

Application to blade stiffened plates and hat stiffened curved panels showed that GAs can be applied successfully to more complex problems. For the blade stiffened plate, the lessons learned from the single-plate investigations paid off by leading to designs that were substantially lighter than previously found. For the stiffened curved panel, response surface approximation helped to reduce computational cost.

Acknowledgement

This research was supported by NASA grant NAG1-168.

References

Ball, N.R., Sargent, P.M., and Ige, D.O., "Genetic Algorithm Representations for Laminate Layups", *Artificial Intelligence in Engineering,* Vol.8, No.2, 1993, pp. 99-108.

Bushnell, D., "PANDA2 Program for Minimum Weight Design of Stiffened, Composite, Locally Buckled Panels", *Computers & Structures*, Vol.25, No.4, 1987, pp. 469-605.

Callahan, K.J., Weeks, G.E., "Optimum Design of Composite Laminates Using Genetic Algorithms", *Composite Engineering*, Vol.2, No.3, 1992, pp. 149-160.

Harrison, P.N., Le Riche, R., and Haftka, R.T., "Design of Stiffened Composite Panels by Genetic Algorithm and Response Surface Approximations," to be published.

Kogiso, N., Watson, L.T., Gürdal, Z., and Haftka, R.T., "Genetic Algorithms with Local Improvement for Composite Laminate Design", *Structural Optimization*, Vol.7 (4), pp. 207-218, 1994a.

Kogiso N., Watson L.T., Gürdal Z., Haftka R.T. and Nagendra S., "Minimum Thickness Design of Composite Laminates Subject to Buckling Strength Constraints by Genetic Algorithm", *Mechanics of Composite Materials and Structures,* (to be published), 1994b.

Le Riche R., and Haftka R. T., "Optimization of Laminate Stacking Sequence for Buckling Load Maximization by Genetic Algorithm", *AIAA Journal*, Vol. 31, No. 5, May 1993, pp. 951-970.

Miki, M., "Optimum Design of Fibrous Laminated Composite Plates Subject to Axial Compression", *Proceedings of the 5rd Japan-US Composite Materials Conference,* Tokyo, Technomic Publishing Co., Lancaster, 1979, pp. 1017-1019.

Nagendra, S., Haftka R. T., and Gurdal Z., "Stacking Sequence Optimization of Simply Supported Laminates with Stability and Strain Constraints", *AIAA Journal*, Vol. 30, No. 8, 1992, pp. 2132-2137.

Nagendra, S., Haftka, R.T., and Gurdal, Z., "Design of Blade Stiffened Composite Panel by a Genetic Algorithm," *Proceedings of the 34th Structures, Structural Dynamics and Material Conference*, La Jolla, CA, April 19-21, 1993, pp. 2418-2436.

Nagendra, S., Jestin, D., Gurdal, Z., Haftka, R.T., and Watson, L.T., "Improved Genetic Algorithm for the Design of Stiffened Composite Panels," to be published, 1994.

Richardson, J.T., Palmer, M.R., Liepins, G. and Hilliard, M., "Some Guidelines For Genetic Algorithms with Penalty Functions," *Proceedings of the Third International Conference on Genetic Algorithms*, George Mason Univ., Morgan Kaufmann Publishers, June 4-7, 1989, pp. 191-197.

Shin, Y.S., Haftka, R.T., Watson, L.T., and Plaut, R.T., "Design of Laminated Plates for Maximum Buckling Load", *Journal of Composite Materials*, Vol.23,1989, pp. 348-369.

Stroud, W.J., and Anderson, M.S., "PASCO: Structural Panel Analysis and Sizing Code. Capabilities and Analytical Foundations," NASA TM 80801, November 1981.

Watabe, H., and Okino, N., "A Sudy on Genetic Shape Design", *Proceedings of the Fifth International Conference on Genetic Algorithms*, Univ. of Illinois at Urbana-Champaign, Morgan Kaufmann Publishers, July 17-21, 1993, pp. 445-450.

Foundations of Genetic Algorithms • 2, Whitley, L.D. Editor, Morgan Kaufmann Publishers, San Mateo, California, 1993.

A Genetic Algorithm for Structural Design Optimization

W M Jenkins
Department of Civil Engineering
University of Leeds
Leeds, UK

INTRODUCTION

The genetic algorithm (GA) is proving useful in solving engineering problems requiring a powerful and robust mechanism for searching large design spaces. The algorithm processes a population of individuals (designs) generation-by-generation operating on a set of coded values of data. Relatively fit individuals are combined to produce new individuals. The maximum fitness in the population tends to increase, rapidly at first, until the process reaches an ultimate maximum fitness. This may or may not be the optimum situation. Although the GA is powerful in operation, when applied in the context of structural design, consideration needs to be given to methods of improving the efficiency and speed of operation. The main purpose of this paper is to describe the form in which the algorithm is used and a combinatorial space reduction heuristic as an enhancement.

1 A SPECIFICATION FOR AN ENGINEERING DESIGN OPTIMIZER

A method of optimization suitable for structural design should.

i) be capable of handling highly constrained situations,
ii) be robust and reliable,
iii) not assume that useful gradients exist in the data,
iv) place minimal restrictions on the form of data,
v) be versatile,
vi) have a large scale capability,
vii) have a simple user interface,
viii) be amenable to self-adaptive controls and
ix) be capable of parallel processing to achieve high speed of operation.

The GA is admirably suited for this purpose particularly since it is a naturally parallel algorithm and moreover possesses an implicit parallelism even when used in a serial mode (Goldberg [1],[2],[3]).

2 DESIGN VARIABLES AND FITNESS EVALUATION

Whilst smoothly varying data is frequently found in design, this is not an acceptable generalization and discrete data handling is more realistic, however, a discrete data formulation leads to very large numbers of possible combinations in all but the simplest of design situations. For an engineering design of even modest size, the combinatorial product can be in excess of 1 E100. Searching a large combinatorial space requires special methods. One of the purposes of this paper will be to present a method by which the combinatorial space may be progressively reduced during the processing.

Having decided that design data will be discrete, we must also accept the need to represent

a) single numbers, eg areas of cross-sections,
b) arrays of numbers. eg section properties,
c) non-numeric data, eg yes/no.

In the GA all such data are processed in a general coded form.

In the context of the GA the objective function in an optimization is expressed in terms of a 'fitness' in the form,

$$\text{Fitness} = F = C - y \qquad (1)$$

where $y = f(\mathbf{x})$ is the objective function, $\mathbf{x}$ is a vector of design variables and C is a constant large enough to avoid the production of negative fitnesses. The GA will try to maximize the fitness and thus produce a minimum of the objective function.

3 REPRESENTING INDIVIDUAL DESIGNS

A single complete design in the GA is represented by a bit string of appropriate length. Each variable is represented by a partial bit string and, when concatenated, these constitute a complete design representation (a 'chromosome' in the GA). The situation is shown in Fig 1.

It is well known (Goldberg [1]) that the performance of the GA is enhanced by the formation of 'building blocks' - certain low order, high fitness individuals with short defining lengths. A bit string with certain locations fixed (1 or 0) and the remaining locations free (wild card or don't care) is termed a 'schema'. In the operations of the GA, some schemata will survive and some will not

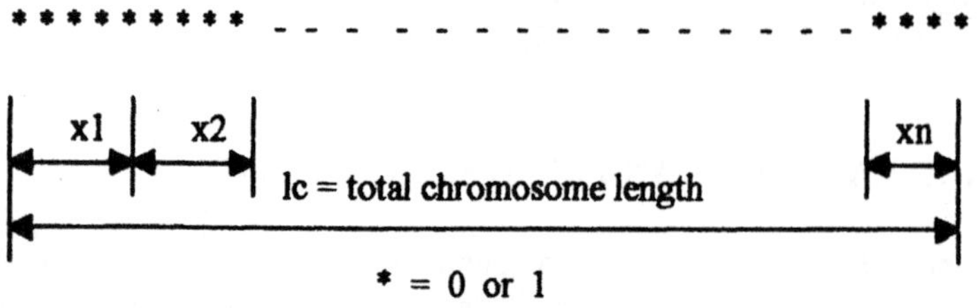

Figure 1

Coded bit string representing an individual design

survive. If the conditions for survival of schemata are analysed it is found that they should exhibit certain properties,

i) should possess above average fitness,
ii) must survive mutation
iii) should be of low 'order' (ie have low specificity of fixed locations)
iv) be of short 'defining length' α. where the defining length is the distance between the outermost defining positions. The probability of survival is a function of the defining length, the total length of the string (L) and the probability of crossover Pc.

$$Ps = 1 - Pc * \alpha / (L - 1) \qquad (2)$$

Such schemata are termed 'building blocks' since they are increasingly reproduced in the population. In structural design we shall be dealing generally with large numbers of design variables, let us say 100 or more. If for example we want to represent something of the order of 60 coded values of data for each variable then a bit string of suitable length would have 6 bits for each variable. This would provide 64 (2^6) coded values for each variable· The total string length would therefore be 600 bits and even longer strings will arise in practical design of large structures, particularly where topological variables are to be represented.. It is difficult to predict the operations of building blocks in such long strings. Furthermore it is unlikely that single point crossover would be effective, clearly some form of multi-point crossover is needed. String partitioning is called for in these circumstances. An intuitively promising approach is to apply these operators to the partial strings representing the individual variables themselves. Now, although we can apply a measure of fitness to a complete string representing all the variables, we cannot do so when we look at a partial string representing one of the variables. However if we nevertheless apply the same fitness measure to all partial strings in an individual we might make progress as the bad strings are progressively replaced by better ones. We could expect that, although the schema survival requirements would not in general be satisfied in respect of high fitness in the early stages of processing, good values would tend to emerge as the processing continues. Once the process was well under way, one could expect to see building block behaviour in operation.

Figure 2 shows a trace on parameter value selections for two variables in a set of 20. After some wide fluctuations in the early generations, the variables settle down quite quickly to their optimum values. A chromosome length of 5 was used for all variables. The variable x_{17} settles on the optimum value (1) at about generation 30. The variable x_1 seems to settle on the near-optimum value of 30 but moves to the optimum (32) towards the end of the processing.

Figure 2 - parameter value selection record

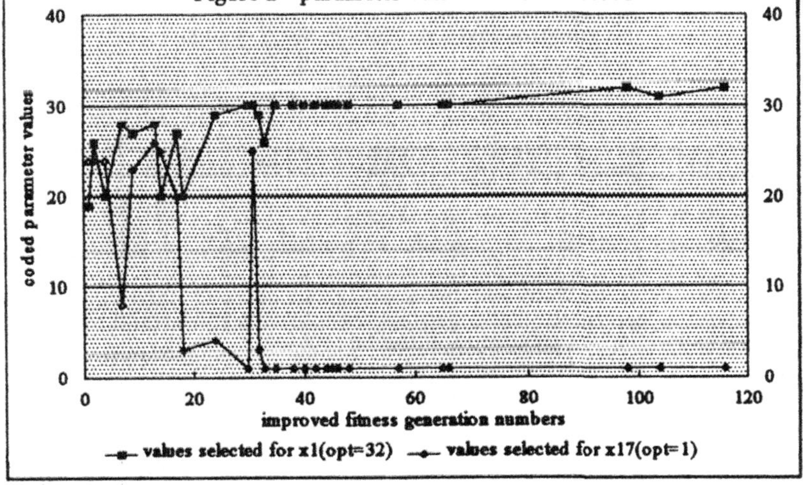

4 SELECTION FOR RECOMBINATION

A straight replacement of two parent strings by two child strings is used. The method of selection of the parents is by stochastic remainder. For each individual the ratio, (fitness / average fitness) is formed. The selection weighting attached to the individual is first the integer part of the ratio increased probabalistically using the fractional part as probability.

Figure 3 shows some alternative approaches to parent selection for recombination. Method A represents simple crossover in concatenated strings and method B represents crossover in sub-strings (variable-by-variable). Method C differs from method B in that parents are selected separately for each variable so we have a multiple parent situation. A comparison of results achieved using methods B and C is shown in figure 4. The chromosome lengths for the 20 variables were taken as [4 4 4 4 4 5 5 5 5 5 5 5 5 5 5 3 3 3 3 3] with the decoded values of the variables equal to the coded values. A population size of 50 was used with pco=1 and pmo=0.005. It is seen that method C reaches the optimum result much more quickly than method B. In the work now described, the selection of parents is carried out individually for each variable so the resulting two children are generated in a multiple parent situation.

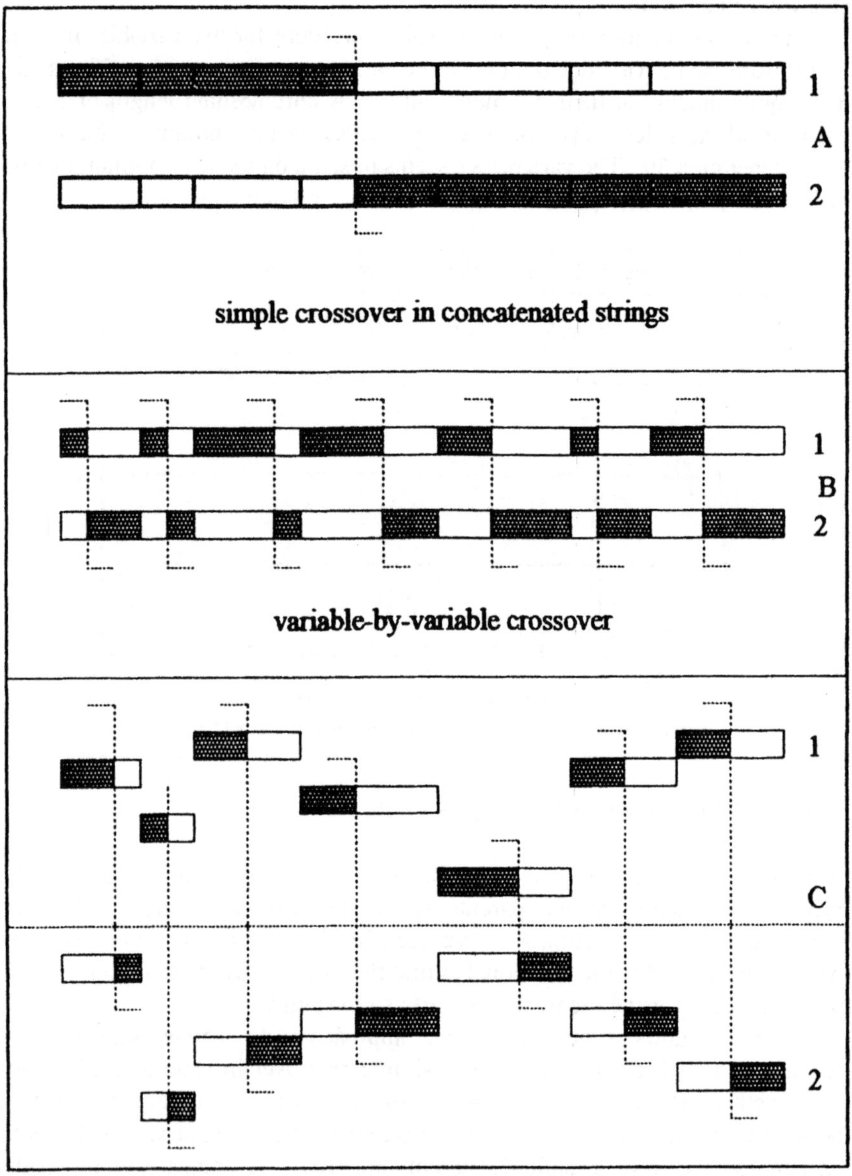

Fig 3
multi-parent crossover in partial strings

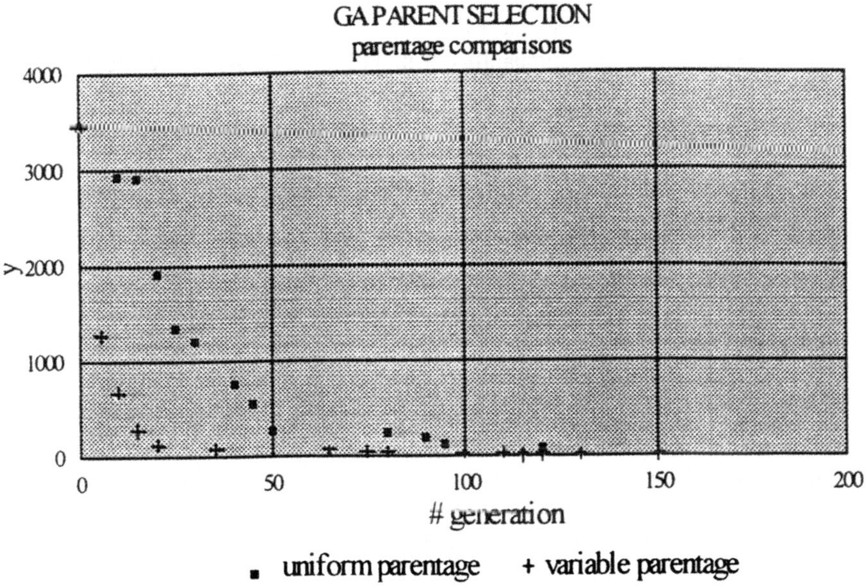

$$y = x_1 x_6 x_{11} x_{16} + x_2 x_7 x_{12} x_{17} + ... + x_5 x_{10} x_{15} x_{20} \; ; \qquad y_{min} = 5$$

Fig 4
comparison of parent selection methods

5 FITNESS SCALING, THE BASIC OPERATORS AND ADAPTIVITY

A linear scaling of fitnesses is carried out in the manner recommended by Goldberg [1]. Relatively high fitnesses are increased and low fitnesses reduced keeping the same average for raw and scaled fitnesses.

The basic GA operators are Pc for crossover and Pm for mutation. It is accepted that in general, Pc should be high and Pm low (Davis [5]). An improvement in operation was obtained when these controls were adapted, variable by variable, according to,

$$Pc[i] = Pco + (1 - Pco) * r \tag{3}$$

$$Pm[i] = Pmo *(1 - r) \tag{4}$$

where	Pc[i]	=	probability of crossover in the string representing variable i
	Pm[i]	=	probability of mutation in the string representing variable i
	Pco, Pmo	=	initial values of control probabilities
	r	=	generation / maximum generation

It is noted that the crossover probability adapts to unity and the mutation probability adapts to zero.

6 APPLICATION OF DESIGN CONSTRAINTS - PENALTY FUNCTIONS

Structural design is usually very heavily constrained. Stresses, displacements, geometry and many other aspects of a structure are subjected to constraints. The application of constraints is perhaps the most difficult area in structural optimization. In the penalty method, the fitness of an individual design is reduced when constraints are violated. From equation (1) we see that the effectiveness of a penalty relates to the maximum value of fitness C rather than to the actual value of the objective function $y = f(\mathbf{x})$. The value of C is to some extent arbitrary but it must be big enough to avoid the production of negative fitnesses. If the penalties are not matched to the values of y encountered, then some heavily penalized individuals can remain in the population. It will help in getting suitable penalty levels if we try to distinguish between significant and insignificant levels of constraint violation. Suppose we have a deflection constraint such that a particular deflection should not be greater than 25 mm. In practice there will be a level of tolerance; perhaps we would apply only a very small penalty to a displacement of 26 mm but a really significant penalty to a displacement of 30 mm. We can build this idea into a penalty function as follows,

$$\text{Penalty} = P = k\,(d)^m \qquad (5)$$

where, d = q/a for **q** > a or a/q for **a** > q
and, a = actual value of the constraint.
q = allowable value of the constraint.

Now we define,

Po = penalty level negligible in comparison with C (eqn (1))
P1 = lowest significant penalty level
do = negligible penalty violation
d1 = lowest level of significant violation.

Substituting in equation (5) we obtain,

$$m = Log(Po/P1) / log(do/ d1) \quad (6)$$

$$\text{and} \quad k = Po / (do)^m \quad (7)$$

As an example, suppose C = 100000 and we decide to take do = 1.01 (ie a 1% violation), d1 = 1.20 (a 20 % violation), Po = C/100, P1 = C/10 then, m = 13.2 and k = 890

A suitable penalty function would then be,

$$P = 900 * (d)^{13} \quad (8)$$

High values of m and low values of k leads to small penalties for minor violations and rapidly increasing penalties as the extent of violation increases. In this case a 39% violation (d=1.39) would apply a penalty equal to C and effectively remove the individual. In multi-constraint design, the constaints k and m are sized separately for each constraint.

7 A COMBINATORIAL SPACE REDUCTION HEURISTIC

7.1 Introduction and the basis of the record

The selection of individuals for processing into the next generation depends on fitness, relatively fit individuals are more likely to be used. Even when operating with comparatively fit individuals, the operations of crossover and mutation cause the GA to visit and re-visit regions of low fitness in the combinatorial space. Such excursions waste the time of the algorithm and it would increase the economy of processing if these regions could be identified and removed. If the algorithm is paused at some particular generation, it will not be possible to identify good and bad parameter values based only on the current data. Although highly fit individuals can be identified, some of the variables will have good (ie near optimum) values and some will have poor values. We cannot make a judgement on this based purely on a current state. If however we keep a record of the parameter values selected we should expect trends to be evident when sufficient processing has been carried out. The record can be more sharply focussed if we record the selections made by the algorithm for those individuals near the maximum fitness positively, and negatively for those individuals with fitnesses near the minimum.

The basis of the record (PSELS = parameter value selections) is shown in Figure 5. The parameter values selected for a fraction (δ_u) of the above average fitness individuals are recorded by adding 1 to the existing record in respect of each parameter value selected. Similarly, the record is updated in respect of the individuals with fitnesses lying in the lower fraction (δ_l) by subtracting 1 from each parameter value selected.

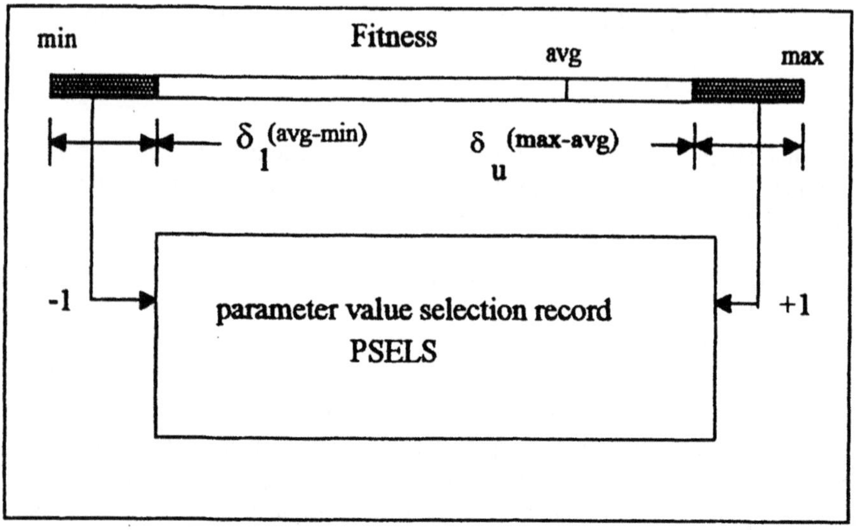

Fig 5
Parameter value selection record

As the processing continues generation-by-generation, the record is expected to indicate any trends in selection patterns and preferred parameter values may become evident. We consider how we might assess the record in order to identify trends.

How might we use the information recorded in PSELS? We shall look first at the principles involved and then examine the process in more detail. At the end of processing in each generation we look for two features in the record. Suppose we consider typical variable x_i. We look first for the parameter value showing the maximum number of selections (MSEL) to see whether this has reached a threshold value λ. We then sort the selections from MSEL downwards and examine the shape of the resulting frequency distribution. If the shape is satisfactory we proceed to condense the parameter value list by retaining the best 50% and removing the rest. This is conveniently achieved by reducing the chromosome length for this variable by one. So a chromosome length of 5 (32 values) will change to 4 (16 values). Eventually the chromosome length will be reduced to 1 (2 values) and, given satisfactory conditions, will then converge to a length 0 ie a single value. If this happens to all the variables then the combinatorial product will reduce to 1 and the process has converged on a single set of values, hopefully at or near the optimum. This is the basic approach to the condensation heuristic, we shall now go on to examine the process in more detail.

7.2 The heuristic in detail

7.2.1 The first threshold

The procedure for updating the selection record is shown in fig. 5. The controls δ_u and δ_l determine the fractions to be recorded at the maximum and minimum ends of the fitness range. As an additional cautionary measure, we decide to adapt these fractions as follows,

$$\delta_u = \delta_{uo} * (avg/max)^2 \qquad (9)$$

and,

$$\delta_l = \delta_{lo} * (1 - (avg/max)^2) \qquad (10)$$

where, δ_{uo} and δ_{lo} are values chosen by the user; avg and max are the current values of average and maximum fitness in the population. In the early generations of a run, all fitnesses can be assumed to be low compared with what is to come. It is therefore appropriate to record a substantial number of the lowest fitness individuals and few if any of the higher fitnesses. In the later stages of the run, all fitnesses will be comparatively high and it will be appropriate to record most of the high fitnesses and few if any of the lower fitnesses. There is some doubt about the use of the ratio (avg/max) in these circumstances since the ratio rises very quickly towards 0.9 and higher, indeed a very high value of (avg/max) is used as a terminating condition for the algorithm. In passing we might mention that the use of a simple ramped adaptation based on the ratio (gen/maxgen) does not improve the performance.

Having decided how we will screen the entries to the selection record made for each variable we now consider how this information will be used. We look first purely at the numbers of selections made and disregard any aspect of distribution. Suppose we stop the process at some particular generation and examine the parameter values selected for variable x_i. If the parameter value showing the maximum number of selections has been selected say λ times then, if λ is big enough, we may have confidence to reduce the chromosome length for this variable by one, retain the best 50% of the values and discard the rest. If for example, lc = 4 for a particular variable, then we have 16 parameter values available to the algorithm. If a condensation is applied, the eight values showing the greatest numbers of selections are retained and the remaining eight values are discarded. Now it is hoped that the optimum value for the variable in question, lies within the retained fraction, if not then it is lost to the process and we shall not reach a global optimum. We may, however, reach a good (near optimum) result nevertheless.

We consider now how we might arrive at a suitable value for the threshold λ. It might be necessary to experiment with trial values of λ but suppose we start with a value of 50. We need to observe that as the chromosome length reduces we shall need to increase λ in order to maintain the same level of significance since the algorithm is now selecting values from a shortened list. The rule for revising λ is,

$$\lambda = \lambda_0 * (2^{lco})/(2^{lc[i]}) \tag{11}$$

where, λ_0 = starting value of λ
lco = initial value of lc for variable i
lc[i] = new value of lc for variable i.

After the first condensation has taken place the value of λ is raised from 50 to 100.

If λ is the only threshold used in the heuristic then we take no account of the distributions of parameter value selections encountered. We can see two extremes, the first where one value predominates completely, no other parameter value being selected. This would be an 'ideal' case. The other extreme occurs when the selection distribution is exactly uniform, ie the number of selections is equal for all parameter values. Clearly we should look at the *distributions* of selecting numbers if we are to develop an appropriate condensation threshold. We shall do this in the next section. We note that each time the space condensation heuristic is invoked, the corresponding part of the record is re-initialized and the counting re-started. An additional , somewhat crude, control can be helpful. The GA can be allowed to run for a stated number of generations γ_o before the record-keeping is started. Any benefit from this device is not yet evident in the work carried out.

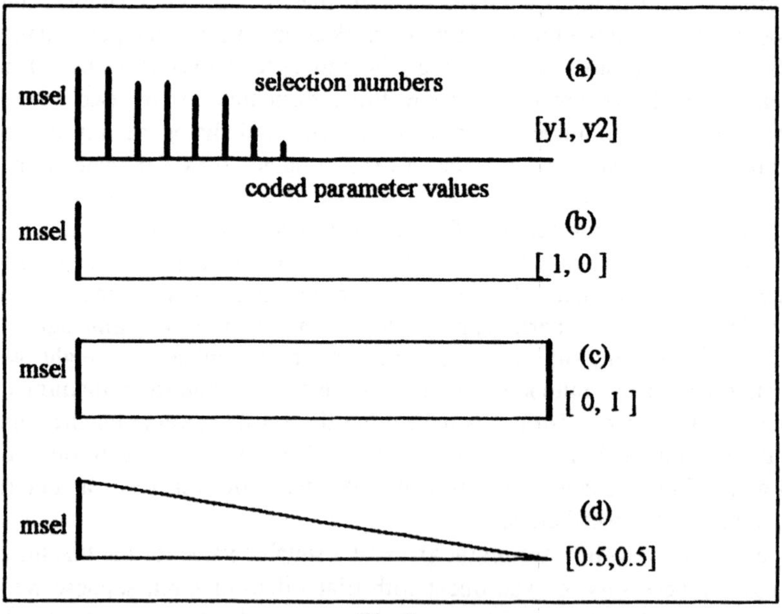

Fig 6
Distributions of parameter value selections showing GRNN outputs

7.2.2 The second threshold

With this threshold we seek to identify the type of distribution of selection numbers which would justify a condensation. We need a classifier and will use a general regression neural network (GRNN) (Wassermann, [16]). Now our parametric data may or may not be monotonically arranged so it would be better to make no assumption about this. Before applying the classifier, we shall carrry out some pre-processing and sort the selection numbers into decreasing numerical order. The corresponding decoded parameter values will be rearranged in the same order. Suppose that the result is a distribution as shown in figure 6 (a) with msel being the maximum number of selections recorded at the current stage. The ordinates in our distribution diagram are simply the numbers of selections made of each of the coded parameter values for the current variable.

The GRNN will provide a binary output $[y_1 , y_2]$ where $y_1 + y_2 = 1$. A distribution as in fig 6 (b) will produce [1 , 0] whereas that in (c) will produce [0 , 1] and (d) will produce [0.5,0.5]. Distribution (b) is the 'ideal' distribution already mentioned where the same parameter value is consistently selected for the variable. Distribution (c) is the uniform selection distribution. If we use these two distributions as training vectors for the GRNN then we observe that they are independent of the abscissae in the data to be presented to the classifier. Now we can improve the action of the classifier if we have regard to the abscissae since we can introduce the influence of positional relationships of the decoded parameter values. Close decoded values should have small relative absicssae, since they are mutually supportive in the classification process, whereas distant decoded values should be spaced further apart. If we are to represent real decoded values of the variables, and hence a non-uniform interval for the abscissae, then the forms of the training functions must be changed so that they are capable of distinguishing the abscissae in the data presented to the classifier. It is clear from these considerations that we should use *differences* in the decoded values of the variables as abscissae (X) when plotting a distribution. The following form has been used,

$$X_i = \sum_{i=1}^{j} ABS(x_{i+1} - x_i); \quad j=2,3 \ldots nv \tag{12}$$

where. $X_1 =$ 0;

$x_1 =$ decoded parameter value corresponding to the maximum selection,

$x_i =$ decoded parameter value corresponding to the i th item in the sorted list of selection numbers.

nv = number of parameter values available for selection (= 2^{lc})

The values X are then normalized by division by the maximum value (X_{nv}) and the resulting abscissae range from 0 to 1. Similarly the ordinates are normalized by division by msel.

A result of this is that parameter values numerically 'near' to the preferred value would have small abscissae near the origin. If the numbers of selections recorded for these 'near' values are similar to the maximum. then the form of the training function must be capable of recognizing this. A suitable function for the training vector u_1 is therefore that shown in figure 7,

$$u_1 = EXP(-100 * (X[i])^3) \qquad (13)$$

where it will be seen that a cluster of popular, near parameter values will guide the heuristic towards a condensation. Some consideration also needs to be given to the shape of the training function u_2.

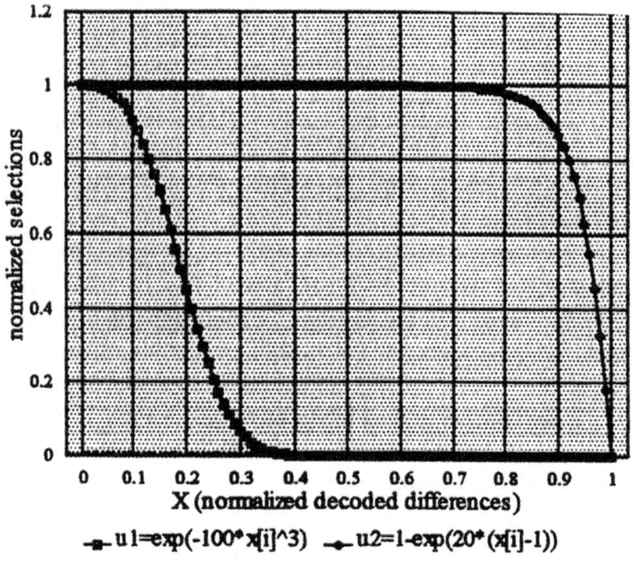

Fig 7
Training vectors for GRNN classifier

We need to provide for the situation where a cluster of values remote from the maximum might effectively trigger the heuristic. We would not wish this to happen so should therefore choose a function such as u_2, fig 5, which will ensure that a remote cluster will deter the heuristic. The function used is,

$$u_2 = 1 - EXP(20 * (X[i] -1)) \qquad (14)$$

Clearly the precise forms of these functions can be changed without necessarily affecting the action of the heuristic to a great extent. There is scope for further consideration here.

The heuristic can be 'calibrated' by observing that the ideal distribution, fig 4(b) should satisfy, hence a simple run of the network offering this distribution can be used as a basis for setting thresholds for invoking a condensation. On this basis the results in table 1 were obtained.

		$[y_1, y_2]$			
lc	nv	high quality		low quality	
2	4	0.69	0.31	0.56	0.44
3	8	0.88	0.12	0.68	0.32
4	16	0.99	0.01	0.86	0.14
5	32	1	0	0.98	0.02
6	64	1	0	0.99	0.01

Table 1: GRNN classifier thresholds

7.3 Employing the heuristic

The record of parameter values selections (PSELS) is examined after completing the processing of each generation. The record is examined variable-by-variable. If the first threshold, based purely on the number of selections made for the preferred value, is reached then the selection distribution is tested with the GRNN. If the distribution threshold is reached then a condensation is invoked, the chromosome length for the variable is reduced by one, the best 50% of the values are retained and the rest discarded. The record for the variable is then re-initialized and the process moves onto the next variable.

It might be helpful to review the GA controls developed so far and to consider what values we might use in practice. We distinguish notionally between 'high' and 'low' quality solutions. It should be said that 'high' and 'low' are used more to indicate 'slow' and 'fast' processing rather than implying a guarantee of the quality of the result. We commence with the basic GA operators for crossover and mutation along with population size and maximum number of generations to be processed. In order to simplify the current investigation the probabilities for crossover and mutation were standardized. The crossover probability was set at 1 throughout. Since 1 was the accepted final value, no adaptation was carried out. The mutation probability was given an initial value of 0.005 and linearly adapted to zero at the maximum generation number. The population size was fixed at 50 uniformly. The maximum generation number was taken generally as 2000 it being expected that one of the terminal conditions would apply before this number was reached. The terminating conditions were.

1) gen = maxgen (2000)
2) combinatorial product = 1
3) average fitness/maximum fitness >0.99999
4) no individual with positive fitness in the population.

Condition 3) indicates convergence (not necessarily to the optimum result). Condition 4) can occur in heavily constrained circumstances and can be handled by easing some or all of the constraints in the early stages of the processing as a temporary measure.

The controls just described may be regarded as 'fixed' for the purposes of this exercise. We turn now to the controls to be applied as options. We have already observed that it is possible to plan for 'high' or 'low' quality processing. The controls considered are,

γ_o = the minimum generation number for commencing record keeping
λ_o = the first threshold for the GRNN = number of selections of preferred value
δ_{uo} = upper fraction of individuals for positive recording,
δ_{lo} = lower fraction of individuals for negative recording
y_l = GRNN output (distribution threshold).

Suggested typical values of the controls are shown in Table 2. It should be emphasized that considerable exploratory work is yet to be carried out before firm recommendations can be made on this matter.

control	Low quality	High quality
γ_o	0	200
δ_{uo}	0.5	0.1
δ_{lo}	0.1	0.2
λ_o	50	100
y_l (lc = 4)	0.86	0.99

Table 2: Suggested typical control values

8 ILLUSTRATIVE EXAMPLES

8.1 A 'deceptive' problem

In the development of efficient forms of the algorithm considerable attention has been directed to its performance in 'deceptive' situations. These are situations where the algorithm is attracted towards sub-optimal values of the variables whilst the optimal values occupy more remote positions in the list of parameter values. One such situation has been studied, Goldberg et al [18], Mahfoud &

Goldberg [19] and it might be useful to see what the current approach with the space condensation heuristic can make of these circumstances. A three-bit string was used in the study, the coded and decoded values are given in table 3.

As in the original work by Goldberg, Korb & Deb, the problem is taken to be the maximization of the sum of ten variables taken from the decoded set. The optimum result is therefore 300 and it is noticed that the optimum value of 30 for each variable is countered by a near-optimum value, 28 at the remote end of the list. It is to be expected that the algorithm will be attracted to the 000 string end of the list thus producing a sub-optimal result. The results obtained from 5 runs of the GA with and without the heuristic are shown in table 4. The population size was 50 throughout, pco = 1, pmo = 0.005. With the heuristic active, $\gamma_0 = 0$; $\lambda_0 = 50$; $\delta_{uo} = 0.05$; $\delta_{lo} = 0.15$.

string	coded	de-coded
000	1	28
001	2	26
010	3	22
011	4	0
100	5	14
101	6	0
110	7	0
111	8	30

Table 3
A deceptive coding of a three-bit string

run	without heuristic			with heuristic			
	ymax	#gen	y	ymax	#gen	nnconds	y
1	282	21	1@30+9@28	300	106	20	10@30
2	286	14	3@30+7@28	300	99	20	10@30
3	284	26	2@30+8@28	300	72	10	10@30
4	288	31	4@30+6@28	300	89	13	10@30
5	286	15	3@30+7@28	298	78	30	9@30+1@28

Table 4
Results from deceptive problem with ten variables

Without the heuristic, the best run (4) finds only 4 of the 10 optimum values (30). With the heuristic invoked, the GA finds the optimum in four of the five runs and in the fifth run, 9 of the 10 optimum values are found. In all cases the maximum generation number was set at 1000. In the runs without the heuristic the GA quickly got into a 'rut' with no progress being made after a comparatively low generation number, nevertheless the algorithm was allowed to proceed to generation 1000. Clearly only a chance mutation would jerk the algorithm out of its slumber. With the heuristic operating it is much easier to prescribe terminating conditions so the runs were much shorter. In column 7 of table 4 we have the number of neural network condensations carried out at the corresponding generation number. With 10, three-bit strings, 30 condensations would reduce the combinatorial product to 1. We note that the original combinatorial product is 8^{10} = 1.07 E9.

This problem was extended to 100 variables to see what the heuristic might achieve. The combinatorial product is now 8^{100} = 2.04 E90 and the optimum result 3000. With the same initial control values, the results in Table 5 were obtained.

Run	without heuristic		with heuristic (low quality)		
	ymax	#gen	ymax	#gen	nn conds
1	(2802) 2808	(200) 979	2944	440	200
2	2802	341	2944	407	293
3	2810	598	2948	408	294
4	2816	512	2966	433	194
5	2814	639	2946	369	200

Table 5
Deceptive problem with 100 variables

Here 300 condensations would lead to a unit combinatorial product. All runs were set for a maximum generation number of 1000. Generally the runs with the heuristic terminated shortly after the best result was reached the terminating condition being, in all cases, a high ratio of average fitness/maximum fitness. The best result without the heuristic is y = 2816 (= 8@30+92@28). The best result with the heuristic is y = 2966 (=83@30+17@28), the heuristic enabling the GA to find 83% of the optimum values along with 17% of the nearest sub-optimal values. The best result without the heuristic contains 8% at the optimum value and 92% at the nearest sub-optimal value.

It is clear that one of the effects of the heuristic is to achieve a progressive more helpful re-coding of the binary strings as these are shortened.

8.2 A structural problem

Some structural optimization studies using the GA have been reported (Jenkins [9],[10],[11],[14]). Here we shall look at a simple problem of finding the minimum volume of material for a trussed arch structure. One symmetrical half of the structure is shown in figure 8. The data file for this exercise is shown in Appendix 1.

The cross-sectional areas of the members (mm^2) were taken from standard tables of square hollow sections (SHS). The optimum sections are shown in bold type. Some 'near-optimum' values occur in the table and these

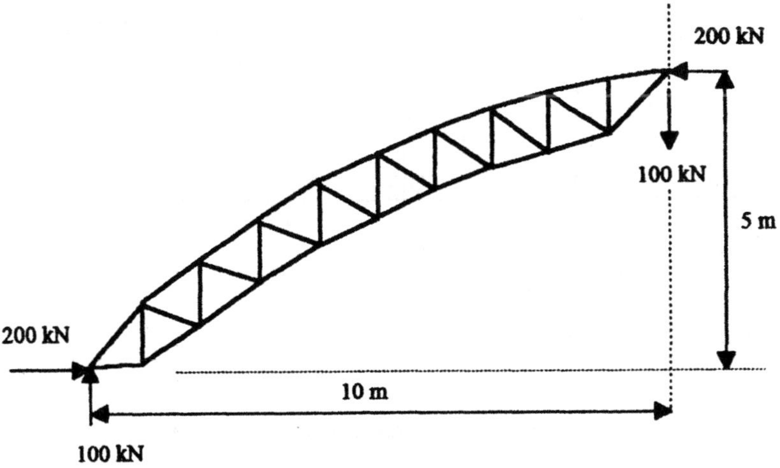

Fig 8
one-half of symmetrical arch

are underlined. The initial chromosome length is taken as 4 (16 discrete values) for each of the 37 members so the initial area data is constructed in an array having 37 rows and 16 columns. A single design constraint was applied to the stress in each member which was taken as 200 N/mm^2 maximum, irrespective of whether the member was in tension or compression. The minimum volume of material in the one-half structure is 27028 cm^3. The results obtained using the genetic algorithm are shown in table 6. The distinction between 'high' and low quality processing was due to the change in GRNN distribution thresholds and the values of δ_{uo}, δ_{lo} and λ_o. All other controls were the same.

In handling the constraints, an option is provided to observe a tolerance. This would generally be the case in design. If constraints are to be applied rigorously then a zero tolerance is selected. The results in table 6 include 0% and 5%

tolerances. It is interesting that when a tolerance is offered the algorithm exploits the situation and can produce an improvement on the optimum result.

To illustrate the operation of the heuristic, the updated list of parameter values at generation #1262 in run 1 is shown in Appendix 2. The bold values are optimal and it is seen that all optimal values remain available to the algorithm at this stage - in fact the global optimum is reached in this generation. We note also that the optimal parameter values generally occupy the preferred positions in the lists.

Run	without heuristic		with heuristic					
			low quality 0% tolerance		high quality 0% tolerance		high quality 5% tolerance	
	#gen	y_{min}	#gen	y_{min}	#gen	y_{min}	#gen	y_{min}
1	998	30344	476	27639	1262	27028	476	26615
2	260	30754	430	27126	1177	27063	529	26807
3	570	30973	432	27140	1214	27028	465	26930
4	458	31155	417	27202	971	27050	401	26703
5	811	31224	427	27140	1128	27111	713	27088
Avg.	619	30890	436	27249	1150	27056	517	26829

$y_{opt} = 27028$; popsize = 50; maxgen = 2000;
low quality settings: $\gamma_0 = 0$; $\lambda_0 = 50$; $\delta_{uo} = 0.1$; $\delta_{lo} = 0.1$
high quality settings : $\gamma_0 = 0$; $\lambda_0 = 10$ $\delta_{uo} = 0.05$ $\delta_{lo} = 0.1$
initial combinatorial product = 3.6 E44

Table 6: Results from trussed arch optimization

9 TOWARDS AN INTEGRATED OPTIMUM DESIGN ENVIRONMENT

The study of optimization methods in isolation from the environment of practical design is unsatisfactory. Insistence on an absolute global optimum is unnecessary, however, we need bench marks to guide us in an assessment of the nearness of a design to the optimum. For this reason regard must be had to the ability of an optimizer to reach a near global optimum - whatever 'near' means! When using the GA as an optimizer, some confidence can be generated by running the algorithm several times from different starting points. Some judgement is needed here. Suppose we run the algorithm ten times from ten different starting points and find that the ten results are within a range of 5% of the best result. Although uncertainty remains, in practical design circumstances, bearing in mind the uncertain nature of some design assumptions, we could probably accept the result.

A serious problem with all structural optimization is the need for re-analysis of the structure during the process. A study of a cable-stayed bridge using the GA (Jenkins [11]) required 500,000 re-analyses of the structure and these were all 'exact'. This was a special case but it high-lights the problem. Generally the structural re-analysis will require more processing time than the optimizing algorithm.

The most promising approach to structural re-analysis lies in the application of a neural network to approximate the re-analysis (Rogers [17]). The NN approach is clearly feasible for integration with a GA optimizer for structures of modest size. It remains to be seen whether the approach can be scaled-up to handle large structures. A possible organizational framework is shown in figure 9.

10 OBSERVATIONS AND CONCLUSIONS

1) The space condensation heuristic assists the GA in finding an improved result and in reducing the processing time needed.

2) Adaptivity of controls also assists the algorithm but appears to be less important than 1).

3) Constraint handling using a simple penalty function is satisfactory and can include a practical tolerance in use. If a non-zero tolerance is specified, the algorithm may exploit this and produce a result beyond the optimum.

4) The form of GA adopted here will handle unsmooth, or even randomly ordered, data.

5) The built-in controls allow the user to choose a 'slow', high quality solution or a 'faster', low quality solution. Some experimentation is needed before making a final choice of controls.

6) The high quality settings reach a virtual optimum in every run but at considerable increase in processing time.

7) The low quality settings give 'near' optimum results sufficient for practical purposes.

8) The purpose of the first heuristic threshold is to avoid premature condensation, the second threshold, based on the distribution shape, is the main decision threshold.

9) The combined use of GA and NN opens up some exiting possibilities for further development.

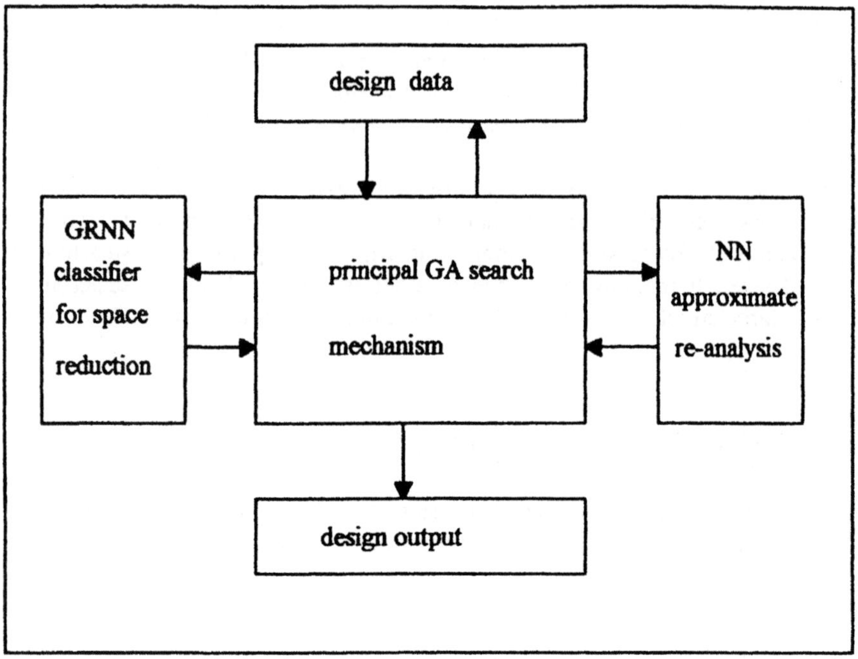

Fig 9
Outline structure for integrated design environment

Acknowledgements

The author expresses his appreciation of the advice and encouragement offered by Professor David E Goldberg. The cooperation of colleagues in the Department of Civil Engineering in the University of Leeds is appreciated as is the interest shown by Professor B H V Topping and colleagues at Heriot-Watt University.

References

1. Goldberg, D.E. (1989(a)) Genetic Algorithms in Search, Optimization and Machine Learning. Adison Wesley
2. Goldberg, D.E. (1989(b)) Sizing Populations for Serial and parallel Genetic Algorithms. Proceedings of the Third International Conference on Genetic Algorithms
3. Goldberg, D.E. (1989(c)) Zen and the art of genetic algorithms. Ibid
4. Maffioli, F. The complexity of combinatorial optimization algorithms and the challenge of heuristics. Chapter 5, Combinatorial Optimization. Eds N Christofides et al. John Wiley & Sons, 1979 pp 107 - 129

5. Davis, L. (1989) Adapting Operator Probabilities in Genetic Algorithms. Proceedings of the Third International Conference on Genetic Algorithms. Morgan Kaufmann Publishers Inc
6. Fogarty, T.C. (1989) Varying the probability of mutation in the genetic algorithm. Ibid
7. Hesser, J., Manner, R. (1991) Towards an optimal mutation probability for genetic algorithms. Lecture Notes in Computer Science 496 - Parallel problem solving from nature (PPSN1) Dortmund FRG October 1990. Springer-Verlag 1991
8. Back, T., Hoffmeister, F., Schwefel, H.P. (1991) A Survey of Evolution Strategies. Proceedings of the fourth International Conference on Genetic Algorithms. Morgan Kaufmann Publishers
9. Jenkins, W.M. (1991(a)) Towards Structural Optimization via the Genetic Algorithm. Computers & Structures Vol 40 No 5 pp 1321-1327
10. Jenkins, W.M. (1991(b)) Structural Optimization with the Genetic Algorithm. The Structural Engineer Vol 69 No 24. December 1991. Institution of Structural Engineers, London
11. Jenkins, W.M. (1992) A Plane Frame Optimum Design Environment based on the genetic algorithm. Journal of Structural Engineering, Proceedings of the American Society of Civil Engineers, Vol 118 No 11 1992, pp 3103-3112
12. Rajeev, S., Krishnamoorthy, C.S. Discrete optimization of structures using genetic algorithms. Journal of Structural Engineering, Proceedings of the American Society of Civil Engineers. Vol 118, No 5, May 1992, pp 1233-1250
13. de Jong, K.A., Spears, W.M. A formal analysis of the role of multi-point crossover in genetic algorithms. Annals of Mathematics and Artificial Intelligence 5 (1992) 1 - 26
14. Jenkins, W.M. (1993) An enhanced genetic algorithm for structural design optimization. 5th International Conference on Civil & Structural Engineering, Heriot-Watt University. CIVIL COMP Press, Edinburgh, August 1993
16. Wasserman, P.D. Advanced Methods in Neural Computing. Van Nostrand Reinhold, New York, 1993
17. Rogers, J.L. Simulating structural analysis with Neural Networks. Journal of Computing in Civil Engineering. ASCE Vol 8 No 2 April 1994
18. Goldberg, D.E., Korb, B., Deb, K Messy genetic algorithms : motivation, analysis and first results. TCGA Report No 89003, May 1989. The Clearing House for genetic algorithms, University of Alabama
19. Mahfoud, S.W., Goldberg, D.E. A genetic algorithm for parallel simulated annealing. Parallel Problem Solving from Nature, 2 1992. Elsevier Science Publishers. B.V

APPENDIX 1
Data file for trussed arch optimization

decoded parameter values (areas mm²) *

372	440	466	568	688	472	**560**	594	728	888	1080	680	722	888	1090	1330
372	440	466	568	688	472	560	594	728	888	1080	680	722	888	1090	1330
372	440	**466**	568	688	472	560	594	728	888	1080	680	722	888	1090	1330
372	440	466	568	688	472	**560**	594	728	888	1080	680	722	888	1090	1330
372	440	466	568	688	472	**560**	594	728	888	1080	680	722	888	1090	1330
372	440	466	568	688	472	560	594	728	888	1080	680	722	888	1090	1330
372	440	466	568	688	472	560	594	728	888	1080	680	722	888	1090	1330
372	440	466	568	688	472	560	594	728	888	1080	680	722	888	1090	1330
372	440	466	568	688	472	**560**	594	728	888	1080	680	722	888	1090	1330
372	440	466	568	688	472	**560**	594	728	888	1080	680	722	888	1090	1330
920	1090	1490	1840	2270	1240	1690	2090	2590	1530	1890	2340	2910	3550	2290	2850
920	1090	1490	1840	2270	1240	**1690**	2090	2590	1530	1890	2340	2910	3550	2290	2850
920	1090	1490	1840	2270	1240	**1690**	2090	2590	1530	1890	2340	2910	3550	2290	2850
920	1090	1490	1840	2270	1240	**1690**	2090	2590	1530	1890	2340	2910	3550	2290	2850
920	1090	**1490**	1840	2270	1240	1690	2090	2590	1530	1890	2340	2910	3550	2290	2850
920	1090	1490	1840	2270	**1240**	1690	2090	2590	1530	1890	2340	2910	3550	2290	2850
920	1090	1490	1840	2270	1240	1690	2090	2590	1530	1890	2340	2910	3550	2290	2850
372	440	466	568	688	472	560	594	728	888	1080	**680**	722	888	1090	1330
372	440	466	568	688	472	**560**	594	728	888	1080	680	722	888	1090	1330
372	**440**	466	568	688	472	560	594	728	888	1080	680	722	888	1090	1330
372	440	466	568	688	472	560	594	728	888	1080	680	722	888	1090	1330
372	440	466	568	688	472	560	594	728	888	1080	680	722	888	1090	1330
372	440	466	568	688	472	560	594	728	888	1080	680	722	888	1090	1330
372	440	466	568	688	472	560	594	728	888	1080	680	722	888	1090	1330
372	440	466	568	688	472	560	594	728	888	1080	680	722	888	1090	1330
372	440	466	568	688	472	560	594	728	888	1080	680	722	888	1090	1330
372	440	466	568	688	472	560	594	728	888	1080	680	722	888	1090	1330
372	440	466	568	688	472	560	594	728	888	1080	680	722	888	1090	1330
372	440	466	568	688	472	560	594	728	888	1080	680	722	888	1090	1330
372	440	466	568	688	472	560	594	728	888	1080	680	722	888	1090	1330
372	440	466	568	688	472	560	594	728	888	1080	680	722	888	1090	1330
372	440	466	568	688	472	560	594	728	888	1080	680	722	888	1090	1330
372	440	466	568	688	472	560	594	728	888	1080	680	722	888	1090	1330
372	440	466	568	688	472	**560**	594	728	888	1080	680	722	888	1090	1330
372	440	466	568	688	472	560	594	728	888	1080	680	722	888	1090	1330
372	440	466	568	688	472	560	594	728	888	1080	**680**	722	888	1090	1330

* optimum values are in bold type, near optimum values are underlined

APPENDIX 2
PARAMETER SELECTION RECORD - PSELS
Updated lists of parameter values at Generation #1262

parameter number

parm\ value/	1	2	3	4	5	6	7	8	9	10	11	12	13	14	15
1	560	372	472	560	560	372	372	372	560	560	920	1490	1690	1690	1690
2	728	594	466	688	594	466	440	472	594	594	2290	1690	1840	1840	1530
3	0	0	722	0	568	0	0	0	0	568	0	1530	3550	1890	0
4	0	0	560	0	680	0	0	0	0	680	0	2590	2590	3550	0
5	0	0	1090	0	0	0	0	0	0	0	0	1090	0	0	0
6	0	0	440	0	0	0	0	0	0	0	0	1240	0	0	0
7	0	0	680	0	0	0	0	0	0	0	0	2910	0	0	0
8	0	0	568	0	0	0	0	0	0	0	0	3550	0	0	0
9	0	0	0	0	0	0	0	0	0	0	0	0	0	0	0
10	0	0	0	0	0	0	0	0	0	0	0	0	0	0	0
11	0	0	0	0	0	0	0	0	0	0	0	0	0	0	0
12	0	0	0	0	0	0	0	0	0	0	0	0	0	0	0
13	0	0	0	0	0	0	0	0	0	0	0	0	0	0	0
14	0	0	0	0	0	0	0	0	0	0	0	0	0	0	0
15	0	0	0	0	0	0	0	0	0	0	0	0	0	0	0
16	0	0	0	0	0	0	0	0	0	0	0	0	0	0	0

parameter number

parm\ value/	16	17	18	19	20	21	22	23	24	25	26	27	28	29	30
1	1530	1240	920	680	560	440	372	372	372	372	372	372	372	372	372
2	1490	1840	1690	722	568	466	466	440	440	440	466	440	472	440	466
3	1690	0	0	0	680	560	440	0	0	0	0	0	440	472	0
4	2270	0	0	0	722	472	472	0	0	0	0	0	1090	680	0
5	0	0	0	0	0	0	594	0	0	0	0	0	0	568	0
6	0	0	0	0	0	0	680	0	0	0	0	0	0	466	0
7	0	0	0	0	0	0	1330	0	0	0	0	0	0	1080	0
8	0	0	0	0	0	0	728	0	0	0	0	0	0	1330	0
9	0	0	0	0	0	0	0	0	0	0	0	0	0	0	0
10	0	0	0	0	0	0	0	0	0	0	0	0	0	0	0
11	0	0	0	0	0	0	0	0	0	0	0	0	0	0	0
12	0	0	0	0	0	0	0	0	0	0	0	0	0	0	0
13	0	0	0	0	0	0	0	0	0	0	0	0	0	0	0
14	0	0	0	0	0	0	0	0	0	0	0	0	0	0	0
15	0	0	0	0	0	0	0	0	0	0	0	0	0	0	0
16	0	0	0	0	0	0	0	0	0	0	0	0	0	0	0

parameter number

parm\ value/	31	32	33	34	35	36	37
1	372	372	372	372	568	372	680
2	440	466	472	440	594	466	688
3	0	0	0	0	560	0	888
4	0	0	0	0	722	0	594
5	0	0	0	0	0	0	0
6	0	0	0	0	0	0	0
7	0	0	0	0	0	0	0
8	0	0	0	0	0	0	0
9	0	0	0	0	0	0	0
10	0	0	0	0	0	0	0
11	0	0	0	0	0	0	0
12	0	0	0	0	0	0	0
13	0	0	0	0	0	0	0
14	0	0	0	0	0	0	0
15	0	0	0	0	0	0	0
16	0	0	0	0	0	0	0

Trussed arch optimization - run 1
popsize = 50 ; maxgen = 2000
pco=1; pmo=0.005
optimum objective = 27028 reached in this generation.
Number of condensations = 93/148
'high' quality processing
0 % constraint tolerance

Bold parameter values are optimum.
0 indicates removal of value

GENETIC ALGORITHMS IN OPTIMAL DESIGN OF CIVIL ENGINEERING STRUCTURES

V.K. Koumousis

Institute of Structural Analysis & Aseismic Research,
National Technical University of Athens,
Zografou Campus, 157 73 Athens, Greece

Abstract. Two distinct structural design problems of combinatorial nature, are presented treated by genetic algorithms. The first problem refers to the topology optimization of plane trusses focusing on representation and analysis methods suitable for genetic algorithms, and the other to the multi-criterion optimal detailing of Reinforced Concrete (R/C) beams. The genetic algorithm is based on a roulette-wheel reproduction scheme, single point crossover and constant or variable mutation schemes. An elitist strategy is also used that passes the best designs of a generation to the next generation. The optimal design of the truss topology problem is based on a genetic algorithm for the layout problem and a fully stressed design of the emerging configurations. The analysis and optimization are performed on the basis of the recursive formulation of the redundant force method. For the detailing problem, the method decides the design on the basis of a multi-criterion objective that represents a compromise between a minimum weight design, a maximum uniformity and the minimum number of bars for a group of members. By varying the weighting factors, designs with different characteristics result. The anchorage lengths are taken into account and the bars are cut at appropriate regions. Various parameters of the genetic algorithm are considered which render the method as an efficient optimization method for discrete structural design problems.

1. Topology Optimization of Plane Trusses

A statically determinate truss (SDT) having N nodes must have 2N-3 members, while the fully connected truss of N nodes will have N(N-1)/2 members. The design space of the problem may contain statically determinate configurations and statically indeterminate ones ranging from degree 1 to A=N(N-1)/2-2N+3. Among the different configurations that can be generated, there may be cases where a node is connected with only one member resulting an unstable structure. In addition, there may be other configurations that correspond to mechanisms not of finite mobility but of infinitesimal one, i.e. to kinematically unstable structures. In the literature, there exist several methods employed to identify these configurations from a kinematical point of view. Most of these methods have a special interest to tensegrity structures and generally to statically and kinematically indeterminate structures [1], [2], [3]. In the proposed approach we rely on a constructive method that by definition eliminates these configurations, provided that the reference truss belongs to the class of simple statically indeterninate trusses, i.e. trusses that define hierarchies, containing configurations from statically determinate ones to the indeterminacy of the reference truss, that involve all nodes of the initial ground structure. Thus, an identifier will be employed to check constructively the kinematical stability of these trusses. The process of eliminating not only members but also nodes, which is common in most topology optimization problems, requires a broader identifier of kinematical stability, and seems more proper to go together with a process that can add nodes to the structure, that is a problem addressed in innovative design.

1.1 First coding method

Frequently, the operations used in genetic algorithms do not have a direct structural significance. Both processes of crossover and mutation can lead to complications that affect the efficiency of the genetic algorithm and therefore relays greately on the successful representation of the problem and the transitions schemes imposed by genetic operations. For example, during a crossover operation, two parents may exchange parts of their chromosomal information and produce structures that may not sustain loads, i.e. mechanisms, or kinematically unstable structures. This can be checked either immediately and resolved by giving the chance to the same parents to produce "healthy offsprings" or to check the complete population and add at the end, members from randomly selected parents, until the complete population of a specific size is formed. Another option is to carry on the process having a variable population size.

Having a fully connected structure of M members, or a reference truss, we label all the members and we assign a binary string of length M full of units to represent this truss, i.e. the ground structure of our problem. All the other possible configurations may have zeros and ones in different positions. A zero meaning that the member is absent while a unit means that the member is present. The same coding was used also by Hajela [4] to solve the more general truss topology problem following a two stage approach based on stiffness method. This coding is not a complete one, in the sense that cannot cover all binary strings from all zeros to all units, because at least 2N-3 units must be in the binary string to represent at least a statically determinate problem.

Considering this binary string coding, the reproduction scheme of the genetic algorithm will suggest the trusses that appear in a population. The part of population that corresponds to kinematically unstable structures must be removed. This infant mortality of the population can be checked in two ways. First, the number of ones in the string is counted. If is less than 2N-3 we exclude these cases from the population. If the number of units is more than or equal to 2N-3 then the identifier of kinematical stability is invoked to check the kinematical stability. Moreover, the indeterminacy of the truss, and the profile of the hierarchy of the population are determined. This corresponds to all trusses of the population arranged in levels of different indeterminacy, from the statically determinate ones to the max indeterminate truss that may appear in a population, or the ground structure.

Crossover can also be controlled to produce healthy structures. By randomly selecting two parents and the digit to perform a single point crossover, the number of ones are counted in the four different parts of the chromosomes of the two parents, i.e. m1, m2 for the first and m3, m4 for the second parent. If m1+m4, and/or m3+m2 is less than 2N-3 this corresponds to a mechanism, while in the opposite case we produce two off springs of m1+m4-2N+3, m3+m2-2N+3 indeterminacy, provided that the two off springs will be checked by the identifier for kinematical stability. Moreover, the offspring are placed directly in the proper level in the hierarchy because they are of known indeterminacy.

This population is decoded, analysed and optimized as an hierarchy of structures and the fitness values are returned to the genetic algorithm to continue the iterations of statistics, reproduction crossover and mutation until we reach a near optimal solution depending on the stop criterion.

1.2 Second Coding Method

The design space can be further organized in groups that have structural significance, related to the method adopted for the analysis and optimization.

Having a statically indeterminate truss (SIT), four times statically indeterminate, with a list of redundants the axial forces of members [6,8,11,14], starting from a statically determinate truss (SDT) we may determine a number of alternative configurations of different indeterminacy up to the given ground structure. In Fig. 1, the 16 alternatives are presented which are linked in 24 different ways. In general, the number of alternative configurations are 2^n, linked by n! edges, where n is the indeterminacy of the plane truss.

Both the SDT and the ground structure are kinematically stable, while the SDT contains all the nodes of the ground structure and thus, all the intermediate configurations are kinematically stable. The alternative configurations are arranged according to the scheme 1-4-6-4-1. For a truss of degree n, it turns out that the alternative configurations are arranged in a pattern following the numbers of Pascal

Triangle presented in Fig. 3. The main property of these numbers is that are equal to the sum of the two numbers placed above them.

Another observation is that for two statically determinate structures, that differ only in one redundant, the lay out of the design space will contain some common configurations. In general, the total number of different configurations, corresponding to k statically determinate configurations that differ between them by only one redundant, is $2^n + (k-1)2^{n-1}$ as can be seen in Fig. 2 for the case of two different SDTs, that differ in the redundants 6 and 3. Moreover, the arrangement of these configurations in classes of the same indeterminacy is determined by adding the relevant numbers of the Pascal triangle that correspond to n and (k-1) times the (n-1) Pascal numbers, i.e. 2-7-9-5-1 for the case of two SDTs of Fig. 2.

The above observations offer the basis for a more efficient representation of the design space within the context of genetic algorithms. The entire problem can be approached in two phases. The first phase refers to the populations of SDTs generated randomly and checked for their kinematical stability. In cases where the number of nodes N is relatively big, as compared to the indeterminacy of the ground structure n, the possible SDT configurations are of the order of $m!/m_d!/n!$ where m_d=2N-3 and $m=m_d+n$. For example for N=30 and n=5, the number of SDTs is of the order of 6,471,002, whereas for N=30 and n=10 the number of SDTs is of the order of 107,518,930,000. For every kinematically stable SDT there exist 2^n additional statically indeterminate configurations. These configurations can be easily investigated using a genetic algorithm.

Following this concept, a two stage approach is proposed that generates randomly a population of statically determinate trusses which are kinematically stable and then for every SDT another GA solves for the near optimum structure of the hierarchy. The attempt to reduce further the groups of configurations that correspond to a number of SDTs that differ by only one redundant leads to complications. The representation of the first scheme follows an incomplete binary string representation with a chromosome length equal to the number of bars of the ground structure. The total number of zeros in the string will be n corresponding to the degree of indeterminacy of the truss. The second GA follows a complete binary coding of the 2^n configurations that have a one to one correspondence with the alphabet of different trusses arranged in a unique way once the list of redundants is specified and ordered. A set of complete paths of the multi-stage graph is determined, probably by adding some configurations to prepare the hierarchies of trusses for the analysis and optimization on the basis of the recursive force method that will be discussed in the sequel.

In the second scheme, crossover and mutation are used in the standard way, while, in the first scheme, the first population of SDTs are generated randomly and then are checked for kinematical stability. In this scheme, the crossover probability is set to a small number approaching zero, while a bipolar mutation is introduced. The mutated configurations are checked for kinematical stability. By introducing this particular mutation scheme, the bipolar mutation, for every mutation, decided on the basis of a probability of mutation, an inverse mutation is performed. In this way, if a zero is mutated to one in some place then a one is changed to zero in another randomly selected place and vice versa. This operation preserves the indeterminacy of the mutated structure, but does not necessarily guarantee its kinematical stability which has to be checked independently. Moreover, the list of mutations, determined by the genetic algorithm, is used to determine the equilibrium equations needed for the solution of the mutated problem based on Henneberg method.

1.3 Binary string representation of Henneberg method

If we subtract the binary strings representing two statically determinate configurations digit by digit we get:

Truss (a)	1 1 1 0 1 1 1 1 1 1 1 1 1 0 1 1 0
Truss (b)	1 1 1 1 1 1 1 1 1 1 1 0 1 1 0 0 1
Transition	0 0 0-1 0 0 0 0 0 0 1 0-1 1 1-1

In the last string, "1" can be interpreted as adding a bar, while "-1" as removing a bar. Thus, for this example to transform truss (a) to truss (b) we have to add bars 11,14,15 and remove bars 4,13,16. It can be observed also that the number of elements in both lists is three thus, their difference is zero which means that the indeterminacy of the structure is not altered. To perform this transformation in pairs, while preserving the indeterminacy, we must combine these two lists in pairs with one element from each list. To preserve the structural significance of this operation we have to rely upon the identifier of kinematical stability of the resulting trusses. In cases where the initial and final trusses are kinematically stable structures there will always be a path linking the two trusses.

In cases where two trusses can be generated by replacement of one bar, then we may establish an appropriate equilibrium equation specifying that the axial force in the removed position is zero due to the external load and unit loads at the position of the added bar times the unknown axial force. This corresponds to Henneberg method and proves that quite easily we may solve for a new configuration at a very low cost, taking advantage of the fact that we already know the axial forces due to unit forces in all unconnected directions, as a characteristic of the employed analysis method.

1.4 Analysis and Optimization

The analysis and optimization of a population is a major problem that affects significantly the computational cost of the entire method. For sizing and shape problems it turns out that the computational cost of applying GAs in many cases is worse than conventional optimization methods [5]. The use of genetic algorithms in an efficient local search for the layout of structures appears more promising [6]. To improve this, either approximate analysis must be employed [Grierson], or special methods that solve and optimize an entire hierarchy of structures must be adopted. In this work we adopt the recursive force method [7], which is further extended to solve and optimize an hierarchy of structures.

The computational complexity of the recursive formulation of the redundant force method, i.e. the number of multiplications and additions needed for the solution suits better hierarchies of problems. The main result is that with the computational cost of the solution of the ground structure we obtain the solution of an hierarchy of structures of all indeterminacy from a statically determinate one up to the ground structure.

1.5 Recursive Force Method

Consider a statically indeterminate plane truss consisting of M members, connected to N nodes in a plane, which are supported by R reaction members, in a way that 2N < M+R. The degree of indeterminacy n = M+R-2N expresses the number of redundants needed to be specified. The loads of a truss are assumed applied at the nodes. The equations of equilibrium are stated assuming that the effect of the deformation on the internal forces is negligible. Moreover, the compatibility of deformations is enforced along the direction of the redundants, which are assumed acting along the directions of their undeformed configuration.

The redundant force method relaxes the indeterminate structure at certain points, introducing appropriate redundant forces, and imposing the requirements of compatibility in the direction of the selected redundants. The principle of superposition, based on the assumption of linear response of the structure, allows the simultaneous enforcement of the compatibility requirements on the structure, and thus, establishes the equivalence of the indeterminate problem with the relaxed one. This approach corresponds to a simultaneous up-lift of the relaxed structure to its true state [12], [13].

The enforcement of the compatibility requirements can also be established on the basis of a step by step up-lift of the structure. The compatibility requirements are applied recursively not on the primary, statically determinate, structure but on the (n-1) indeterminate structure. To determine this (n-1) state, the ((n-1)-1) state is needed and so on, until the process reaches a statically determinate state. Then, it can start the calculations backwards establishing the required forces. This method, when applied manually, leads to difficulties having to deal with many different forces and displacements at different structures. For this reason it was mainly used in cases where the (n-1) state was a known state. Today, the ability of incorporating more advanced data structures and recursive processes overcomes these shortcomings leading to a robust analysis method.

It can be noticed that the number of required parameters of the redundant force method and the recursive force method are the same. Thus, for a given list of redundants, first the recursive approach hierarchically decomposes the problem into simpler ones, terminating at a statically determinate structure. Then, going upwards, enforces the compatibility requirements recursively calculating also the required forces at all levels of the hierarchy. Therefore, the analysis of an n-degree statically indeterminate truss, under a specific loading condition, can be performed by solving recursively n groups of single compatibility equations (in total n(n+1)/2 single equations) , instead of a system of n linear algebraic equations. The selection of the redundants can be made either in the beginning, as in the standard redundant force method, or at every step concerning only one redundant at a time. Thus, for a structure that is n times statically indeterminate, by selecting one redundant, the axial forces can be expressed as:

$$N_{i,q}^{n} = N_{i,q}^{n-1} + N_{i,n}^{n-1} X_{n,q} \quad i=1,\ldots,m \tag{1}$$

and the compatibility requirement along the direction of the selected redundant as:

$$D^{n}_{n,q} = D^{n-1}_{n,q} + F^{n-1}_{n,n} X_{n,q} = 0 \tag{2}$$

where the superscript indicates the structure, while the first subscript refers to the redundant along which the displacement is calculated, due to the loading specified by the second subscript. $F^{n-1}_{n,n}$ is the flexibility coefficient of the (n-1) structure in the direction of the redundant X_n due to a unit redundant X_n. To determine the axial forces that appear in relation (1), the solution of the following two problems is required on the ((n-1)-1)=(n-2) redundant structure.

$$\begin{aligned} N^{n-1}_{i,q} &= N^{n-2}_{i,q} + N^{n-2}_{i,n-1} X_{n-1,q} \\ N^{n-1}_{i,n} &= N^{n-2}_{i,n} + N^{n-2}_{i,n-1} X_{n-1,n} \end{aligned} \tag{3}$$

which must satisfy the following compatibility equations:

$$\begin{aligned} D^{n-1}_{n-1,q} &= D^{n-2}_{n-1,q} + F^{n-2}_{n-1,n-1} X_{n-1,q} = 0 \\ D^{n-1}_{n-1,n} &= D^{n-2}_{n-1,n} + F^{n-2}_{n-1,n-1} X_{n-1,n} = 0 \end{aligned} \tag{4}$$

This process is repeated n times. At the jth level, (j+1) new problems must be stated with their corresponding j equilibrium and compatibility equations. Finally, at the nth step, two sets of n equations are determined as expressed in relations (5) and (6). At this stage, the resulting structure is a statically determinate truss, which can be solved for the (n+1) unknown sets of axial forces using only the equilibrium equations.

$$\begin{aligned} N^{1}_{i,q} &= N^{0}_{i,q} + N^{0}_{i,1} X_{1,q} \\ N^{1}_{i,n} &= N^{0}_{i,n} + N^{0}_{i,1} X_{1,n} \\ &\cdots \\ N^{1}_{i,3} &= N^{0}_{i,3} + N^{0}_{i,1} X_{1,3} \\ N^{1}_{i,2} &= N^{0}_{i,2} + N^{0}_{i,1} X_{1,2} \end{aligned} \tag{5}$$

$$\begin{aligned} D^{1}_{1,q} &= D^{0}_{1,q} + F^{0}_{1,1} X_{1,q} = 0 \\ D^{1}_{1,n} &= D^{0}_{1,n} + F^{0}_{1,1} X_{1,n} = 0 \\ &\cdots \\ D^{1}_{1,3} &= D^{0}_{1,3} + F^{0}_{1,1} X_{1,3} = 0 \\ D^{1}_{1,2} &= D^{0}_{1,2} + F^{0}_{1,1} X_{1,2} = 0 \end{aligned} \tag{6}$$

Thus, by applying the unit load theorem (n+1) times the displacements and the flexibility coefficient, that appear in the compatibility equations (6), for the first up-lift of the truss, can be calculated. Subsequently, all the axial forces of structure #1 can be calculated using relations (5) to establish the base for the second up-lift. The second up-lift requires (n) applications of the unit load theorem. The process elevates up to the final enforcement of compatibility equation (2) and the calculation of the axial forces of the truss using relation (1). The total number of applications of the unit load theorem is n(n+1)/2 which is the same needed for the evaluation of the coefficients of the nxn symmetric flexibility matrix of the standard redundant force method. The difference is that these coefficients are evaluated on different substructures of the hierarchy and not on the primary structure as in the standard force method.

1.6 Fully-Stressed Design of an Hierarchy of Truss

Force method, applied to structural optimization, usually leads to smaller problems as compared to displacement method [8]÷[11]. Following the recursive force method, the compatibility requirements are imposed one at a time. The information from the previous stage concerns the axial forces for all different loading cases and the axial forces due to unit redundant force of that stage. The resulting axial forces will have the following form:

$$N_{i,q}^{j+1} = N_{i,q}^{j} + N_{i,j+1}^{j} X_{j+1,q} \qquad i=1,...m;\ q=1,...,L \qquad (7)$$

where i is the member, q is the loading case and j, j+1 are the previous and current stages.

The optimality criterion of fully stressed design implies that the areas of the cross sections must be determined from the following relation:

$$A_i^{j+1} = \max\{|N_{i,q}^{j+1}|\}/\sigma \qquad q = 1,2,\ldots,L$$

or

$$A_i^{j+1} = \max\{|N_{i,1}^{j} + N_{i,j+1}^{j} X_{j+1,1}|, |N_{i,2}^{j} + N_{i,j+1}^{j} X_{j+1,2}|, \ldots, |N_{i,L}^{j} + N_{i,j+1}^{j} X_{j+1,L}|\}/\sigma$$
$$i=1,...,m \qquad (8)$$

where σ is the maximum allowable stress, for simplicity taken the same in tension and compression. Using this expression for the cross sectional areas, a system of L compatibility equations is formed as follows:

$$D_{j+1}^{j+1} = D_{j+1,q}^{j} + F_{j+1,j+1}^{j} X_{j+1,q} = 0 \qquad q=1,...,L \qquad (9)$$

where $D_{j+1,q}^{j}, F_{j+1,j+1}^{j}$ are determined using the unit load theorem.

The above system of L nonlinear equations is solved using a fixed point method. The values of $X_{j+1,q}$ that satisfy the system of equations are determined iteratively using the following relations, together with relations (7) and (8).

$$X_{j+1,q} = -\frac{D_{j+1,q}^{j}}{F_{j+1,j+1}^{j}} \quad \text{or} \quad X_{j+1,q} = -\frac{\sum_{i=1}^{n} \frac{N_{i,q}^{j} N_{i,j+1}^{j} l_i}{EA_i^{j+1}}}{\sum_{i=1}^{n} \frac{(N_{i,j+1}^{j})^2 l_i}{EA_i^{j+1}}}, \quad q=1,...,L \qquad (10)$$

where l_i is the length of the member i, and E is Young's modulus. Having these solutions, the axial forces are calculated from relation (7), and the corresponding values of the cross sectional areas from relation (8).

If we do not update the axial forces in relation (7), by inforcing all the compatibility equations starting from the statically determinate truss, we end up with an approximate solution that corresponds to a force approximation.

The axial forces due to unit redundants correspond to the adjoint loads for the stress constraints, thus, the analytical sensitivities for the hierarchy of one degree problems can be calculated using the adjoint method, and can be used also in a rigorous mathematical programming formulation of the minimum weight design of an hierarchy of trusses.

1.7 Implementation of the Genetic Algorithm

The above methodologies have been programmed by linking together several modules developed in different languages that suit better the particular problem. Thus, the genetic algorithm has been implemented in C language, while the analysis and optimization, based on the recursive force method, in Prolog language. Moreover, the identifier of the kinematical stability of the generated configurations is again programmed in Prolog. At the moment we are not in a position to report final results, but the estimated computing time is expected to be quite reasonable.

The basic parameters of a simple genetic algorithm are the population size of the generation, the probability of crossover and the probability of mutation. By varying these parameters, the behaviour of the algorithm is altered. To maintain the robustness of the algorithm, it is important to assign appropriate values for these parameters. Usually, the population size and the probability of mutation are related. With larger population size the probability of mutation is smaller. In this problem we use Population sizes (Pop) that vary with the number of nodes of the truss. We use 30-50 members per ten nodes. During the process if we don't observe a change towards the optimum we alter the probability of mutation. Another way of improving the performance of the algorithm is to pass some of the best members of the generation to the next generation.

The proposed method, based on the recursive force method performs the analysis and optimization of hierarchies of structures providing the optimal solution by moving along a physical path, i.e. enforcing gradually the compatibility equations. The method iterates on small systems of nonlinear compatibility equations recursively from a statically determinate state to subsequent indeterminate states. Elimination of a member occurs only when the axial forces in all loading cases are zero. The genetic algorithm, that improves a population of designs, is in alignment with the proposed methodology.

2. Multi-criterion Optimal Detailing of R/C Beams

(Co-author S.J.Arsenis)

A genetic algorithm is used for the optimal detailed design of reinforced concrete members of multi-storey buildings. The method synthesizes the information from the cross-sectional level, to the member-level and finally to a group of members where the detailed design of the whole group is decided. For many cases the number of alternative solutions are of the order of several millions. For example, with two layers and three regions considered in every span of a continuous beam, the design space of the problem is of the order of $2(8)^{3n}$, where n is the number of spans of a continuous beam and 8 represents the alternative solutions in every region. For these design spaces enumeration methods lead to expensive solutions. A genetic algorithm is adopted for the solution of the problem. The method is applied to problems having large design spaces and near optimum solutions are found in reasonable computing time. The genetic algorithm is based on a roulette-wheel reproduction scheme, single point crossover and constant or variable mutation schemes. An elitist strategy is also used that passes the best designs of a generation to the next generation. The method decides the detailed design on the basis of a multi-criterion objective that represents a compromise between a minimum weight design, a maximum uniformity and the minimum number of bars for a group of members. By varying the weighting factors, designs with different characteristics result. The anchorage lengths are taken into account and the bars are cut at appropriate regions. The performance of the system is illustrated with a number of examples. Various parameters of the genetic algorithm are considered.

2.1 Detailed Design of R/C Beams

The design of reinforced concrete beams is performed first by solving a sizing optimization problem that determines the dimensions of the cross section and the required reinforcement in cm^2. The detailed design, i.e. the selection of the set of bars of certain diameter and length, and the exact placement of this reinforcement, is performed in two distinct phases. In the first phase the cross-sections are examined independently, while in the second level, related cross-sections are combined to design the entire member, or group of members.

2.2 Cross-section design

The exact geometry of a continuous beam is determined by consulting the geometrical database to get all the information concerning the continuous beams, including beam lengths and column widths and the connectivity of the structure, etc. Every span of a continuous beam, is separated into a number of regions. By default three regions are determined, two support regions, and a middle region, each having a geometrical and nominal length. The nominal length corresponds to the length of the reinforcing bars in the region. The nominal lengths of the two support regions are equal to their geometrical lengths, while the middle region by default is considered from column face to column face.

Every region is assigned a specific cross-section whose shape depends on the geometry of the structural element and the loading conditions. In each section two layers of reinforcement are considered, the lower and the upper layer. In every region the required reinforcement in cm^2 is converted into different sets of bars, each set consisting of bars of the same diameter. All different diameters available in the market are considered.

Moreover, the maximum and the minimum shearing forces and the torsional and bending moments are determined, as well as the reinforcement requirements, as they are deduced by the code requirements. All possible combinations of the loading cases are checked, and the reinforcement requirements are finalized for every cross-section.. The calculation of the required single or double steel reinforcement in a beam cross-section is obtained iteratively using the equivalence of the internal and external forces in the cross section and code requirements. For beams of orthogonal cross-section, the cases of single reinforcement and double reinforcement are treated separately. Also for T beam sections, two distinct cases are considered for single and double reinforcement. For these calculations, the relevant tables are used in the form of matrices. This process may alter the dimensions of the orthogonal or T section and thus the solution may be determined iteratively. The entire design process is programmed in Prolog language [14].

The calculated shear and torsion reinforcement is checked against the requirements of the code (min-max, percentage of reinforcement, width of section, etc.). At this stage, the requirements are expressed in cm^2. At the next stage, a recursive predicate, undertakes the task to transform the proposed requirements from cm^2 to a number of different solutions having the same bar diameter. Thus, several pairs of the form: "Number of bars, Diameter of bars" are produced. The algorithm checks for the appropriate spacing of the bars in one layer according to the code requirements and rejects the solutions that do not satisfy these requirements.

After the completion of this stage, all the results are saved in the form of relational database, easily accessible in later stages.

2.3 Member design

In this level, the information from the cross-sectional level is combined to determine the design of the members, or groups of members. Thus, the reinforcement requirements in the consecutive cross-sections of a support are unified by considering the extreme situation for both sections.

The solutions determined, by the proposed method, comply with the requirements of the Greek Code for concrete structures. According to this code, the reinforcement at different regions of a beam must obey certain relations which are described in Fig. 4.

Then, the beams are designed as complete continuous beams by taking into account the necessary anchorage lengths for the reinforcing bars, the length of the bars and the proper positions for cutting the bars [14].

2.4 Multi-criterion Optimal Design

The method of objective weighting that represents the most usual substitute method for vector optimization problems is employed [15]. The utility function or preference function which is independent of the individual minima for positive values of the weights, is determined as a linear combination of the individual criteria $f_1, f_2, \ldots, f_m$ together with the corresponding weighting factors $w_1, w_2, \ldots, w_m$:
where it is usually assumed that

$$0 \le w_j \le 1, \quad \sum_{j=1}^{m} w_j = 1 \tag{12}$$

The Pareto optimal set can be obtained by varying the weights w_j in the utility function. This strategy is not the only one used to formulate a multi-criterion objective. In the literature [16] the method has been criticized, whereas other methods like distance functions, trade-off methods, min-max formulations etc. have been used [15].

2.5 Conflict and Scaling of Criteria

The criteria used in this work are three. The minimum weight of reinforcement in the members, the maximum uniformity of the reinforcing bars, i.e. the property of having bars of the same diameter in different consecutive regions and different consecutive beams, and the criterion of minimum number of bars in the cross-section, which enhances the constructability of the member and can control indirectly the development of cracks. In addition to the above criteria, other criteria such as ductility and/or a direct crack control may be added.

These three criteria do not directly oppose each other. "Minimum weight" tends to assign small diameters trying to get closer to the calculated cm^2 of reinforcement. But as it tends to assign different diameters to every region, it increases the length of the bars by adding the appropriate anchorage lengths. Thus, the minimum weight criterion is penalized by the extra anchorage length, leading partially to more uniform solutions. On the other hand the "minimum number of bars" criterion works similarly to the uniformity criterion tending to provide big diameters. Therefore, the conflict between the criteria can be characterized as a mild one, where both uniformity and the number of bars contradict the minimum weight criterion.

Due to the limited search, scaling of the possible solutions is performed on an absolute scale as follows:

a) For the minimum weight criterion, the maximum value is calculated by adding all the heaviest solutions in every region multiplied by the longest possible length for the region plus the anchorage lengths, while the minimum is calculated by adding the lighter solutions multiplied by the smallest nominal length in every region plus the minimum required number of anchorage lengths. This means the in the scale 0-100, 100 is rarely achieved and the best solutions are in the range of 85/100.

b) The uniformity criterion is graded using the following relation:

$$U=100-(N-1)*(100/ND) \tag{13}$$

where N is the number of different diameters used, and ND is the total number of available bar diameters.

c) Scaling of the third criterion is straightforward and is provided by the following relation:

$$NB = 100 \cdot \frac{C - Max}{Min - Max} \tag{14}$$

where

$$Max = \sum_{i=1}^{K} \max Nb_i, \qquad Min = \sum_{i=1}^{K} \min Nb_i, \tag{15}$$

Nb_i is the number of bars in region i, K is the total number of regions and C is the actual number of bars in all regions.

The above scaling of the criteria in the range 0-100 will be combined to form the preference function, using relations (11) and (12) in discrete form, that as a single objective will control the detailed design of the reinforced concrete members.

2.6 Basic Parameters of the Genetic Algorithm

By using eight (2^3)different solutions per region, we assign a binary string of chromosome length 9n, where n is the number of beams of a continuous beam to represent the particular detailed design of a R/C beam. The Population size (Pop) varied with the number of spans. We used 30-50 members per span. The probability of crossover we used was (P_C) 0.6. The mutation probability we considered, also varies according to the number of spans. Mutation is important in the evolution of the method, because through mutation new members enter the population bringing new blood to the generation. In this way, the initial population which might have been very far from a satisfactory solution can adapt itself towards the optimum solution. On the other hand, mutation tends to disorganize the convergence of the problem and this makes the selection of this parameter, together with the population size, very crucial for the overall performance of the genetic algorithm. For this reason we tried several scenarios of adapting the probability of mutation. During the process if we don't observe a change towards the optimum we alter the probability of mutation. The schemes adopted are the following: for constant probability of mutation in all generations, the probability of mutation varies with the number of spans. For a beam of a single span, P_m=0.01, for a two span beam, P_m=0.02, for a three span beam, P_m=0.04, for a four span beam P_m=0.08, and so on. For a variable scheme, the initial probability is assigned as before. Then, if the algorithm is not improving for N/3 generations, where N is the maximum number of generations used as an on line stop criterion, then P_m is divided by 3, and in case of no improvement again is lowered in the same way. This scheme gave much better results in most of the cases that was applied. Another way of improving the performance of the algorithm is to pass some of the best members of the generation to the next generation.

The structural design problem was programmed in Prolog that accommodates relational databases, whereas the genetic algorithm was developed in C language. Linking of the different modules at the present stage was indirect, but at the end the final program will link the modules directly. This is expected to increase the performance of the proposed method considerably.

The selected designs of a continuous beam can be drawn on the screen for every span [9]. The different objects of the drawings are converted from Prolog predicates to "dxf" files that can be seen and plotted from any drawing package like Autocad, Clariscad, etc. to produce the final detailed drawings.

2.7 Numerical Results

As an example we consider a building that has 10 beams in every floor and 8 columns. A continuous beam of the first floor consists of the beams 1501, 1502 and 1503. In Table 2 all the proposed solutions for every region are presented. These solutions are determined prior to the initiation of the optimal selection using the genetic algorithm.

It can be seen that the proposed solutions are too many. Therefore, complete enumeration for all the proposed solutions would take too long to be performed. For comparison reasons we established this solution using an exhaustive search. It took several days on a 486 DX2/66 PC with 8Mb RAM. The number of alternative solutions examined was 16,003,008. The solutions were represented in the form of a list of the following form:

[Number of Beam, Number of Region, Number of Bars, Diameter in mm ...]

The best list, resulted from the exhaustive search, was the following:

BEST LIST : [1503,3,2,24,1503,2,3,24,1503,1,2,24,
1502,3,2,24,1502,2,4,24,1502,1,2,24,
1501,3,2,24,1501,2,3,24,1501,1,2,24]

and the total grade of this list was 89.5%, determined for weighting factors 40/30/30 for weight, uniformity, and number of bars criteria respectively. This solution had as single criterion grades 73.75%, 100%,100%, for weight, uniformity, and number of bars criteria respectively.

The best solution was attained in 37 generations, among which the last ten are with no improvement and act as an on-line stopping criterion. The total grade of the near optimum solution is 88.28%.

Therefore, the genetic algorithm searched in total 37 * 152 = 5624 solutions which correspond to the 0.018 % of the entire design space. The CPU time, on the same computer, was less than 4 minutes to determine a solution that corresponds to 98.7% of the absolute optimum.

Table 2. Alternative solutions for a three span beam

Lower Layer of Reinforcement																
Beam	Reg.	Sol	#	Ø	#	Ø	#	Ø	#	Ø	#	Ø	#	Ø	#	Ø
1501	1	7	2	24	2	22	2	20	3	18	4	16	4	14	6	12
1501	2	5	3	24	4	22	4	20	5	18	6	16				
1501	3	7	2	24	2	22	2	20	3	18	3	16	4	14	6	12
1502	1	7	2	24	2	22	2	20	3	18	4	16	4	14	6	12
1502	2	3	4	24	5	22	6	20								
1502	3	7	2	24	2	22	2	20	3	18	3	16	4	14	6	12
1503	1	7	2	24	2	22	2	20	3	18	4	16	4	14	6	12
1503	2	5	3	24	4	22	4	20	5	18	7	16				
1503	3	7	2	24	2	22	2	20	3	18	3	16	4	14	6	12
Upper Layer of Reinforcement																
1501	1	6	3	24	3	22	4	20	5	18	6	16	7	14		
1501	2	7	2	24	3	22	3	20	4	18	5	16	6	14	8	12
1501	3	6	3	24	3	22	4	20	5	18	6	16	7	14		
1502	1	6	3	24	3	22	4	20	5	18	6	16	7	14		
1502	2	7	2	24	3	22	3	20	4	18	5	16	6	14	8	12
1502	3	6	3	24	3	22	4	20	5	18	6	16	7	14		
1503	1	6	3	24	3	22	4	20	5	18	6	16	7	14		
1503	2	7	2	24	3	22	3	20	4	18	5	16	6	14	8	12
1503	3	6	3	24	3	22	4	20	4	18	5	16	7	14		

Two different scenarios for the probability of mutation were used, the one with constant probability of mutation, and the other with a variable one, following the scheme described previously. The results are presented in Fig. 5, where we can observe that the variable scenario had an accelerating effect. In Fig. 6 the detailed drawing of a three span beam is presented.

2.8 Concluding Remarks

A genetic algorithm is proposed for detailing reinforced concrete members of multi-storey buildings. The method decides the detailed design on the basis of a multi-criterion objective that represents a compromise between a minimum weight design, a maximum uniformity and the minimum number of bars for a group of members. These criteria control the design in a satisfactory way. By varying the weighting factors, designs of different characteristics result. The anchorage lengths are taken into account and the bars are cut at appropriate regions. Due to the combinatorial nature of the problem in big design spaces of the order of several millions result. A genetic algorithm is adopted to search for the near optimum solution. The parameters of the genetic algorithm are customized to the needs of the problem and satisfying solutions are determined by searching less 1% of the design space via a genetic algorithm. The resulting near optimum solutions correspond to more than 97% of the global solutions for a wide range values of the parameters of the genetic algorithm.

REFERENCES

1. Pellegrino S. and Callandine C.R. "Matrix Analysis of Statically and Kinematically Indeterminate Frameworks", Int. J. Solids Structures, Vol. 22, No. 4., pp.409-428, 1986.
2. Salerno G. "How to Recognize the Order of Infinitesimal Mechanisms: A Numerical Approach", Int. J. for Numerical Methods in Engineering, Vol. 35, pp. 1351-1395, 1992.
3. Horton J.D. (1987) "A polynomial time algorithm to find the shortest cycle basis of a graph", SIAM J. Comput., 16, pp. 358-366.
4. Hajela P., Lee E., Lin C.-Y., (1992) "Genetic Algorithms in Structural Topology Optimization", M.P. Bendsoe and C.A. Mota Soares (eds.), Topology Design of Structures, 117-133. NATO ASI Series E: Applied Sciences - Vol.227.
5. Rajeev S. and Krishnamoorthy, (1992) "Discrete Optimization of Structures using Genetic Algorithms", ASCE, Journal of Structural Engineering, Vol. 118, No.5, May, pp. 1233-1250.8. Reinschmidt K. F., Russell A.D. "Applications of Linear Programming in Structural Layout and Optimization", Computers & Structures,Vol. 4, 1974, pp. 855-869.
6. Koumousis V.K., Georgiou P.G, (1994) "Genetic algorithms in Discrete Optimization of Steel Truss Roofs", ASCE Journal of Computing in Civil Engineering, July, Vol. 8,No. 3, July, pp. 309-325.
7. Koumousis V.K., "Statically Indeterminate Trusses Programmed in Logic", 3rd International Conference on the Applications of Artificial Intelligence in Structural Engineering, Civil Comp-AI, August 1993 Edinburgh.
8. Reinschmidt K. F., Russell A.D. "Applications of Linear Programming in Structural Layout and Optimization", Computers & Structures,Vol. 4, 1974, pp. 855-869.

9. Farsi B., Schmit L.A, " Minimum Weight Design of Stress Limited Trusses", J. Struct. Div., ASCE, Vol. 100, no. ST1, January 1974, pp. 97-107.
10. Sheu C. Y, Schmit L. A., "Minimum Weight Design of Elastic Redundant Trusses under Multiple Loading Conditions", AIAA J., Vol. 10, no. 2, February 1972, pp. 155-162.
11. Kirsch U. "Optimum Structural Design", Mc Graw Hill Inc. 1981, pp.292-296.
12 Argyris J.H., and Kesley S. "Matrix Force Method of Structural Analysis with Applications to Aircraft Wings", Wissenschaften Gessellschaft Luftfahrt Jahrbuch 78-98 ,1956.
13. Kaveh A. "Recent developments in the force method of structural analysis", Applied Mechanics Reviews vol. 45, no. 9, September 1992.
14. Koumousis V.K., Arsenis P.G, Vasiloglou V.B, (1993) "Detailed Design of R/C Buildings using Logic Programming", 3nd Int.Conf. on the Applications of Artificial Intelligence in Civil and Structural Engineering, Edinburgh, UK.
15. Koski, J. "Defectiveness of Weighting Method in Multi-criterion Optimization of Structures", Comm. in Applied Numerical Methods, Vol. 1, 1985, pp. 333-337.
16. Eschenauer H., Koski J., Osyczka A., (Eds.) "Multi-criteria Design Optimization, Procedures and Applications" Springer-Verlag 1990.

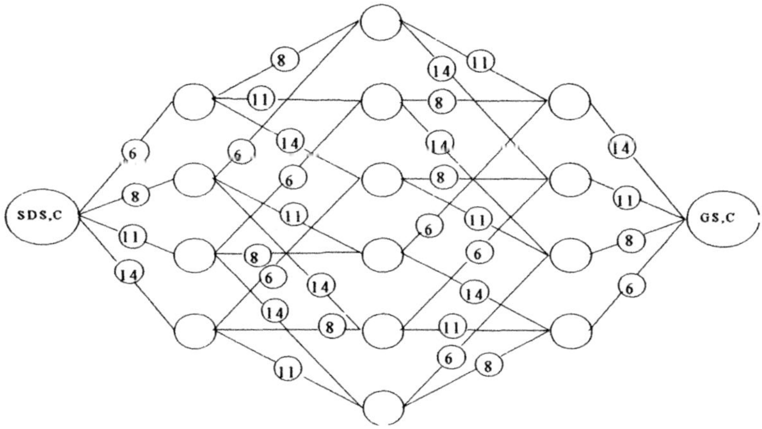

Fig. 1 All different configurations from a specific SDT to a four degree SIT.

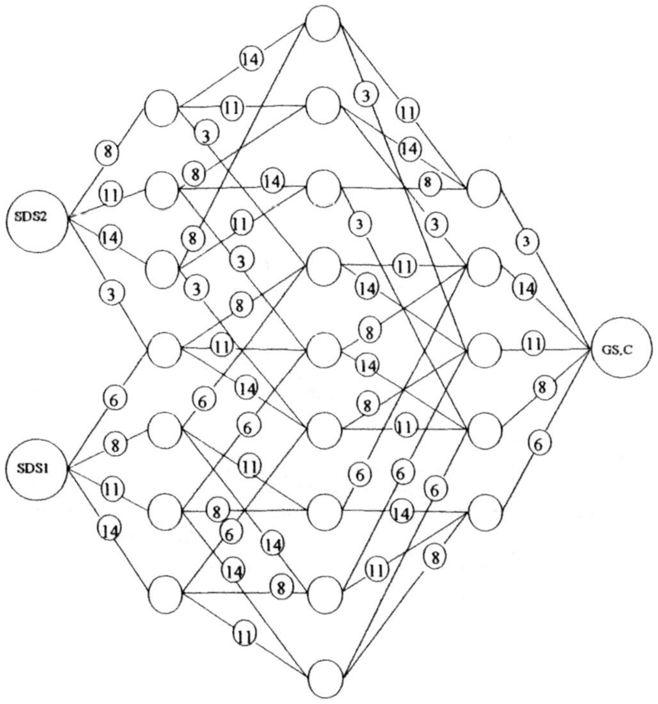

Fig. 2 Design space from two SDTs that differ in one redundant to a fourth order SIT

1
1 1
1 2 1
1 3 3 1
1 4 6 4 1
1 5 10 10 5 1
1 6 15 20 15 6 1
1 7 21 35 35 21 7 1

Fig. 3 Pascal triangle

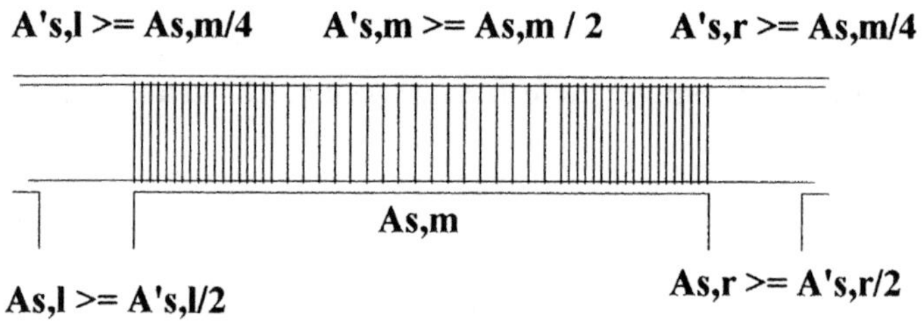

Fig. 4 Required reinforcement in different regions

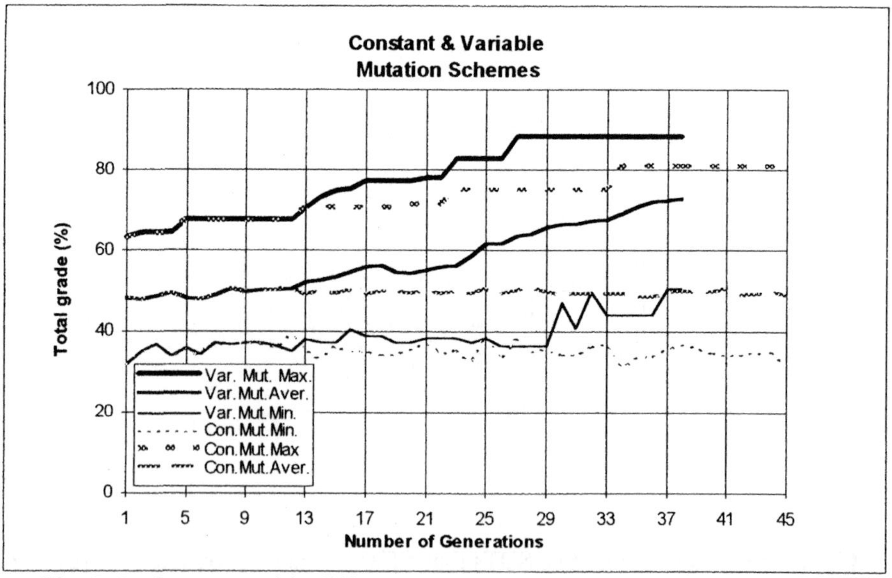

Fig. 5. Performance with different mutation schemes.

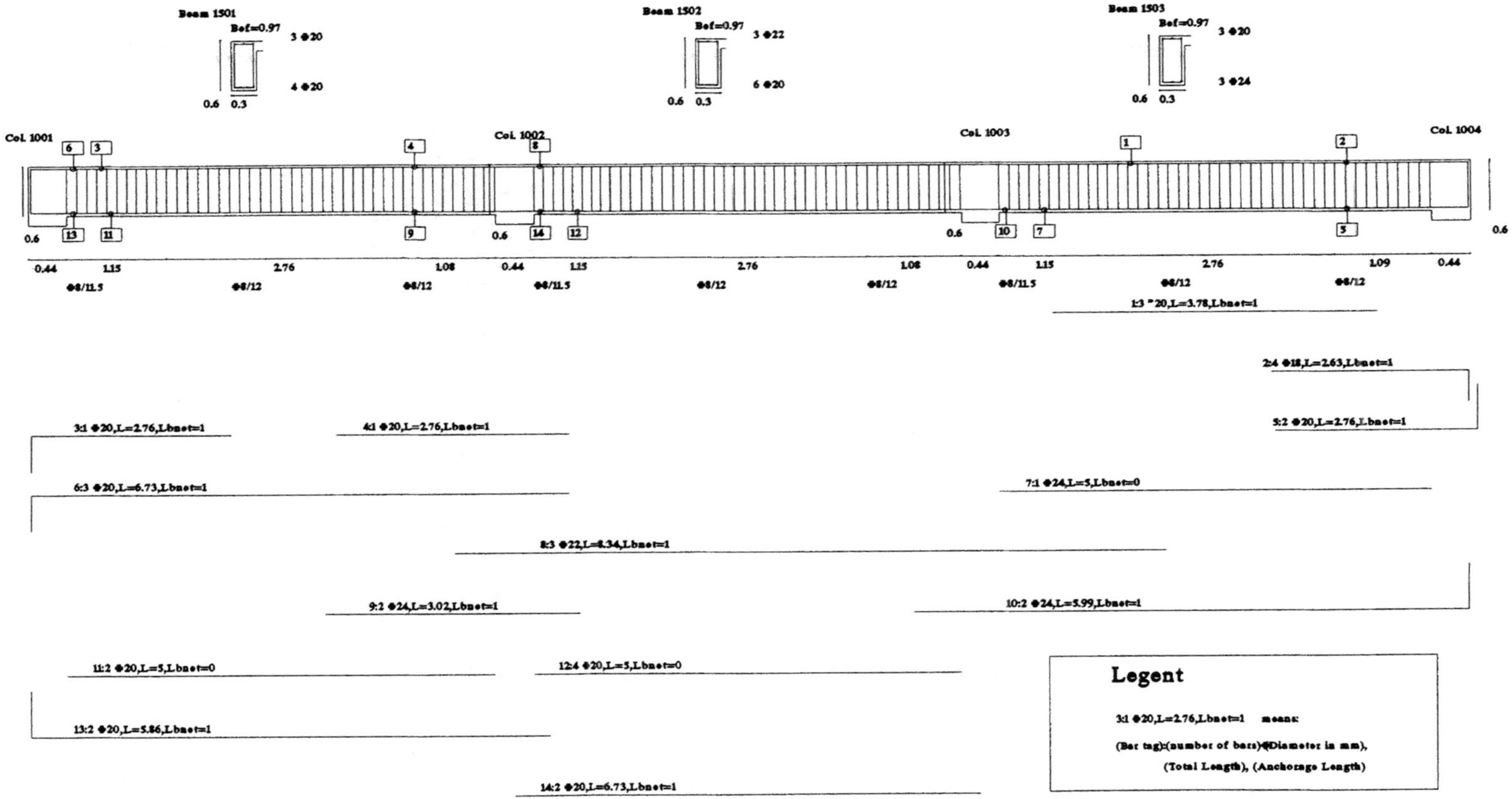

Fig. 6 Detailed drawing of a continuous beam

Genetic Search Based Hybrid Multimodal Optimization

C.-Y. Lin, Y.-J. Yang, and J.-Y. Liou

Department of Mechanical Engineering, National Taiwan Institute of Technology, Taipei, Taiwan 106, Republic of China

Abstract. A genetic search based hybrid multimodal optimization strategy is proposed. This work includes the use of techniques in sharing-enhanced genetic algorithms to identify regions where designs cluster in a multimodal design space. After every period of predetermined generations in a genetic search, the design located in the center of the identified cluster with the highest population density is used as an initial design to start a gradient-based local optimization. Cluster identification techniques in genetic algorithms help the fast determination of regions containing relative optima and provide good initial designs for subsequent local searches. Hybrid optimization strategies based on genetic algorithms is tested in an illustrative multimodal optimization problem.

Keywords. Hybrid optimization, genetic algorithms, nonlinear programming

1 Introduction

Since John Holland published his famous book, *Adaptation in natural and artificial systems* [1], in 1975, genetic algorithms have attracted many researchers and have been intensively investigated in the past few years on areas such as machine learning, and optimization. Based on rules of natural selection, genetic algorithms are robust and powerful optimization tools in solving design problems with a nonconvex, disjoint design space and with discontinued design variables [2,3]. Genetic search based methods can be used as effective optimization tools to globally search for near optimal designs in a design problem with multiple relative optima. A population of finite design candidates in a regular genetic search will eventually converge to a single solution regardless of the fact that the chance to obtain the true optimum is not significant. By implementing the sharing strategy proposed by Goldberg and Richardson [4], genetic searches can locate multiple solutions for problems with a multimodal design space. For best results of creating multiple design clusters in a multimodal design space, precise values of sharing radii need to be defined in advance. Cluster identification techniques can automatically identify an approximate radius for each extreme-containing region where a number of designs cluster. Mating restrictions proposed by Deb and

Goldberg [5] increase the convergence rate for designs clustering around each extreme value. Cluster identification techniques ensure correct mating restrictions with which designs performing crossover are indeed located in a common region containing an relative optimum. Generally speaking, although a genetic search utilizing cluster identification techniques provides a good opportunity of locating the global optimum if a sufficiently large design population is used, the computational cost can be prohibitively high. For an efficient multimodal optimization, a hybrid optimization strategy is proposed, which involves the use of genetic search for near optimum solutions and the use of gradient-based local search techniques for accurate optima.

2. Genetic Search for Multiple Optima

Genetic algorithms provide a robust tool to solve an optimization problem with a multimodal design space for a global solution. However, a traditional genetic search based method will always converge to a single solution unless the search is terminated at a predetermined generation, regardless of this solution being a true global optimum or not. Sharing function implementation, aided by mating restriction, provides a special power which enables the population of designs to pursue multiple relative optima in a same genetic evolution. The sharing function implementation [4] is a breakthrough for locating multiple relative optima in a multimodal design space by using a genetic search. The sharing function approach is based on a concept of shared resources among distinct subsets of population. Each such subset converges to one relative optimum, and in doing so, maximize its payoff. The principle of sharing is implemented by degrading the fitness of each design in proportion to the number of designs located in its neighborhood through the use of sharing functions. The extent of sharing is controlled by a sharing radius σ_{sh}.

The raw fitness value of each design is decreased due to the presence of a large number of designs in its vicinity. The sharing function implementation will prevent all candidate designs from converging to a single optimum, and help locating multiple optima in a genetic search.

3. Cluster Identification Techniques

In the beginning of the genetic search, the design density (number of designs divided by a finite part of the design space) is roughly averaged for each part of the design space. As designs gradually converge to a number of extreme regions due to the effects of sharing function implementation, design densities will increase in those crowded areas where design clusters form. The crowdness function $C(X)$ for a given design X is defined as the degree of design crowding in its neighborhood and the mathematical expression of the crowdness function is as follows:

$$C(X) = \sum_{i=n}^{m} \left(\frac{N_i}{R_i} \right) \left(\frac{L}{R_i} - 1 \right) / N \tag{1}$$

where L: the possible largest normalized distance between two designs in the design space.
m: total number of divisions along L.
n: initial number of divisions of L.
R_i: distance measured from the design X, and $R_i = (i/m)L$.
N_i: number of designs located within R_i measured from X.
N: total number of designs in the design space.

In Equation 1, (N_i/R_i) represents the design density inside the hypersphere with a radius R_i. The effect of the design density will be magnified by a distance ratio $(L/R_i\text{-}1)$ that will be equal to zero as R_i=L and become infinity as R_i=0. The division by N serves to take away the effect of the population size. The term $(N_i/R_i)(L/R_i\text{-}1)/N$ has to be summed up for each i increment to account for the effect of design concentration. Figure 1(a), and 1(b) show the crowdness function contour for a design space containing 3 clusters, one in the upper middle and two in the lower corners, with a population size of 324, and 144, respectively. By inspecting the value of the crowdness function for each design in Figure 1(a) and 1(b), it is clear to see that a design with a relative maximum crowdness function value compared to its neighboring designs is located in or near the center of a design cluster. The above observation may not be apparent in another situation where a design group has not yet form a significant cluster.

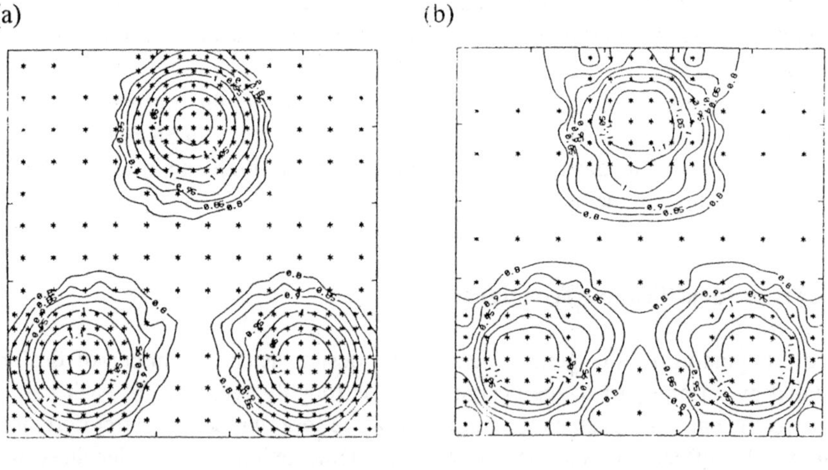

Fig. 1. Crowdness function distribution for (a) 324, and (b) 144 designs with 3 clusters.

A crowdness value can be used as a reference to determine if a design cluster is formed. More specifically, a crowdness function for each design is calculated and a systematic procedure is then used to identify if any design cluster is created. Once a design cluster is located, the design with the highest crowdness value in the region will be considered as the center of the cluster and specific criteria will define the radius of the cluster.

Through several numerical experiments on design spaces with different physical dimensions and different number of population sizes, it is found that the highest crowdness function value for evenly-distributed designs is a value around 1.62. This significant value C_{even}=1.62 is then selected as a reference index in the cluster identification procedure. After the crowdness function value of each design is computed, the following two steps will be applied for cluster identification.

Step 1. Identify a cluster center: Find all designs with a crowdness function value greater than $1.5C_{even}$. Continue the following procedure on each of qualified designs one after one starting from the highest-crowdness design. Proceed to Step 2 if a design is not located inside any existing cluster. If a design is located in one of existing clusters, proceed to Step 2 only if its crowdness value is greater than the highest crowdness value in this cluster when it was first determined. Otherwise, the process for this design is fathomed.

Step 2. Determine the radius: Identify the design with the highest crowdness value and align all other designs in a sequence according to the distance from the highest-crowdness design. Designs with shorter distance are placed near the design with the highest crowdness. The following two criteria will then be used to check designs along the line one after one to determine if a cluster is created.

Criterion I: The crowdness of the design being considered is greater than the previous one AND the crowdness of the previous design is smaller than the average crowdness of the current generation multiplied by 0.9.

Criterion II: The crowdness of the design being considered is smaller than $0.78C_{even}$.

If any one of the above two criteria is satisfied, the design with a highest crowdness value is defined as the center of the cluster, and the distance from this highest-crowdness design to the design qualified any above criterion is defined as the radius of the hyperspherical cluster. The second part of Criterion I ensures the acceptance of small perturbation of crowdness function values before reaching a significant relative minimum. Criterion II safeguards the determination of a solely existing cluster in a unimodal design space. Once the center and radius of a cluster are defined, all designs reside inside the hypersphere will disappear from the list containing designs with crowdness value greater than $1.5C_{even}$. Continue to Step 1 until the above list is empty.

Another impact of the identified clusters on the genetic search is that with the precise information of the cluster, the best design inside each cluster may be recovered after a generation if no better design occurs and the previous best design disappears due to gene-exchanging actions. Simply speaking, the elitist strategy can now be applied in each of identified clusters to speed up the convergence rate of multiple extremes.

4. Hybrid Multimodal Optimization

The proposed hybrid optimization strategy comprises a complete genetic search and several gradient-based local optimization. The genetic search with sharing implementation and cluster identification techniques will be able to pursue several relative optima at the same time. For every *N* generations in the genetic search, the design with highest crowdness function value in an identified cluster will serve as an initial design for a subsequent gradient-based local search. The accurate local optimum thus obtained can be used as the center of the extreme-containing region, and the radius of the original cluster is used as the radius of the extreme-containing region. This newly identified extreme-containing hypersphere will be cut off from the original design space. All designs originally located in the new cut-off area will be moved into randomly-selecteded positions in the remained design space. Since the defined design space for the genetic search will be reduced every *N* generations if a new relative optimum is successfully discovered, the design density will continuously increase so that the power to explore the entire design space is also increased. The whole hybrid optimization can be terminated after a predetermined maximum number of function evaluations is reached or other more sophisticated stopping criteria can be employed. The schematic flowchart of the hybrid optimization used in this work is shown in Figure 2.

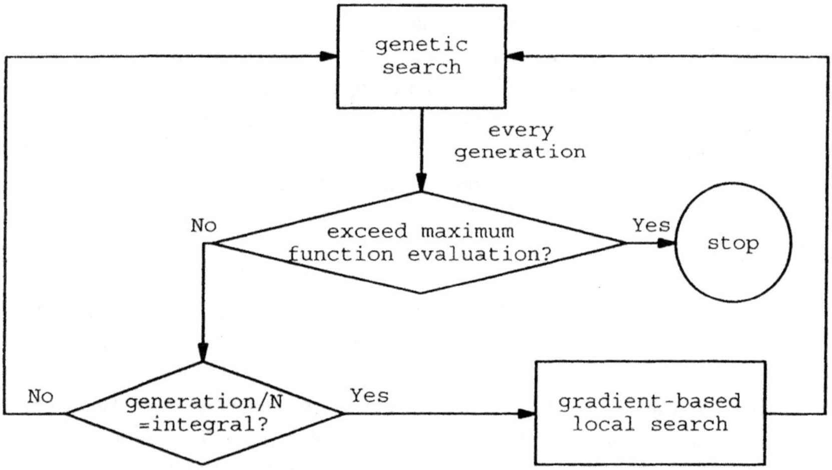

Fig. 2. Schematic flowchart of hybrid optimization

5. Illustrative Example

The illustrative example in this section was solved by using a multiple-strategy genetic search program EVOLVE [6] and integrated gradient-based technique, BFGS method which is one of the variable metric methods.

The example is to minimize an exponential spike/bump function which has four relative minima and these four extreme-containing regions have different shapes sizes, and depths as can be seen in Figure 3.

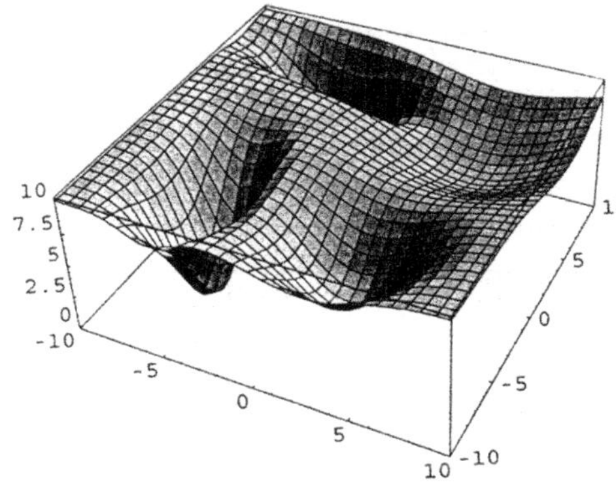

Fig. 3. A 3-dimensional diagram of the exponential spike function

The upper-right region contains a local minimum of near 5.0, the upper-left region contains a local minimum of near 2.6, the lower-right corner contains a local minimum of near 3.9, and the lower-left corner contains a global minimum of near 0.0. Two design variables are both bounded between -10.0 and 10.0. Each variable is represented by a 11 bit binary substring and the precision of each variable is also 0.01. Population size is selected to as 30. Probability of crossover and mutation is selected as 0.6 and 0.001, respectively. In the first part of the experiment for testing cluster identification techniques, all genetic searches are terminated after 1800 function evaluations. The genetic search with cluster identification techniques is tested against three genetic searches with a pre-determined sharing radius, 2.0, 4.0, and 6.0. Table 1 lists the best designs located in each of four extreme-containing regions after the 15th, 30th, 45th, and 60th generation of the genetic search. It can be seen that much better designs are located in a shorter period of time by using cluster identification techniques in most regions. Figures 4(a) and 4(b) illustrate clusters identified at the 15th, and 30th generation, respectively. It can be seen that four clusters have been identified in an early stage of the genetic search and then adjust smoothly along the evolution. The hybrid optimization as described in Fig. 2 is then performed with N=5, 10, and 20, and the genetic search will be terminated after all four relative optima are identified. Results obtained by hybrid optimization are listed in Table 2 where results from the previous genetic search with cluster identification techniques are also included for comparison. It is clear that hybrid optimization can be effectively used to locate precise multiple optima in an efficient manner. Although the hybrid optimization with the smallest N value gains solutions of simple problem, it is suggested to choose a larger N value for more complicated problems.

Table 1. Best results obtained by different genetic search strategies

	σ_{sh}=1.0	σ_{sh}=2.0	σ_{sh}=3.0	Cluster
Upper-left	9.540	*	8.822	2.773
Upper-right	*	6.015	5.723	4.998
Lower-left	-0.043	-0.046	-0.046	-0.042
Lower-right	4.292	4.253	4.334	3.885

* represents no solutions in the region

(a) (b)

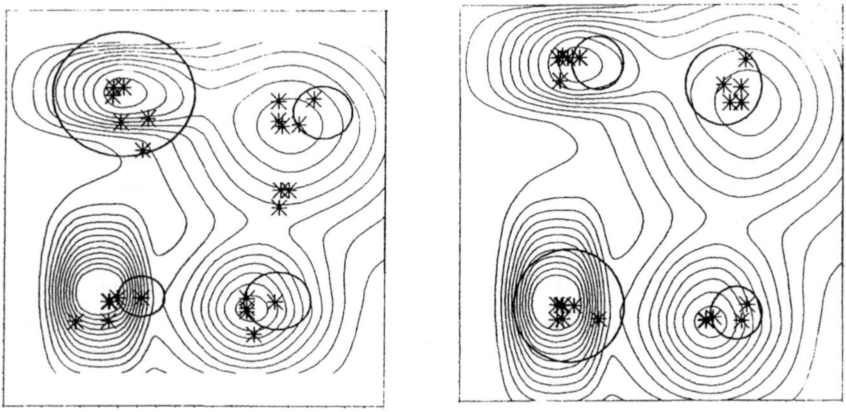

Fig. 4. Clusters distribution after (a) 15th, and (b) 30th generations

Table 2. Comparison of results of hybrid optimization

	GA with C.I.	Hybrid optimization		
		N=5	N=10	N=20
Upper-left	2.773	2.589	2.588	2.588
Upper-right	4.998	4.998	4.998	4.998
Lower-left	-0.042	-0.045	-0.046	-0.046
Lower-right	3.885	3.879	3.879	3.879
Function evaluations	1800	847	1430	2603

6. Concluding Remarks

Genetic search based hybrid optimization can be effectively used as multimodal optimization tools for solving problems for global and other relative optima. This approach combines advantages provided by robust genetic algorithms based multimodal optimization and efficient gradient-based local search. Cluster identification techniques effectively assist sharing function implementation in locating multiple minima in multimodal optimization problems. More self-adjusting strategies integrating genetic searches and gradient-based searches are feasible and currently under investigation.

Acknowledgments

This research was supported by the National Science Council of the Republic of China under grant number NSC83-0401-E-011-039.

References

1. Holland, J.H.: Adaptation in natural and artificial systems. The University of Michigan Press, Ann Arbor (1975)
2. Hajela, P.: Genetic search-an approach to the nonconvex optimization problem. AIAA Journal, 26(7), 1205-1210 (1990)
3. Lin, C.-Y., Hajela, P.: Genetic algorithms in structural optimization problems with discrete and integer design variables. Journal of Engineering Optimization. 19(3), 309-327 (1992)
4. Goldberg, D.E., Richardson, J.,: Genetic algorithms with sharing for multimodal function optimization. In: Proceedings of the 2nd International Conference on Genetic Algorithms, 41-49, Cambridge (1987)
5. Deb, K., Goldberg, D.E.,: An investigation of niche and species formation in genetic function optimization. In: Proceedings of the 3rd International Conference on Genetic Algorithms, 42-50, George Mason University (1989)
6. Lin, C.-Y., Hajela, P.,: EVOLVE: a genetic search based optimization code with multiple strategies. In: Proceedings of the 3rd International Conference on Computer Aided Optimum Design of Structures, 639-654, Zaragoza, Spain (1993)

The Development of Generic Adaptive Search Tools for Engineering Design

I. C. Parmee and M. J. Denham

Plymouth Engineering Design Centre, School of Computing, University of Plymouth, Plymouth, UK.
email: ian@cis.plym.ac.uk

1 Introduction

The Plymouth Engineering Design Centre (PEDC) at the University of Plymouth is one of seven Engineering Design Centres that have been established within the UK by the Engineering and Physical Science Research Council. The primary aims of the PEDC are to carry out fundamental research into the application of the Genetic Algorithm [1,2,3] and related Adaptive Search (AS) techniques to engineering design. The objective is to make these techniques more accessible to designers via computer based support tools and by the provision of generic strategies and methodologies for their adoption and integration into the design process. Concurrently we wish to inform the engineering design community of the potential of these techniques and assist in their introduction in design education and training.

Three areas of Adaptive Search research are evident at the Centre:

- fundamental research concerning the theory and dynamics of the Genetic Algorithm
- research concerning the integration of Adaptive Search techniques with current design practice
- the application of the Genetic Algorithm and related AS techniques to complex design problems

The Centre is collaborating with major industrial groups within the electronic, mechanical and civil engineering sectors. The complementary nature of each of these areas is resulting in the overall development of generic techniques.

The paper describes the nature of the various design problems that are currently under investigation. These include:

- the design of large scale hydropower systems
- the design of FIR digital filters
- the design of blade cooling hole geometries for gas turbine engines

The utilisation of a structured genetic algorithm both in hydropower system design and the FIR digital filter work is described. In addition to these design areas the Centre has recently established collaborative projects in:

- thermal cycle design for power generation
- the optimum design of precast concrete components

Adaptive Search techniques such as the Genetic Algorithm (GA) are seen, in the main, as powerful optimisation algorithms that are capable of negotiating complex design spaces avoiding local optima to eventually converge upon a global optimum solution. However they can also be utilised as global search algorithms that can provide a number of diverse high fitness solutions to a design problem. The paper concentrates upon this aspect and describes several methodologies that can provide a high level of decision support by allowing the designer to utilise both global and local search to locate and investigate specific regions of the design space.

The first technique utilises a hybrid GA which incorporates a standard clustering algorithm. High fitness areas of the overall search space are rapidly identified by the GA and these areas are further defined by the clustering algorithm. A number of independent searches are then initiated within the high fitness regions. This technique is extended to address the problem of design sensitivity by investigating the high fitness regions to locate robust clusters that can satisfy some preset sensitivity criteria.

In some cases the engineer will not require the characteristic diverse search of the evolutionary algorithms. Perhaps the objective is to determine the parameter combination that relates to a specific solution. This approach is taken by the second technique which utilises a multi-criteria approach to define specific solutions / regions within the design space. This technique is also utilised to locate parameter combinations which define feasible solutions within a heavily constrained space. The overall objective of the work is to develop a flexible adaptive design approach that provides the engineer with a selection of tools to enable the successful negotiation of design spaces, no matter how complex, performing either cursory examination or detailed search of selected regions.

The fundamental GA research has included the development of a tool for the extraction and analysis of those building blocks that are more regularly sampled during the optimisation process. This has led to the empirical testing of the building block hypothesis which indicates that the GA does not necessarily arrive at its final solution through the combination and recombination of highly fit building blocks. Random mutation should have a similar effect upon building block improvement.

The effect of crossover is described by the building block hypothesis as combining and recombining building blocks for sampling. An alternative description of the action of crossover is proposed. The action of crossover may be described as a geometric rotation and scaling of the parameter sets associated with a pair of chromosomes about their midpoint. The rotational effect of crossover is apparent provided that there is a linear mapping from the chromosome to the parameter set. This description allows a prediction of GA performance and can offer an improved performance in some cases.

The paper also discusses the introduction of intelligence into the search process to enable the GA in particular to better negotiate those design spaces characterised by features such as:

- uncertain and incomplete design models;

- non-linear, discontinuous objective functions;
- non-linear, constraints;
- noise in the objective function owing to model inaccuracies;
- critical, stringent and conflicting design objectives.

The extent to which an adaptive search procedure is able to deal with these features could be said to be a measure of its intelligence.

2 Collaborative Projects

Initial research concerning the Genetic Algorithm at the University involved the generation of an optimal chamber shape for a novel, low-head hydropower device [4] and the application of the technique to the design of FIR digital filters [5].

Current research at the Centre involves a number of collaborative projects investigating the integration of the Genetic Algorithm with several differing areas of engineering design. The highly collaborative nature of the applied aspects of the research ensures that due consideration is given to the requirements of current engineering design practice. Having established the feasibility of the basic application of the GA to engineering design a more fundamental approach involving an analysis of the individual design processes is also taking place in addition to fundamental research investigating the evolutionary process and the development of more efficient hybrid techniques. The complementary nature of the applied and fundamental aspects of the research is essential for the successful development of a generic methodology.

2.1 The Design of Computationally Efficient FIR Digital Filters

In recent times there has been a move to reduce the computational complexity of finite impulse response (FIR) filters, especially when the filters are applied in VLSI technology at high sample rates or when using gate array technology. These filters can be roughly split into two categories; those with a reduced number of multipliers [6,7,8,9] , and those which are truly multiplierless [10,11,12,13], where the coefficients are made up from combinations of power of two coefficients. In this case the multiplication processes are modified not only to a simpler but also faster bit shift and add operations. The filters designed here fall into this category.

The primitive cascade method generates a frequency response, which meets a desired frequency specification, by cascading together very simple sections which alone have a very simple structure and frequency response. The impulse response of the overall filter, *h(n)* , can therefore be obtained by convolving the individual impulse responses, $H_i(z)$, of the sections:-

$$H(z) = \prod_{i=1}^{P} [H_i(z)]^{n_i} \qquad n_i = 0,1,2 \ldots \qquad (1)$$

where n_i is the number of times primitive i is used. A library of suitable primitives is made up from six different sections. The set of primitives is a reduced set containing no sections with on-circle zeros.

The objective is therefore to select the P primitives, from the reduced set, for a specific maximum cascade length, m, choosing each primitive type and its corresponding delay integer value and power of two coefficient whilst minimising hardware needed, i.e. reducing the number of adders a_i and delays d_i, within the circuit, in order that the filter frequency response falls within or very close to the desired specification.

This can be successfully done using integer linear programming for low-order filters [13], $m<10$, but for medium order filter designs, $m>10$, integer programming becomes impractical since computation time rises exponentially with filter order. The increase in the size of the problem can best be understood if the size of the search space is considered. For a value of $m=15$ different primitives arranged in any order and with a raw library size of 687 different primitives, a search space of the order 10^{30} has to be searched, rising to 10^{38} for $m=20$.

A variant GA, called the *structured* GA (sGA) [14], has been found to work better than a normal GA for synthesising medium order primitive cascade filters. The sGA has a multi-level parameter combination structure where genes can either be active or passive. Higher level genes control or switch (on/off) lower level genes. Three levels of genes are used for the filter design, filter primitive type, filter delay and filter coefficient (if applicable); highest level to lowest level respectively. Results from this form of GA encoding are encouraging.

A measure of the performance of the filters designed by the structured GA is achieved by a comparison against classical transversal FIR filters designed using the equiripple (minimax) algorithm for the same filter specification. Comparison shows that typically a 2:1 advantage in both chip area and clock rate is achieved by the primitive cascade method [15]. The use of a structured GA allows effective search for answers in a domain that is otherwise practically inaccessible. However, due to the stochastic nature of the GA operation and the size and complexity of the search space only sub-optimal results are being found, not the desired global optimum.

Recent work has moved on to look at the practical aspects of the filters designed by the sGA. In practical implementations Rabiner [16] states that it is of the utmost importance to maximise the output signal to noise ratio for cascade FIR filters, such as the filters designed here. In order to achieve this goal the ordering of the sections, wordlength and scaling between sections needs to be considered. The filters designed are implemented using fixed point format. If this is the case then the scaling factors between sections are scaled to satisfy the l_p- norm criteria:-

$$\|h\|_p = \left[\sum_{n=0}^{\infty} |h(n)|^p \right]^{\frac{1}{p}} = 1 \qquad p=1,2,... \tag{2}$$

where $h(n)$ is the scaled section impulse response. In order to keep the risk of overflow to a minimum and since the scaling factor is rounded to the nearest power of two value ,l_1-norm is chosen. Removing scaling from the problem we are left with how to order the sections whilst minimising the overall wordlength of the cascade of primitives.

A simple algorithm developed from sources [16,17] was initially tested. This works on the principle that the action of scaling and rounding at the input of each filter generates a roundoff noise source of variance $Q^2/12$, where Q is the quantisation level. The noise sources generated within each section which would appear at the output are generally negligible and are therefore ignored for simplicity. The output noise variance from the ith section to the output can be defined as

$$\sigma_{no}^2(i) = \frac{Q^2}{12}\sum_k g_i^2(k) \tag{3}$$

where g_i is the impulse response from the ith section to the output. From this definition a classical algorithm presented in the literature [16,17] can be expressed : "Starting with $i=N_s$, permanently build into position i the section which with the sections already built in, results in the smallest possible value for $\sum_k g_i^2(k)$, where N_s is the number of sections in the cascade."

Tests using the above algorithm on several filters gave results that are reasonably good, such that the integrity of the frequency response was maintained to a certain degree. However, because the same method produces the same order of sections irrespective of wordlength, a better method of ordering the sections is necessary.

The problem of ordering sections for a fixed wordlength is a similar problem to the well known Travelling Salesman Problem (TSP), Michaelwicz [18]. The GA has satisfactorily solved this type of combinatorial problem. It is therefore possible that the ordering of the primitives using a combinatorial GA with its adapted operators will overcome the approximations made by the classical algorithm. Indeed, the GA would optimise orderings at specific wordlengths. The objective function for such a GA would be the optimal ordering of the sections whilst maintaining the frequency response for a fixed wordlength.

For all the filters optimised using the TSP type GA, at different wordlengths, it was found that the orderings delivered out performed the classical algorithm in terms of frequency response disruption. It was also noted that the optimal order for different wordlengths for the same filter were different. Curves plotted with error versus wordlength also showed that not only did the minimum error drop with increasing wordlength as would be expected since Q is increasing, but also there was an optimal wordlength, where the error generally dropped to zero and then rose again. As a result current work is concentrating on using the GA not only to find the optimal ordering of the sections but also to minimise the wordlengths within each section of the primitive cascade. This problem has moved into the area of multi-objective optimisation. Initial results look promising [19].

2.2 The Concrete Arch Dam

Unlike other dams which rely upon self-weight to resist the hydrostatic forces generated by the impounded reservoir the Concrete Arch relies upon its shape alone. The designer implements horizontal and vertical curvatures that transmit thrust to the dam's foundation and abutments in a manner which ensures that uniform compressive stresses and limited tensile stresses are generated within the structure. The constituent curves may take elliptical, circular or polynomial forms. Alternatively they may be based upon a logarithmic spiral. An overall dam geometry will incorporate some if not all of these forms. In addition, a number of horizontal curves of differing radii may be combined at any particular elevation and the radii will vary with depth to better accommodate increasing hydrostatic force.

Work has been concentrated upon a three-centred geometrical configuration which utilizes three separate plan curves at each elevation. The designer must first establish vertical curvatures into which the horizontal, three-centred curves can be integrated to form the smooth internal and external faces of the dam (known respectively as the intrados and extrados). Initially, a reference plane is established and a reference cylinder defined within the valley. The intrados and extrados curves can then be established from secondary and tertiary coordinate systems. To ensure that the horizontal curvatures vary in a satisfactory manner with depth their radii are defined by the lines of centres the slope or curvatures of which can be varied in order that the increasing hydrostatic forces are better transmitted via the dam curvature into the abutments.

A Genetic Algorithm has been integrated with a flexible computer model of the three-centred geometry [20,21]. The model at this stage incorporates circular curves only. Typical variable parameters include the X and Y coordinates of the centres of curvatures of both the intrados and extrados vertical curves and the gradients / curves of the lines of centres. Variability of the parameters has been suitably constrained to ensure that feasible geometries are generated. The objective function is to reduce the dam volume whilst minimizing stress variation within the structure. Simple stress analysis techniques have been utilized during the development stage to provide a measure of the relative fitness of each parameter combination. A typical evolution of the crown cantilever is shown in figure 2. The GA has also been integrated with AVS visualization software which produces three dimensional images, plans and elevations of the best configuration of each generation.

A Structured Genetic Algorithm [14] is currently being introduced at a higher level of the design process. During the initial stages of the design of a large scale hydropower system the engineer can be faced with a number of possible sites each possessing significant potential due to differing individual characteristics. The software under development allows the simultaneous GA processing of five differing hydropower sites. Variables for each site include:

- Dam Type and alignment
- Dam Height
- Characteristics of Headrace / Tailrace tunnels
  ie Length, diameter, lining etc
- Powerhouse location
- Flow rate for power generation
- Spillway type
- Material type and availability

The objective function is to maximise power available at minimum cost. The overall objective of the project is to develop a design tool which provides the engineer with a high level of decision support and significantly reduces design lead time by rapidly processing design options at this preliminary design stage.

The Structured Genetic Algorithm allows parameters defining the characteristics of each site to be encoded in the same string. Controlling genes within the string then determine the site and dam type and 'switch on' the relevant parameters which are subsequently passed to the fitness function. Other values remain dormant within the string unless activated in a later population as a result of crossover or random mutation.

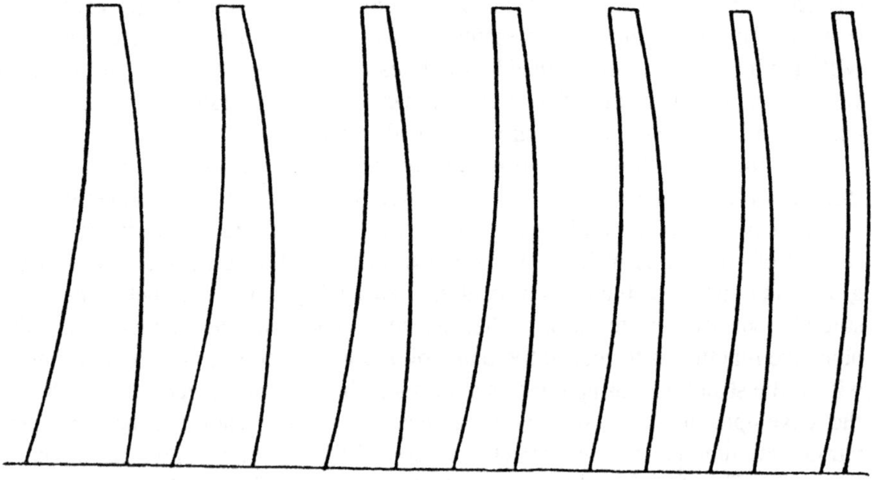

Figure 1: Typical Evolution of the Crown Cantilever

2.3 The Design of Gas Turbine Engine Components with Rolls Royce, plc.

The GA has been applied to two areas of gas turbine design namely the design of blade cooling hole geometries and the design of the turbine annulus. Application has been at the preliminary design stage in both cases.

The cooling hole geometry problem has involved the development of a simple, one-dimensional, steady-state model of a cooling passage situated at the leading edge of a high pressure gas turbine blade. The model represents a purely convective radial passage and provides a basis for extension to cover multiple passages and the inclusion of more comprehensive features and inter-dependencies. A constant cross-section is assumed from passage inlet to exit and the thermal design section is located at the aerofoil mid-span.

Typical variables manipulated by the GA include the coolant temperature at the passage inlet (TCI), the ratio of the passage inlet pressure to passage outlet pressure (RAT) and the ratio of the cooling passage perimeter to the effective gas-side perimeter of the blade (SGC). Various configurations (IGEOM) to promote mixing within the coolant passage have been included and these introduce variation to a heat transfer coefficient factor and coefficient of discharge. In order to introduce a variation in blade material and to investigate the significance of the application of insulative coatings, wall thickness (DTH) and thermal conductivity (KW) have also been included as independent variables. The primary objective function has been to minimize the coolant flow rate.

In an unconstrained form the search space defined by this initial model presents little difficulty to the Genetic Algorithm. Although a degree of discontinuity is created by the varying internal configurations of the cooling passage, appropriate model coding ensures that such obstacles are easily negotiated by a simple GA. A global minimum cooling flow rate can be achieved within thirty five generations. The situation significantly changes however with the introduction of metal temperature and cooling hole area constraints. The utilisation of appropriate penalty functions to draw the GA search into feasible space distorts the fitness surface creating local minima and destroying the intermediate structures required to establish a path to the feasible region / regions. A near optimal solution can still be achieved however although at a significantly increased computational expense (150 to 200 generations).

In this initial form the model is providing a practical fitness function that relates real engineering objectives to the search capabilities of the Genetic Algorithm at an acceptable level of computational expense. This allows extensive experimentation to take place which benefits from design-oriented analysis. The model provides a compromise between the rapid processing of some mathematical function with little relation to any physical problem and the computationally intensive dynamic analysis that provides definitive information but severely restricts experimentation due to associated run time. The initial model has been utilised in this manner for the development of the generic GA design strategies described in the following sections.

The development of the cooling hole model is continuing. Film cooling passages have recently been included and flight cycle considerations are now being taken into account. The intention is to develop a GA-oriented preliminary design model that will provide the engineer with a robust optimal starting point for the detailed design process.

3 Global Search

With regards to the utilisation of Adaptive Search techniques for engineering design the diverse search characteristics of the evolutionary algorithms offer a level of decision support that cannot readily be achieved using point-to-point techniques such as Simulated Annealing or Tabu Search. The Genetic Algorithm initially populates the design space with a number of points equal to the number of parameter combinations within each population. Continuation of the search in those regions surrounding each point depends upon their relative fitness ie there is a greater probability of those parameter sets defining points in the better regions being reproduced into subsequent generations than those sets lying in regions characterised by poor performance. In early generations a parallel search of these better regions develops. Current research at the PEDC is indicating that these natural clustering tendencies can be reinforced and that the introduction of standard clustering techniques after a set number of generations will increase their definition (22). To illustrate this figure 2 shows the results of the introduction of clustering (23) after fifty generations of the GA manipulation of a preliminary, steady state design model describing the cooling hole geometry of a gas turbine blade. Three clusters of good solutions have been identified and have been plotted in two dimensions, initial coolant temperature (TCI) and thermal conductivity (KW), of the six dimensional problem. Each of the three clusters defines a different region of the design space. Bearing in mind that thermal

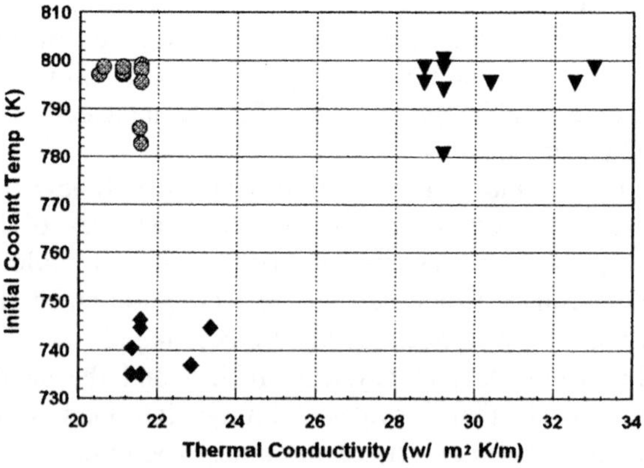

Figure 2: Clusters in Two Dimensions

conductivity defines material type and initial coolant temperature can vary between 700 and 800°K, the designer has a choice of continuing the search in any of the following regions:

- mid-range values of TCI, high thermal conductivity
- upper-range values of TCI, low thermal conductivity
- upper-range values of TCI, high thermal conductivity

At the initial design stage therefore, the engineer can take into account directions being proposed in related areas of the overall turbine design and any preliminary design decisions. An appropriate region can then be selected and a further search within that specific region can be implemented.

4 Design Sensitivity

When searching for a global optimum solution using a mathematical simulation of the system under design consideration must be given to the engineering significance of that solution. In order that the global optimum represents a true engineering optimal design the model must take into account a range of criteria that are at best difficult to describe in mathematical terms. Such criteria may include buildability, site/problem specifics, manufacturing constraint and customer requirement. Another aspect is the sensitivity of the solution to mild parameter perturbation.

The Genetic Algorithm samples widely varying areas of the design space with little regard to the gradient characteristics of each region. Good solutions may lie in regions characterised by steep gradients or discontinuity and the related parameter sets will be highly sensitive to slight perturbation which may cause significant solution degradation. A better approach is to avoid such regions and concentrate effort in those areas that offer sufficiently robust solutions that satisfy some sensitivity criteria.

The sensitivity criteria could be included as a second objective within the fitness function in order to ensure the rejection of non-robust solutions. However the inclusion of such a criteria would involve the following steps:

- the systematic perturbation between some upper and lower bound of each parameter in turn for each parameter set.
- a call to the mathematical model to ascertain the change in performance caused by each perturbation.
- a means of measurement of the variation in performance and thus the sensitivity of each set.

If we therefore assume a six-dimensional problem (ie six parameters) and a population size of fifty parameter sets the overall analysis of the relative fitness of one population will involve 50×3^6 (36450) calls to the mathematical model. The computational expense of such an exercise would not be acceptable and an alternative methodology is therefore required.

Using clustering techniques after a preset number of generations the average sensitivity of each cluster can be ascertained by testing each parameter

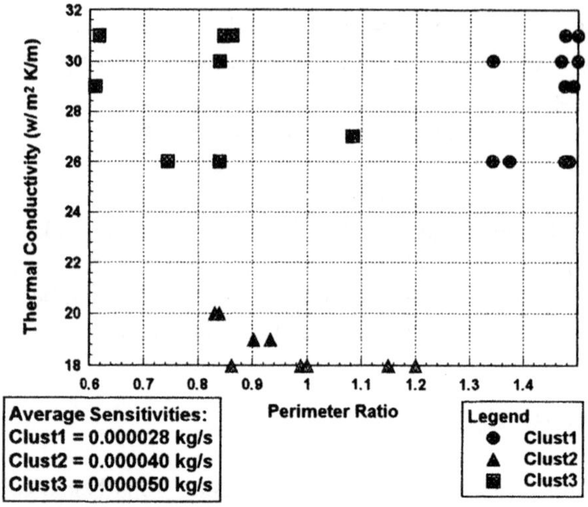

Figure 3: Clustering with Relative Sensitivities

combination in the manner described above. The design engineer can then decide (depending upon some preset sensitivity criteria) in which region to concentrate the search effort and can discretize a further investigatory run of the Algorithm accordingly. This approach allows the engineer, at little computational expense, to eliminate those areas of high sensitivity during the early stages of the search and to continue in those regions that will provide a sufficiently robust optimal design.

The average sensitivity of the parameter sets contained within each of the three clusters of figure 3 illustrate the feasibility of this approach

5 Constrained Search

The diverse search characteristics of the GA can be utilised to provide a high level of decision support for the design engineer as described in previous sections. However, the efficiency of the GA search can be significantly reduced in heavily constrained design spaces. Constraints can be effectively removed from the design space by the introduction of appropriate penalty functions. This allows the GA to search across the entire space but the fitness of each solution is penalised in proportion to any violation of a constraint boundary. The search will eventually be concentrated within feasible regions of the design space. In complex constrained spaces, however, convergence can be extremely difficult and it is still possible for the search process to converge upon a high fitness, non-feasible solution due to the constant rejection of the intermediate structures required to establish a path to a feasible region. A number of more complex methodologies are available from the literature [24].

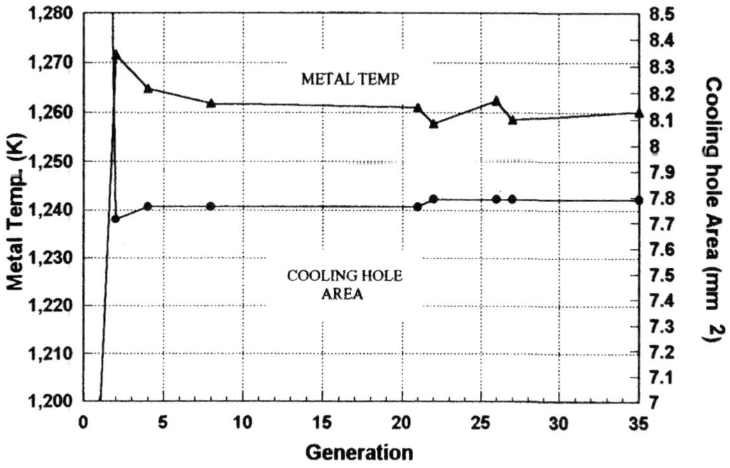

Figure 4: VEGA Search for Utopian Values

An alternative approach utilising a modified VEGA [25] is under development at the PEDC. Using the directed search capabilities of VEGA it is possible to identify parameter combinations that define feasible points. For instance, both the values of metal temperature and cooling hole area returned from the turbine cooling hole model must lie within appropriate constraints. By selecting utopian values for both TW and AC that lie within these constraints and minimising coolant flow rate, VEGA will identify a relatively high fitness, feasible parameter combination. Convergence upon the Utopian Values of TW=1260°K and AC=7.8mm^2 is illustrated in figure 4. Having defined a point within a feasible region a secondary search can take place to establish the bounds of a feasible subset and an optimal design solution can be achieved from this subset by initiating a tertiary search within it.

5.1 Creating a Feasible Subset

Let us assume that the following constraints apply:

$1210° K < TW < 1250° K$; $7.6mm^2 < AC < 8.6mm^2$

The results from a VEGA run searching for TW = 1230°K, AC = 8.0mm^2 , CFR = minimum; define the following Utopian Values for the five variable parameters: TCI=703.91° K; RAT=1.385; SGC=1.172; KW=29.25 DTH=1.61mm

It is now possible to introduce ten further parameters (LB_1 ....LB_5; UB_1 ....UB_5), where LB=lower bound and UB=upper bound, which define the upper and lower bounds of any variation of TCI, RAT, SGC, KW and DTH about their Utopian Values. This is achieved by discretising the ranges between each of the Utopian parameter values and their respective initial upper and lower limits ie those limits that describe the overall search space.

Table 1: Preliminary Directed Search Results

Constraints: 1210° K<TW<1250°K and 7.6mm²<AC<8.6mm²

Hypercube Bounds:

	Vertex Testing			Mid-Point Testing			Overall Search Space	
	UB	**UT**	**LB**	**UB**	**UT**	**LB**	**UB**	**LB**
TCI (°K):	743.28	703.9	701.75	752.4	703.9	700.0	800.00	700.00
RAT:	1.50	1.39	1.34	1.57	1.39	1.34	1.60	1.05
SGC:	1.191	1.172	1.156	1.207	1.172	1.141	1.50	0.50
KW:	31.43	29.25	21.67	31.50	29.25	20.93	33.00	18.00
DTH (mm):	2.0	1.61	0.94	2.31	1.61	0.88	2.5	0.75

Violation Measures:

	Vertex Testing	Mid-Point Testing	Overall Search Space
%age violation:	5.3%	18.8%	93%
Max. TW (°K):	1261.63	1271.34	1402.00
Min. TW (°K):	1202.182	1194.17	1062.56
AMV TW(°K):	2.36	4.49	46.47
Max AC (mm²):	8.66	8.81	18.18
Min AC (mm²):	7.40	7.40	5.47
AMV AC (mm²):	0.028	0.079	0.0043

AMV = Average Magnitude of Violation
UT = Utopian Value as defined by Vega.

A secondary GA search for the optimal upper and lower bounds of the five parameters can now be initiated. The objective function is to minimise the number of non-feasible solutions returned from these runs whilst maximising the range of each parameter within their respective initial limits. In the first instance the hypercube vertices were considered to be suitable points at which to test the degree of constraint violation of each of the GA generated hypercubes. To do this it is necessary to initiate thirty-two runs of the mathematical model per parameter set, one for each of the combinations of LB_n and UB_n which define the hypercube vertices. The relative fitness of the hypercubes is determined from the degree of constraint violation and the magnitude of the allowable parameter variation. This results in conflicting criteria as violation is minimised by drawing in the hypercube vertices whilst increasing parameter variation is achieved by forcing the vertices apart.

The obvious disadvantage of this approach is the computational expense incurred by the thirty-two calls to the fitness function for each hypercube. To reduce the number of calls from 2^n to $2n$ (n = no of parameters) mid-point testing has been introduced. This involves the testing of the mid-points of each of the hypercube surfaces and requires ten calls to the fitness function per parameter set. It is apparent that mid-point testing is likely to result in a higher degree of constraint violation. It is also possible that the topology of the feasible region may be such that vertex testing results in the generation of a highly restricted subset. Initial results have been very encouraging [25]. The results shown in Table 1 compare the two approaches. Constraint violation within the GA generated hypercubes is shown as a percentage of the solutions returned from an exhaustive search of the overall search space.

6 Fundamental GA Research

Fundamental research concerning the operational characteristics of the GA is at present concentrated upon three areas of interest :

- An analysis of the static building block hypothesis
- An analysis of the action of crossover
- The development of a generic search space visualisation tool

The first of these areas involved the extraction of building blocks [3] from the populations of the GA during the optimisation process [27]. These short, low order schemata were then assigned a fitness equal to the mean fitness of a large random sample ($o(10^3)$) of chromosomes which match them. This measure approximates the static fitness of the building block. The 100 most sampled building blocks were then combined with each other to measure the effect upon fitness. A control experiment involved the addition of 100 randomly generated building blocks to each of these highly sampled building blocks and measuring the fitness gain in the same fashion. Figure 5 illustrates the results of this analysis upon the function f5 from De Jong's test suite [28].

This result is typical of the study - namely that the schemata resulting from the combination of two highly sampled building blocks are not of significantly different static fitness to the better of their constituent pairs of building blocks.

The action of crossover is traditionally described as constructing solutions through the combination of building blocks. However, if we examine the effect of crossover upon the parameter sets associated with the chromosomes we notice that provided the mapping from chromosome to parameter set is linear, the action of crossover closely resembles a rotation [29]. Figure 6 illustrates the unique solutions generated with single point crossover on the parent chromosomes 10101101 and 01010100. Each chromosome maps to two 4 bit binary integers which are then translated into decimal integers.

$$10101101 \rightarrow 1010,1101 \rightarrow (10,13)$$
$$01010100 \rightarrow 0101,0100 \rightarrow (5,4)$$

From the above figure we find that the effect of crossover is a rotation and scaling about the midpoint of the parent parameter sets. This is true for a variety of crossover types and chromosomal representations. From this description of crossover we may deduce an accurate lower bound upon the expected performance (in terms of the mean fitness of the population) of a GA, albeit ignoring the effect

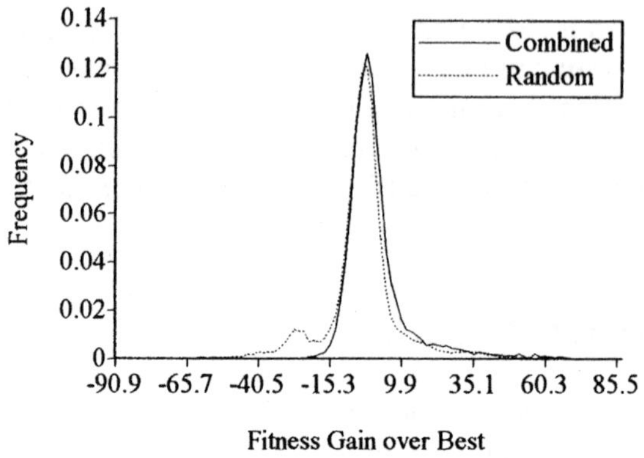

Figure 5: Fitness gain through building block combination for function f5

of mutation, upon the simple function $f(\mathbf{X}) = |\mathbf{X}|^2$. This lower bound is given by the inequality:

$$\langle f(P_{t+1}) \rangle \geq f(P_t) + \left(1 - \frac{p_c}{4}\right) \frac{\sigma_t^2}{f(P_t)} \tag{4}$$

where $f(P_t)$ is the mean fitness of the population at generation t, p_c is the crossover probability and s^2_t is the variance of fitness at generation t. Figure 7 illustrates this lower bound upon the average mean fitness of 500 runs of a GA with two point crossover at a rate of 0.6. Current research includes the extension of this model from predicting one generation ahead to predicting many generations ahead.

This description of crossover has led to the development of an alternative crossover operator for real coded GAs. The parent chromosomes are rotated

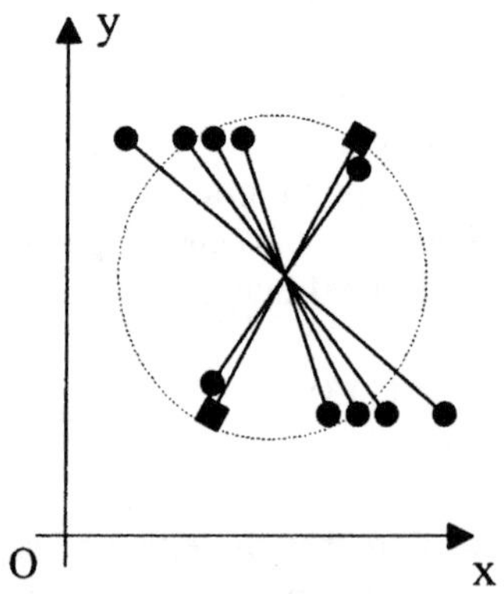

Figure 6: The Rotational Effect of Crossover

through a random angle about their midpoint to create the offspring chromosomes. This rotational crossover operator has performed well in comparison to traditional real coded GA crossover operators upon a number of test functions.

The visualisation research currently involves the search for a qualitative, rather than quantitative, general technique. The difficulty of representing highly dimensional surfaces may be circumvented by generating one or more two dimensional surfaces which provide a qualitative representation of the original multidimensional surface. From a suitable representation the designer should be able to determine whether the surface is noisy, discontinuous, multimodal or smooth. This information will help the designer in the selection of a suitable optimisation technique for the problem or for the tuning of the control parameters of the GA. It is intended that the visualisation technique will work in tandem with the GA, using the solutions generated by the GA to create the representation of the surface. This will reduce the computational expense associated with visualisation. Techniques such as neural networks are currently under consideration for the generation of this qualitative representation.

7 Introducing Intelligence into the Search Process

The techniques described in previous sections for dealing with problems which are characteristically "hard" for the GA and other adaptive search methods are examples of ways in which more "intelligence" can be introduced into the search process in order to cope with complexity in the search space. In general complex search spaces can be characterised by a number of features including: uncertain and incomplete design models; non-linear, discontinuous objective functions; non-linear, discontinuous constraints; noise in the objective function owing to

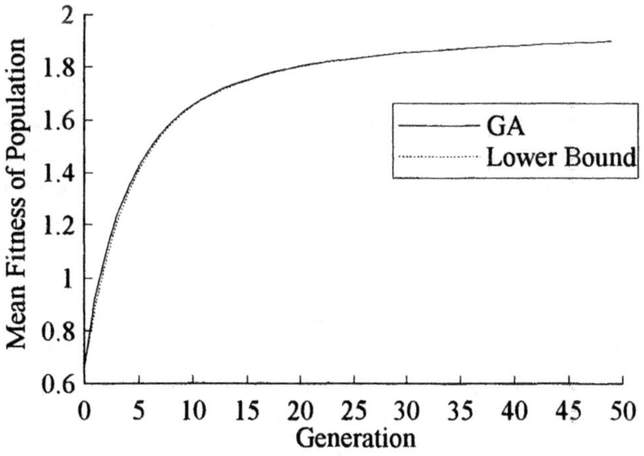

Figure 7: Expected Lower Bound of Performance

model inaccuracies; critical, stringent and conflicting design objectives. The extent to which an adaptive search procedure is able to deal with these features could be said to be a measure of its intelligence.

Of course, the intelligence could be exercised during the search by the human designer, using clues provided by the computational search method, together with domain knowledge both of the design method and of the design task, to guide in the selection of appropriate search directions/regions. The use of clustering techniques together with engineering insight is an example of this. Alternatively, the computational search process itself could be provided with a higher degree of machine intelligence in order to deal with complexity more effectively and efficiently. In many cases the dimensionality of the problem will preclude a purely human approach owing to the difficulties of visualisation. On the other hand it is difficult to conceive how all the intuition and sophisticated knowledge which the human designer brings to a problem might be captured in a purely machine-based approach to intelligent search. Most likely is a combination of the two approaches.

7.1 Fundamental abilities and mechanisms of intelligent search

An intelligent search process, whether human-based, machine-based or both, must possess the following abilities:

- selection - the ability to selectively interpret and process observations of the search space in order to cut down on the potentially enormous number of possible observations, by selecting those observations which are most effective in the search process, at the same time not discarding useful knowledge; an example of this is the detection of, and focussing attention, on feasible regions of a heavily constrained search space;
- learning - the ability to maintain knowledge of the search space for use in the planning and selection of future search actions, implying the existence of a memory of the search space and of past search actions; an example of this is the maintenance of a memory of those search actions which resulted in feasible solutions in order to assist in the location of feasible regions;
- adaption - the ability to detect and make use of novel information carried in the observations of the search space, in order to make the search more efficient by modifying the search strategy to take account of this information; there are several examples of adaption of the search strategy parameters, eg mutation rates, in the GA literature, but few, if any, appear to make explicit use of the detection of novel features in the search space to guide strategy adaption.

The diagram in Figure 8 shows how these abilities can be seen as fundamental mechanisms in a dynamic system representation of the search process. The selection mechanism identifies and interprets a subset of the3 observations for retention in memory and use in the planning and generation of future search actions. The observations may consist of both the search actions and evaluations of their fitness. The memory and working memory mechanisms retain long-term

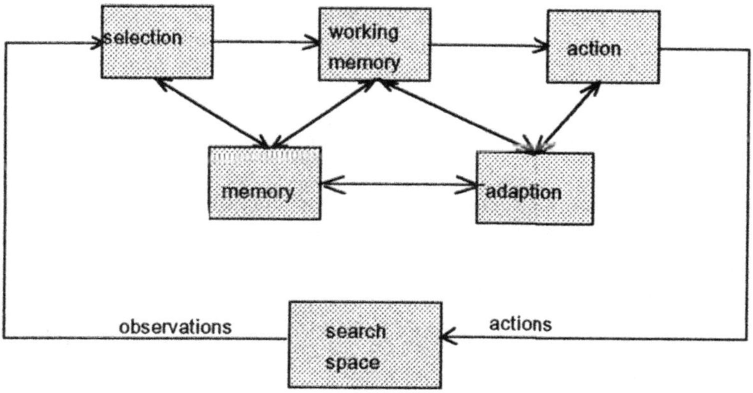

Figure 8: The Intelligent Search Process

knowledge of the search space and short-term information necessary for immediate selection of appropriate search actions respectively. The adaption mechanism embodies the necessary interplay between memory and observations which ensures that detection of new features in the search space causes appropriate responses in the search strategy. The action mechanism plans and selects new actions, ie locations in the search space, based on the selected subset of observations held in working memory.

Of course, the GA and other search methods already have these abilities to a greater or lesser extent. The Darwinian evolution process on which the GA is based has diversity and selection as two of its fundamental requirements and one can easily map the GA onto the dynamic representation of the search process shown in Figure 8. However one can also use this overall representation to identify ways in which the GA itself might be enhanced to make it more effective in searching complex spaces. Learning and memory mechanisms in the GA are implicit in the way in which the next generation is created from the last, hopefully making use of "good" past observations of the search space, embodied perhaps in the fundamental "building blocks" which are preferentially maintained in each successive population. The way in which the next generation is created could perhaps be improved if long-term memory of the search space was used to process the current observations to create a better perception of the space and a more effective set of search actions. This might be seen as analogous to the ability of certain members of a population to actively enhance their chances of survival and hence the passage of their genes to the next generation by making the most effective use of their knowledge and skills during their lifetime, based on their ability to learn and use memory.

We are currently investigating ways in which such explicit use of memory might be implemented in the search process, perhaps using neural network methods to store long-term information about the search space based upon the observations.

8 Adaptive Search Software Development

Experimental adaptive search design software incorporating the concepts and methodologies previously described is under development within the Centre. The objective of this work is to develop a tool which provides the engineer with a degree of control of the search process. This includes the ability to change the nature of the search by selecting an alternative adaptive or deterministic technique.

The software currently incorporates a Genetic Algorithm, Simulated Annealing, Tabu Search, Hooke and Gieves and a Simplex method. The user can integrate a fitness function and initiate any one of these search / optimisation techniques. The clustering algorithm and the sensitivity analysis are also included and it is intended to also integrate multi-criteria techniques based upon Schaffer's VEGA and the directed search technique for constrained spaces as previously described. The implementation of a Structured Genetic Algorithm will also be possible. Suggested user scenarios include:

- Initiate a diverse GA search
- Cluster after *n* generations
- Compare characteristics of each cluster of solutions
- Test for sensitivity
- Select region for further investigation
- If sensitivity is low try numerical methods from various starting points
- Otherwise initiate SA, Tabu or a secondary GA search.

If a specific region is of interest the user will be able to implement a directed search and establlish a hypercube defining a particular search space around a specific point. This technique can also be utilised to partially define feasible regions within heavily constrained spaces. Having defined suitable subsets of the overall search domain any of the search / optimisation techniques available can be utilised for a secondary local search. Alternatively a diverse GA search can be initiated to identify clusters of good solutions within the subset and the search be further localised.

The overall objective is to develop a tool which allows the engineering designer to rapidly scan the available search space identifying regions of interest before isolating those which warrant closer investigation and initiating local searches within them. Initially the tool is primarily aimed at the preliminary design process. It is intended to further develop and refine the software as research progresses and the efficiency of the various techniques improves. New approaches will also be included and the ultimate long term aim must be the development and integration of intelligent processes to enable the engineer to negotiate those regions exhibiting particularly complex characteristics. In the short to medium term however it is expected that the software can be introduced to industry and can be utilised as an effective decision support tool for preliminary design.

9 Summary

Each of the industrial case studies offer a rich design environment which, although amenable to the application of conventional optimization techniques, present such techniques with severe difficulties. The multi-dimensional aspects of the design problems combined with high modality, conflicting criteria and physical constraints create complex fitness landscapes which contain numerous local optima and discontinuity. The powerful, non-linear processing techniques of the Genetic Algorithm are currently negotiating these design spaces in a satisfactory manner. It is expected that further development and the integration of allied techniques plus the inclusion of new techniques arising from the fundamental GA research will result in the establishment of highly interactive preliminary and detailed experimental design tools that are both potent and robust.

In order to develop a designer / GA system that will produce a truly optimum design, a significant amount of research will be necessary to establish methodologies which will allow the successful integration of adaptive search with current design practice. It is intended that the evolutionary-oriented design methodology will be apparent at every level of the overall design process with the engineer providing the control and continuity both within and between individual design areas.

It is the opinion of the author that, in the short to medium term, Adaptive Search has most to offer when utilised during the preliminary stages of the design process. The integration of techniques such as clustering and directed search will allow the engineer to rapidly scan the overall search space before isolating regions of particular interest. These regions can then be further investigated to provide an optimal preliminary design solution. This approach could significantly reduce design lead times and provide innovative alternatives and approaches to particular design problems. In the longer term the integration of some form of machine intelligence to supplement designer knowledge will be required in order to increase the efficiency of the search process. This will allow the GA processing of complex, computationally expensive analytic problems at the detailed design stage within acceptable time and budget constraints. At the moment the GA, with its various operators and operator probability represents a multi-parameter control problem that the engineer must solve to some extent before successful negotiation of a complex design space can be accomplished.

10 Acknowledgements

The research described in the paper is currently funded by the UK Engineering and Physical Science Research Council. Financial support and design expertise for the gas turbine engine work is also being provided by Rolls Royce plc. GEC Plessey Semiconductors of Plymouth and Knight Piesold and Partners, a UK civil engineering consultancy of Ashford, Kent are supplying the necessary design software and/or expertise to support the concrete arch dam and digital filter research respectively. The authors wish to thank the above organizations for their continuing assistance.

References

1. Goldberg D.: Genetic Algorithms in Search, Optimization and Machine Learning. Addison Wesley, 1989.
2. Davis, L.: Handbook of Genetic Algorithms. Van Nostrand Reinhold, 1991.
3. Holland J.: Adaptation in Natural and Artificial Systems. The University of Michigan Press, 1975.
4. Parmee I. C.: Pneumatic Hydropower Devices. PhD thesis, Polytechnic South West, Plymouth, 1990.
5. Suckley D.: Genetic Algorithms in the Design of FIR Filters. IEE Proceedings, Vol 138,No.2, 1991.
6. Adams J.W., Wilson A.N.Jr.: A new approach to FIR digital filters with fewer multipliers and reduced sensitivity. IEEE Trans.,1983, CAS-30, pp.277-283
7. Adams J.W.,Wilson, A.N.,Jr.: A novel approach to the design of efficient FIR digital bandpass filters. IEEE, 1984, ISCAS, pp.28-32
8. Saramaki T., Neuvo Y., and Mitra, S.K.: Design of computationally efficient interpolated FIR filters. IEEE Trans., 1988, CAS-35,(1),pp.70-87
9. Cabezas C.E., and Dinez P.S.R.: FIR filters using interpolated prefilters and equalizers. IEEE Trans. on Circuits and Systems,990, 37,(1) ,pp.17-23
10. Lynn P.A.: FIR digital filters based on difference coefficients:design improvements and software implementation. IEE Proc., 1980,127,(6),pp 253-258
11. Agarwal R.C., and Sudhaker R.: Multiplier-less design of FIR filters. IEEE , 1983, ICASSP,pp.209-212
12. Noonan J.P., and Marquis, D.A.: New algorithm for the design of linear phase filters with +1,-1 and 0 coefficients. Signal Processing,Elsevier Science Publishers B.V., 1989,pp.81-85
13. Wade G., Van-Eetvelt P., and Darwen H.: Synthesis of efficient low-order FIR filters from primitive sections. IEE Proc., 1990,G,(5),pp.367-372
14. Dasgupta D., and McGregor D.R.: A Structured Genetic Algorithm. Research report IKBS-2-91, University of Strathclyde, U.K., Sept. 1991
15. Wade G., Roberts A., Williams G.: Multiplier-less FIR Digital Filter Design Using a Genetic Algorithm. IEE Proc. Visual Image Signal Processing, Vol. 141,No3, June 1994
16. Rabiner L., and Gold B.: Theory and Application of Digital Signal Processing. Prentice-Hall 1975.
17. Chan D. and Rabiner L.: An algorithm for minimising roundoff noise in cascade realisations of FIR digital filters. B.S.T.J., 1973,52,(3),pp 347-385
18. Michalewicz Z.: 'Genetic Algorithms + Data structures = Evolutionary Programs', Springer-Verlag, 1992.
19. Roberts A., Wade G.: Optimisation of Finite Wordlength FIR Filters Using a Genetic Algorithm. Procs. 'Adaptive Computing in Engineering Design and Control', Plymouth Engineering Design Centre, University of Plymouth, 1994.

20. Parmee I. C.: The Concrete Arch Dam - An Evolutionary Model of the Design Process. Procs: International Conference on Neural Nets & Genetic Algorithms, Springer-Verlag Wien, 1993
21. Parmee I. C.: The Genetic Algorithm and Civil Engineering Design. The International Journal of Construction Information Technology, University of Salford, Vol2, No 1, Spring 1994.
22. Jarvis R. A., Patrick E. A.: Clustering using a Similarity Measure based upon Shared near Neighbours. IEEE Transactions on Computers, Vol. C22, No 11, November 1973, pp 1025 - 1034.
23. Parmee I. C.: The Implementation of Adaptive Search Tools to Promote Global Search in Engineering Design. Procs. 3rd IFIP Working Conference on Optimisation-Based Computer-Aided Modelling and Design, Prague, May 24-26th, 1994
24. Michalewicz Z., Janikow C. Z.: Handling Constraints in Genetic Algorithms. Proccs. International Conference on Genetic Algorithms, Morgan Kaufman ,1991.
25. Schaffer J. D.: Multiple Objective Optimization with Vector Evaluated Genetic Algorithms. Procs. International Conference on Genetic Algorithms, 1985, pp 93 - 100.
26. Parmee I. C., Purchase G.: The Development of a Directed Genetic Search Technique for Heavily Constrained Design Spaces. Procs 'Adaptive Computing in Engineering Design and Control', Plymouth Engineering Design Centre, University of Plymouth, Sept.1994
27. Harris, R. & Ellis, C.: An Empirical Investigation of the Schema Theorem and Building Block Hypothesis. Internal report PEDC-01-93. Plymouth Engineering Design Centre, University of Plymouth, 1993.
28. De Jong K.: An Analysis of the Behaviour of a Class of Genetic Adaptive Systems. University of Michigan press, Ann Arbor, 1975.
29. Harris R. & Ellis C.: An Alternative Description of the Action of Crossover. Internal report PEDC-02-93. Plymouth Engineering Design Centre, University of Plymouth, 1993.

Evolution Strategy, its Parallelization and Application to Discrete Optimization Problems

G. Thierauf and J. Cai

Department of Civil Engineering, University of Essen,
45117 Essen, Germany

Abstract: The parallelization of a modified multi-membered evolution strategy for solving structural optimization problems with discrete variables is described. Evolution strategies (ESs) imitate biological evolution and combine the concept of artificial survival of the fittest with evolutionary operators to form a robust search mechanism. An important characteristic of ESs that differs from other conventional optimization algorithms is that instead of a single design point the ESs work simultaneously with a population of design points in the space of variables. This allows for an implementation in a parallel computing environment. In this paper two ways of the parallel implementation of ESs and their realization on two parallel computing environments are presented.

Keywords: Discrete optimization, parallel evolution strategy

1 Introduction

To solve discrete optimization problems several methods have been developed. Most of the methods transform a discrete problem into a sequence of continuous problems which are solved iteratively [2,3,5]. In every iteration a continuous optimization problem must be solved. Two recently developed direct search methods which solve the discrete problems directly are genetic algorithms (GAs) and evolution strategies (ESs). Rajeev et al. [6] presented an artificial genetic approach for discrete optimization of structural design problems. Cai and Thierauf [4] modified the evolution strategies developed by Rechenberg [7] and Schwefel [8] for solving discrete problems.

Instead of a single design point, the ESs work simultaneously with a population of design points in the space of variables. This characteristic allows for an implementation in a parallel computing environment. There are two ways of parallel implementation: a directly parallelized evolution strategy (DPES) and a parallel sub-evolution-strategy (PSES). In the following the application of ESs to solve discrete optimization problems and its parallelization on a Power-Xplorer parallel system and on a workstation-cluster by using

PVM (Parallel Virtual Machine) are described. In the PVM environment each workstation is considered as a processor.

2 The evolution strategies

For convenience of the further discussion the evolution strategies (ESs) will be described briefly. The ESs can be divided into the two-membered evolution strategy (2-ES) and the multi-membered evolution strategy (M-ES)

2.1 The 2-ES

An optimization problem can be formulated in the following form:

$$\begin{array}{lll} \min & f(X) & X = \{x_1, x_2, \ldots, x_n\}^T, \\ \text{subject to}: & g_i(X) \geq 0, & i = 1, 2, \ldots, m. \end{array} \tag{1}$$

For a discrete optimization problem the design variables x_i $(i = 1, 2, \ldots, n)$ can only take the discrete values of a given set R^d.

To solve the optimization problem (1) the 2-ES works in two steps:

Step 1 (mutation):

In the g–th generation a new design-vector is computed from

$$X_O^{(g)} = X_P^{(g)} + Z^{(g)}, \tag{2}$$

where $X_O^{(g)}$ and $X_P^{(g)}$ are the offspring vector and parent vector, respectively; $Z^{(g)} = [z_1^{(g)}, z_2^{(g)}, \ldots, z_n^{(g)}]^T$ is a vector of random change.

For continuous problems the components $z_i^{(g)}$ are random numbers with the normal distribution [8]:

$$p(z_i) = \frac{1}{\sqrt{(2\pi)}\,\sigma_i} \exp(-\frac{(z_i - \xi_i)^2}{2\sigma_i^2}), \tag{3}$$

where ξ_i is the expectation, which should have the value zero, and σ_i^2 is the variance, which should be small.

As a modification of the ES, for discrete problems the components of the random change vector $Z^{(g)}$ should be chosen as follows [4]:

$$z_i^{(g)} = \begin{cases} (\kappa + 1)\delta x_i & \text{for } l \text{ randomly chosen components } (l < n), \\ 0 & \text{for all other } (n - l) \text{ components}, \end{cases} \tag{4}$$

where

$\frac{l}{n}$ ranges from $\frac{1}{10}$ to $\frac{2}{5}$,

δx_i is the current difference between two adjacent discrete values and

κ is an integer random number with Poisson-distribution:

$$p(\kappa) = \frac{(\gamma)^\kappa}{\kappa!} e^{-\gamma} . \tag{5}$$

γ is the expectation and variance of κ, which should range from 0.001 to 0.1.

Step 2 (selection):

The selection chooses the best individual from the parent and the offspring to survive:

$$X_P^{(g+1)} = \begin{cases} X_O^{(g)} & \text{if } g_i(X_O^{(g)}) \geq 0\,,\, i = 1, \ldots, m \\ & \text{and } f(X_O^{(g)}) \leq f(X_P^{(g)})\,; \\ X_P^{(g)} & \text{otherwise .} \end{cases} \tag{6}$$

According to Schwefel's research [8] the search is terminated if the value of the objective functions has been improved by less than a given value ε in the course of $20n$ trials. For discrete problems the value ε can be given as $\varepsilon = 0$.

2.2 The M-ES

In the case of multi-membered ES a population of μ parent vectors and λ offspring vectors will be handled simultaneously.

Step 1 (recombination and mutation):

In the $g-$th generation μ vectors called parent vectors are given. λ offspring vectors are created from the μ parent vectors by means of recombination and mutation. The operator recombination will be described in the flowing.

Step 2 (selection):

There are two variants of M-ES, the $(\mu + \lambda)-$ES and the $(\mu, \lambda)-$ES.

$(\mu + \lambda)-$ES

The best μ individuals are selected from parents and offsprings to form the parents of the next generation.

$(\mu, \lambda)-$ES (with $\lambda > \mu$)

The best μ individuals are selected only from the λ offsprings. The μ parents in the $g-$th generation die out.

In step 1, a temporary parent vector $\tilde{X} = [\tilde{x}_1, \tilde{x}_2, \ldots, \tilde{x}_n]^T$ for every offspring vector is first built by means of recombination. The recombination can be carried out as follows.

For a continuous problem five recombination cases are given in Ref. [10]:

$$\tilde{x}_i = \begin{cases} x_{a,i} \text{ or } x_{b,i} \text{ randomly} & \text{(A)} \\ 1/2(x_{a,i} + x_{b,i}) & \text{(B)} \\ x_{bj,i} & \text{(C)} \\ x_{a,i} \text{ or } x_{bj,i} \text{ randomly} & \text{(D)} \\ 1/2(x_{a,i} + x_{bj,i}) & \text{(E)} \end{cases} \tag{7}$$

where X_a and X_b are two parents randomly chosen from the population, the i-th component $x_{bj,i}$ is chosen randomly from the i-th components of all μ parent vectors.

For a discrete problem the recombination cases (B) and (E) in Eq. (7) are not suitable, because the mean value of two discrete values is usually not a discrete value from the given set. For a discrete problem the following recombinations are suggested in Ref. [4]:

$$\tilde{x}_i = \begin{cases} x_{a,i} \text{ or } x_{b,i} \text{ randomly} & \text{(A)} \\ x_{bj,i} & \text{(B)} \\ x_{a,i} \text{ or } x_{bj,i} \text{ randomly} & \text{(C)} \\ x_{m,i} \text{ or } x_{b,i} \text{ randomly} & \text{(D)} \\ x_{m,i} \text{ or } x_{bj,i} \text{ randomly} & \text{(E)} \end{cases} \tag{8}$$

The vector X_m is the best of the μ parent vectors in the $g-$th generation. In cases (D) and (E) the information from the best parent can be used and a better convergence for many problems can be obtained. From the temporary parent $\tilde{X}$ an offspring can be created in the same way as in the two-membered ES (Eq. (2)).

For discrete problems the following termination criteria are suggested here. The search will be terminated,

a) if the best value of the objective function in the last K_I ($\geq 4n\mu/\lambda$) generations has not been improved or

b) if the mean value of the objective function values of all parent vectors in the last K_{II} ($\geq 2n\mu/\lambda$) generations has been improved by less than a given value ε_b or

c) if the relative difference between the best objective function value and the mean value of objective function values of all parent vectors in the current generation is less than a given value ε_c or *d*) if the ratio μ_b/μ has reached a given value ε_d, where μ_b is the number of the parent vectors in the current generation which have the best objective function value.

For discrete problem the value ε_b can be set to $\varepsilon_b = 0$. For the examples in this paper the values ε_c and ε_d are usually set to $\varepsilon_c = 0.0001$ and $0.5 \leq \varepsilon_d \leq 0.8$.

3 Parallelization of the M-ES

The M-ES works simultaneously with a population of design points in the space of variables. This inherent parallelism allows for an implementation in a parallel computing environment. In the following two ways to parallelize the M-ES, a directly parallelized evolution strategy (DPES) and a parallel sub-evolution-strategy (PSES), are described.

3.1 Directly parallelized evolution strategy (DPES)

In every generation λ offspring vectors will be generated and their objective function and feasibility must be calculated and checked. For a structural optimization problem the check for feasibility is performed by a FE-analysis. The idea of the implementation of DPES is to parallelize the λ FE-analyses on $(n-1)$ processors. In the implementation on a n-processor computer, the n processors will be divided into one master-processor and $(n-1)$ slave-processors. A master job on the master-processor runs the M-ES, generates offsprings and sends them to the slave-processors. The slave jobs on slave-processors check the feasibility of the offsprings and then send the information about the offsprings back to the master job. In the master job μ best individuals will be selected from the complete population as the parent vectors of next generation. A parallel environment works for the communication between the processors and has a role of information transferrer.

If possible, the ratio $\lambda/(n-1)$ should be integer so that the processors share the loading equally. This implementation can be called directly parallelized evolution strategy (DPES). The program flowchart with DPES is shown in Fig. 1.

The first implementation was performed on an architecture with 7 HP 700 workstations under PVM environment. Every workstation has its own CPU and all workstations share only one hard disk. A general purpose finite element program B&B [9] was employed for the FE-analysis on the slave-processors.

As an example, the well-known 10-bar truss problem was considered. Four tests with different numbers of slave-processors have been performed. Unfortunately the results show a low speed-up of real time. Two factors might be

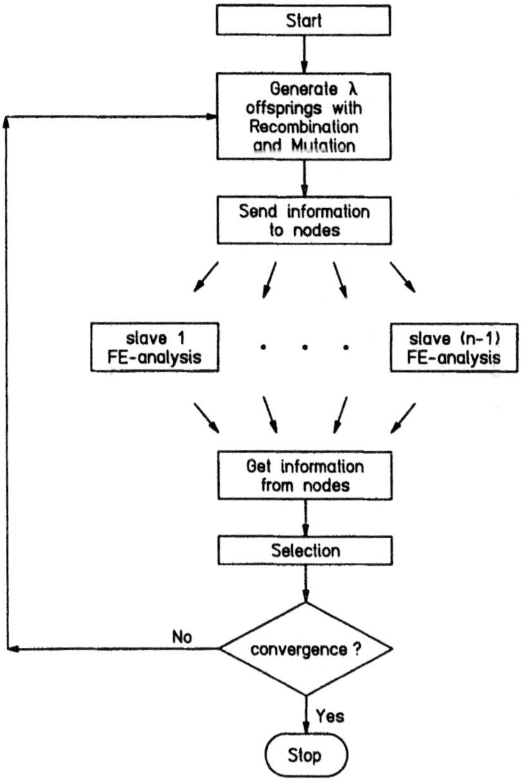

Figure 1: Program flowchart with DPES

considered. First, the frequency of information exchange in this implementation is high, because information exchange must be carried out before and after every FE-analysis. Second, there is no direct connection between the workstations, all information must be transferred under PVM through the network.

The second implementation was performed on a Power-Xplorer system based on PowerPC 601 processors and transputer communication technology. The parallel operating environment is PARIX.

In this implementation two examples, the 10-bar truss problem and a 200-bar truss problem, were considered. For FE-Analysis a special program for truss analysis was employed. For the first example, because a FE-analysis for such a small truss does not take much more computing time than the ES operations and the communications, only a low speed-up has been reached. In the second example a FE-analysis needs much more computing operations than in the first example and the result shows a quasi-linear speed-up.

3.2 Parallel sub-evolution-strategy (PSES)

In structural optimization by using M-ES the probability to achieve the global optimum is proportional to the size of the population [8]. However, in some applications time limitations make it impossible to run the M-ES with a large population. The idea of the PSES is to divide the large population into several small subpopulations and to parallelize the ES.

In the parallel implementation on a n-processor computer, the whole population will be divided into n subpopulations, each processor runs the ES on its own subpopulation (called sub-evolution-strategy or sub-ES). Periodically some good individuals will be selected and copies of them will be sent to one of its neighbors. It also receives copies from its neighbors, which replace the bad individuals in its own subpopulation. This implementation is called parallel sub-evolution-strategy (PSES).

In the realization of the PSES, for convenience of management and control, a further processor will be introduced. A master job on this extra processor plays the role of manager and organizes the information exchange between other n slave-processors. It starts n optimization processes with sub-ES (slave jobs) on the slave-processors, gets and sends information from or to them.

The information exchange can be carried out arbitrarily between the slave jobs or with a determined sequence, for example, from slave job $i-1$ to i and from i to $i+1$. In the case of an arbitrary exchange an optimization process with PSES itself becomes random, because a slave job can be located on different processor when restarting the process, and the optimization process can be badly controlled. Thus, the information exchange in this paper will be carried out in a determined sequence. During a test, if a slave job is finished, the master will redefine a new sequence of information exchange for the slave jobs which are still running.

The exchange of individuals between the subpopulation is the main advantage of PSES over the sequential M-ES and makes it possible that the n subpopulations work with a similar probability to achieve the global optimum like the original large population, but within a much shorter computing time.

The frequency of exchange and the number of exchanged individuals are two adjustable parameters. The frequency and the number should not be too small, otherwise the evolutionary process in every subpopulation will be isolated and the advantage of a large population to get global optimum will be lost. If the frequency and the number are too large, the individuals in a subpopulation will be quickly replaced by the ones from other subpopulations. After several exchanges the n subpopulations become the same or similar and a few of the good individuals dominate over the whole population, which may lead to a premature termination of the search.

The frequency of exchange can be defined as the number of generations. After a certain number of generations (say m) the exchange can be performed.

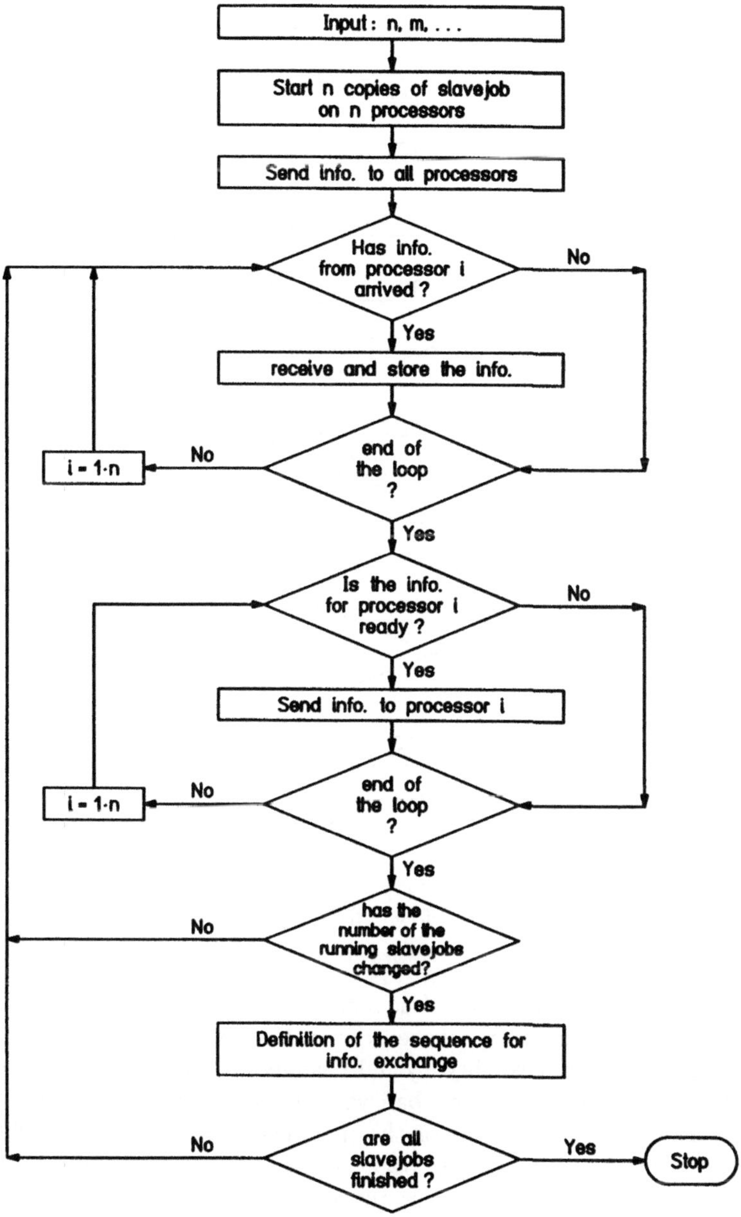

Figure 2: PSES - flowchart of the master program

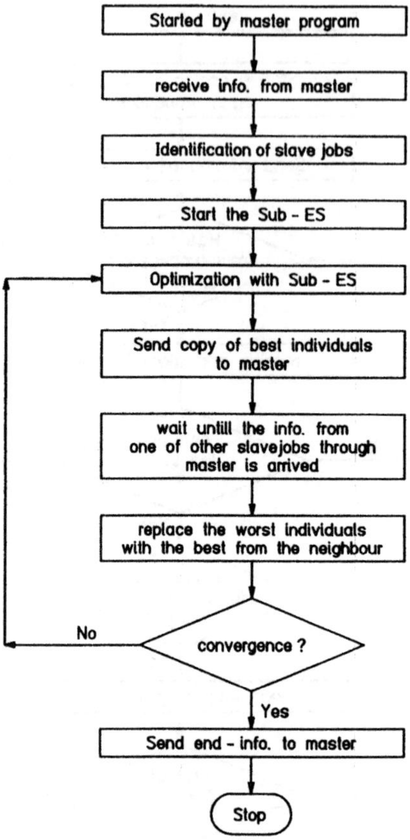

Figure 3: PSES - flowchart of the slave program

In many cases this definition of frequency could lead to long waiting states during the run of PSES. In a generation not all of the λ offsprings should be checked for their feasibility by a FE-analysis, some of them might be eliminated only because of their bad value of the objective function. After m generations the unevenness of the numbers of FE-analyses in different slave jobs causes waiting states, because the FE-analyses take much more computing operations than the operators of ES. For this reason, instead of the number of generations, we define the number of FE-analyses as the frequency of exchange m. According to our experience the frequency m can be chosen as $(0.7 \sim 0.9)\lambda_s K_I$ or $(0.7 \sim 0.9)\lambda_s K_{II}$, where λ_s is the number of the offsprings of a subpopulation.

The number of exchanged individuals n_{exch} depends on the number of parent vectors in a subpopulation. As result of our study, n_{exch} could be

20% to 30% of the number of parents of a subpopulation.

The implementation with PSES has a low frequency of information exchange and is suitable for the parallelization of M-ES on a workstation-cluster. The flowcharts of master and slave program with PSES are shown in Fig. 2 and 3, respectively.

4 Examples

Example 1: Ten-bar truss

As first example we consider the well-known 10-bar truss problem shown in Figure 4. The objective function of the problem is the weight of the structure. The design variables are the cross-sectional areas of the 10 members. The constraints are the member stresses and the vertical displacements of the nodes 2 and 4. The allowable displacement is limited to 2 in. and the allowable stress to ± 25 ksi. The design parameters are $E = 10^4$ ksi, $F = 100$ kips, $\varrho = 0.1$ lb/in^3 and $a = 360$ in. (ksi-units are chosen, for reference only).

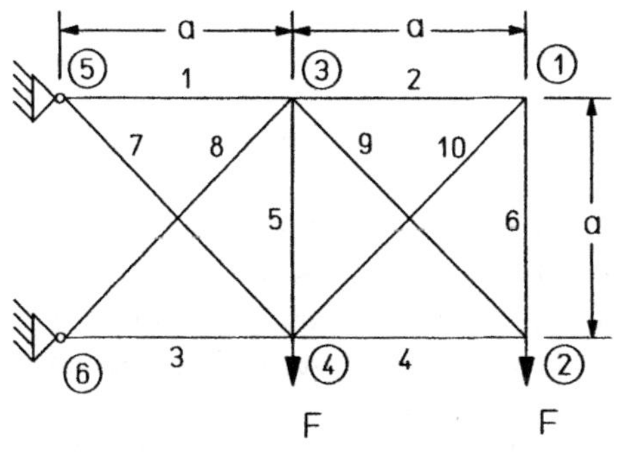

Figure 4: 10-bar truss

The following discretization of the design variables is considered (according to DIN 1028, double angle profiles).

$$S = ((0.100,)\ 0.347,\ 0.440,\ 0.539,\ 0.954,\ 1.081,\ 1.174,\ 1.333,\ 1.488,\ 1.764,\ 2.142,\ 2.697,\ 2.800,\ 3.131,\ 3.565,\ 3.813,\ 4.805,\ 5.952,\ 6.572,\ 7.192,\ 8.525,\ 9.300,\ 10.850,\ 13.330,\ 14.290,\ 17.170,\ 19.180,\ 23.680,\ 28.080,\ 33.700)\ \text{(sq in.)}$$

Bremicker et al [2] solved this problem using a MINSLIP-approach (f_{min} = 5153.90 lb). Cai and Thierauf [3] solved it using a combination of penalty

Table 1: Results for Example 1, sequential M-ES

test	$(\mu+\lambda)$	recom.-case	n_{FE}	n_{ge}	f_{min} (lb)
1	10+10	D	1783	325	5113.4
2	10+10	E	3313	676	5100.3
3	60+60	E	7373	221	5100.3

Table 2: Results for Example 1, DPES - with PVM

test	n_{proc}	real time (s)	discrete solution (lb)
p10-01	1	2472.23	5100.3
p10-02	2	1972.79	5100.3
p10-05	5	1437.49	5100.3
p10-10	10	1453.13	5100.3

algorithm and a sequential quadratic programming code NLPQL, the discrete solution is $f_{min} = 5152.92$ lb. To get this solution, 6 iterations were needed and in every iteration a complete continuous optimization problem had to be solved [3].

In this paper the example was solved with sequential M-ES, DPES and PSES. The results with (10+10)-M-ES and (60+60)-M-ES are shown in Table 1, where n_{FE} means number of FE-analyses and n_{ge} number of generations. Table 2 shows the results of four tests with DPES on a workstation-cluster by using PVM, where n_{proc} means number of slave-processors. Only a low speed-up of real time has been achieved. In the test p10-10, 5 workstations have worked as 10 virtual processors. A similar speed-up is given by the tests on a Power-Xplorer system (Table 3). The real computing times of the implementations of DPES on workstation-cluster and on Power-Xplorer system are quite different. That is because different programs for FE-analysis were used.

Table 3: Example 1, speed-up of DPES - with Power-Xplorer

n_{proc}	1	2	5
time(s)	2.065	1.563	1.391
speed-up	1	1.32	1.48

Table 4: Results for Example 1, PSES

slave job	subpopulation	recom.-case	n_{FE}	n_{ge}	f_{min} (lb)
1	10+10	E	1387	250	5100.3
2	10+10	E	1305	226	5100.3
3	10+10	E	1355	246	5100.3
4	10+10	E	1462	261	5100.3
5	10+10	E	1153	203	5113.4
6	10+10	E	1049	178	5113.4

Table 4 shows the result of a test on a workstation-cluster with PSES in which the whole population of (60+60) is divided into 6 subpopulations. In this test the frequency of exchange was set to $m = 80$ and the number of exchanged individuals $n_{exch} = 2$. The test with PSES reached a good result. Most slave jobs with sub-ES have the same solution as the sequential (60+60)-ES, but much fewer FE-analyses were needed. In consideration of the numbers of FE-analyses the PSES achieves a quasi-linear speed-up.

Example 2: Two hundred-bar plane truss

The second example is a 200-bar plane truss shown in Fig 5. The design parameters are: $E = 30000$ ksi, $\varrho = 0.283$ lb/in^3 (for reference ksi-units are chosen). The objective function of the problem is the weight of the structure. The constraints are the member stresses and the vertical and horizontal displacements of all nodes. The allowable displacement is limited to 0.5 in. and the allowable stress to ± 30 ksi. The members of the truss are divided into 96 groups and the cross-sectional areas of every group are the design variables (Table 5). There are three loading cases

Loading case 1:

One kip acting in positive x direction at nodes 1, 6, 15, 20, 29, 34, 43, 48, 57, 62, 71;

Loading case 2:

10 kips acting in negative y direction at nodes 1, 2, 3, 4, 5, 6, 8, 10, 12, 14, 15, 16, 17, 18, 19, 20, 22, 24, ..., 71, 72, 73, 74, 75;

Loading case 3:

cases 1 and 2 combined.

The same discretization of the design variables in example 1 is considered.

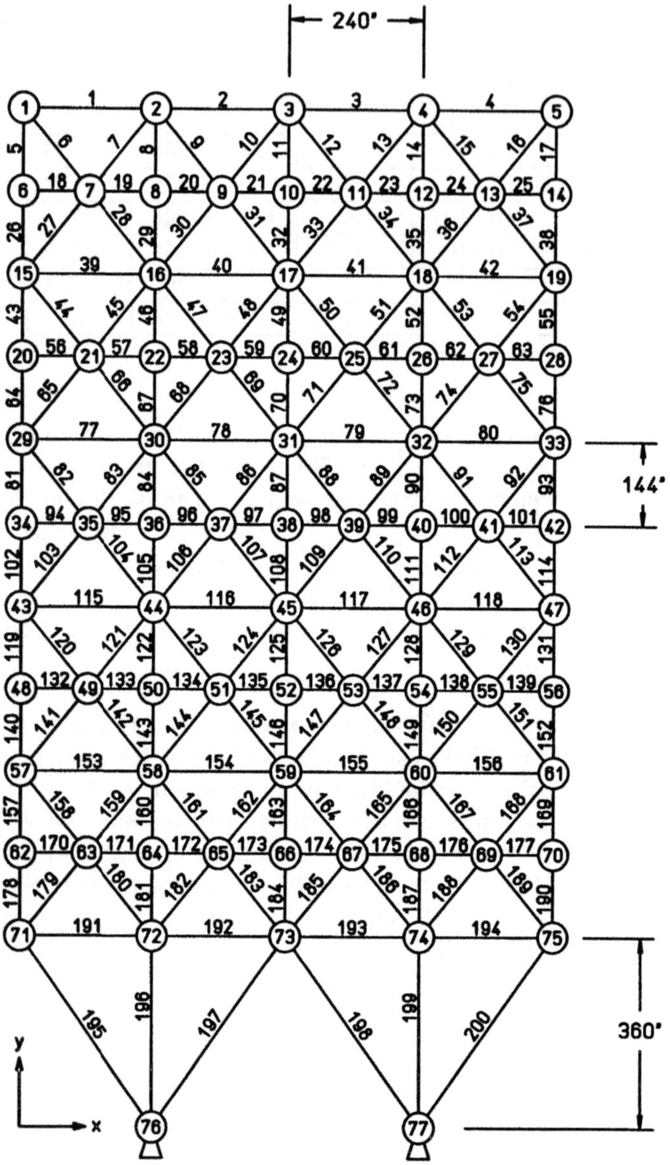

Figure 5: 200-bar truss

Table 5: Design variables in Example 2

No.	member numbers	No.	member numbers
1	1,4	48	97,98
2	2,3	49	102,114
3	5,17	50	103,113
4	6,16	51	104,112
5	7,15	52	105,111
6	8,14	53	106,110
7	9,13	54	107,109
8	10,12	55	108
9	11	56	115,118
10	18,25,56,63,	57	116,117
	94,101,132,139	58	119,131
	170,177	59	120,130
11	19,20,23,24	60	121,129
12	21,22	61	122,128
13	26,38	62	123,127
14	27,37	63	124,126
15	28,36	64	125
16	29,35	65	133,134,137,138
17	30,34	66	135,136
18	31,33	67	140,152
19	32	68	141,151
20	39,42	69	142,150
21	40,41	70	143,149
22	43,55	71	144,148
23	44,54	72	145,147
24	45,53	73	146
25	46,52	74	153,156
26	47,51	75	154,155
27	48,50	76	157,169
28	49	77	158,168
29	57,58,61,62	78	169,167
30	59,60	79	160,166
31	64,76	80	161,165
32	65,75	81	162,164
33	66,74	82	163
34	67,73	83	171,172,175,176
35	68,72	84	173,174
36	69,71	85	178,190
37	70	86	179,189
38	77,80	87	180,188
39	78,79	88	181,187
40	81,93	89	182,186
41	82,92	90	183,185
42	83,91	91	184
43	84,90	92	191,194
44	85,89	93	192,193
45	86,88	94	195,200
46	87	95	196,199
47	95,96,99,100	96	197,198

Table 6: Example 2, speed-up of DPES - with Power-Xplorer

n_{proc}	real time (s)	speed up	discrete solution	
			volume (in^3)	weight (lb)
1	18766.87	1	104730	29641
2	9711.30	1.93	104730	29641
5	4215.44	4.45	104730	29641

Table 7: Results for Example 2, PSES

slave job	sub-population	recom.-case	number of FE-analyses	number of generations	discrete solution	
					volume (in^3)	weight (lb)
1	10+10	E	11261	2299	105105	29745
2	10+10	E	11856	2438	105097	29742
3	10+10	E	12521	2632	105076	29737
4	10+10	E	11451	2306	105097	29742
5	10+10	E	12454	2552	105076	29737
6	10+10	E	10773	2182	105105	29745

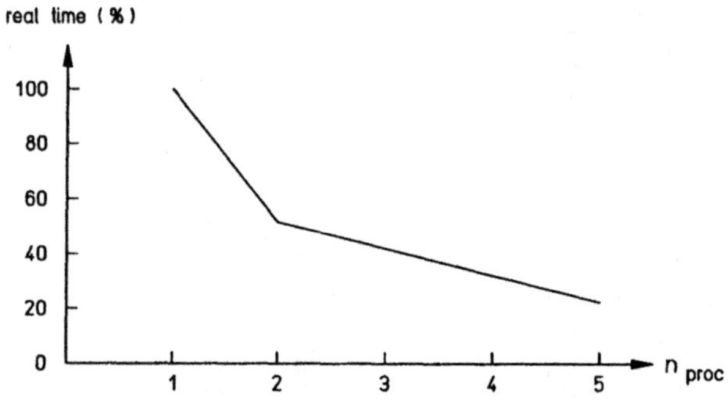

Figure 6: 200-bar truss, speed-up of DPES on a Power-Xplorer system

An optimal solution for a continuous problem was given by Arora et al. [1]: optimal volume 102343 in^3 and weight 28963 lb. In Ref. [4] this problem was solved as a discrete optimization problem by using sequential (20+10)-ES. The optimal volume and weight are 109590 in^3 and 31014 lb, respectively. To solve this problem by using DPES on a Power-Xplorer system and by using PSES on a workstation-cluster two tests were made. By the implementation of DPES a (10+10)-ES was considered. Three tests with different numbers of slave-processors were carried out. Because the FE-analyses for this problem take much more computing operations than the ES-operators and the communications, a quasi-linear speed-up has been achieved (Table 6 and Fig. 6). By the implementation of PSES a population of (60+60) was divided into 6 subpopulations. The number of exchanged individuals was set to $n_{exch} = 2$ and the frequency of exchange was $m = 150$. The results are given in Table 7.

5 Conclusion

The M-ES works on a population of individuals. This inherent parallelism allows for an implementation in a parallel computing environment. The examples show that the directly parallelized evolution strategy (DPES) is more suitable for a parallel computer than for a workstation-cluster. In this case, the parallelization of the M-ES is of advantage only if the FE-analyses take more computing operations than the ES-operators and communications between the processors. For the parallelization of M-ES on a workstation-cluster by using PVM the parallel sub-evolution-strategy (PSES) is a better alternative.

Acknowledgement

The research described in this paper was partially supported by the German Research Foundation (DFG) under grant TH-218/12-1 and by the Ministry of Science and Research of State North Rhine-Westphalia in Germany. The authors would like to thank DFG and the Ministry for their support.

References

1. Arora, J. S. and Govil, A. K., 1977, "An Efficient Method for Optimal Structural Design by Substructuring". *Comp. & Struct.*, Vol. 7, pp. 507-515.

2. Bremicker, M.; Papalambros, P. Y. and Loh, H. T., 1990, "Solution of Mixed-Discrete Structural Optimization Problems with a New Sequential Linearization Algorithm", *Comp. & Struct.*, Vol. 37, No. 4, pp. 451-461.

3. Cai, J. and Thierauf, G., 1993, "Discrete optimization of structures using an improved penalty function method", *Eng. Opt.*, Vol. 21, pp. 293-306.

4. Cai, J. and Thierauf, G., 1993, "Discrete structural optimization using evolution strategies", in *Neural networks and combinatorial optimiza-*

tion in civil and structural engineering, edited by B.H.V. Topping and A.I. Khan, Edinburgh, Civil-Comp Limited, pp 95-100.

5. Fu, J.; Fenton, R. G. and Cleghorn, W. L., 1991 "A Mixed Integer-Discrete-Continuous Programming Method and its Application to Engineering Design Optimization", *Eng. Opt.*, Vol. 17, pp. 263-280.

6. Rajeev, S. and Krishnamoorthy, C. S., 1992, "Discrete Optimization of Structures Using Genetic Algorithms", *J. Struct. Eng.*, ASCE, Vol. 118, No. 5, pp. 1233-1250.

7. Rechenberg, I., 1973, "Evolution Strategy: Optimization of Technical Systems according to the Principles of biological Evolution (in German)", Frommann-Holzboog, Stuttgart.

8. Schwefel, H.-P., 1981, "Numerical Optimization of Computer Models", Wiley & Sons, Chichester.

9. Thierauf, G., 1993, "B&B – Programmsystem zur Berechnung und Bemessung allgemeiner Tragwerke". Universiät GH Essen.

10. Hoffmeister, F. and Baeck, T., 1992, "Genetic Algorithms and Evolution Strategies: Similarities and Differences", Technical Report No. SYS-1/92, University of Dortmund.

Some Results in Optimal Structural Damping Using a Genetic Algorithm

P. Trompette, J.L. Marcelin

Laboratoire Sols, Solides, Structures URA CNRS 1511
BP53 38041 GRENOBLE Cedex 9 FRANCE

Abstract. The possibility of finding the optimal locations of passive dampers in structural engineering design is investigated with the use of a genetic algorithm (GA). Several and different problems of optimal damping of beams or plates are presented. It is demonstrated that the GA is a very attractive and powerful method for this kind of problems in which the objective function has no derivatives and several optimal solutions simultaneously exist.

Keywords. Optimization, structural damping, genetic algorithm

1 Introduction

Structural vibration control is a major design problem for a variety of structures. This control may be approached in several ways as active or passive attenuators, structural damping, etc... In most cases the objective of the designers is to minimize the vibration amplitudes on a wide range of frequencies in order to prevent damage by fatigue. For this purpose, two main processes may be followed by the engineers. One is the use of composite materials which generally exhibit excellent material damping properties one or two orders higher than most common metals, the other is to damp the structure itself by viscoelastic material coatings or to insert at the most efficient locations passive dampers. Each of these solutions has its own advantages, but it may be pointed out that the second way may often be used without any change in the design requirements.

This paper is devoted to the determination of optimal structural damping solutions in several design processes. The topic is concerned with beams, plates and shells. The major problem is to find the best positions for the dampers or the coatings or the stacking sequence of composite materials in order to maximize a modal damping factor. At present these optimization problems have not yet received effective solutions. Indeed, as the design variables which represent the damper positions are not continuous, the objective function has no derivatives and the classic mathematical approaches are invalid. It is shown here, how genetic algorithms offer a new and attractive way for solving this kind of question.

The paper is divided into two sections. Section 2 gives some background on structural or viscous damping, viscoelastic materials, and the effects of viscoelastic constrained coatings are briefly recalled. In section 3, results obtained with GA are reported through several examples but also information is given on the changes performed on genetic algorithms to improve their efficiency when they are used to solve this kind of specific problem.

2 Statement of the Problem

2.1 Damping Representations

In most structures the damping originates from several sources : to the structural material's damping itself may be added the effects of local frictions, air or support dampings... This is the reason why it is generally difficult to quantify the total damping from a mathematical point of view. Its common representations are : viscous, proportional, structural. If the matrices $|M|$ and $|K|$ represent the mass and the stiffness of a structure, the vectors $\{u\}$, $\{\dot{u}\}$, $\{\ddot{u}\}$ the displacements, velocities and accelerations, then the equilibrium equations associated to these kinds of damping are res-pectively :

$$|M|\{\ddot{u}\} + |C|\{\dot{u}\} + |K|\{u\} = \{F(t)\} \quad (1)$$
$$|M|\{\ddot{u}\} + \left(a|M| + b|K|\right)\{\dot{u}\} + |K|\{u\} = \{F(t)\} \quad (2)$$
$$[-w^2|M| + |K_r| + j|K_i|]\{u_0 e^{jwt}\} = .\{F_0 e^{jwt}\} \quad (3)$$

Because of time dependent stress-strain relations, the representation (3) of the structural damping is simple only for periodic excitations.

For such excitations a viscoelastic material behavior may be represented by a complex Young's modulus : $E_v = E_{vr} + jE_{vi}$. This means that the damping introduced by viscoelastic constrained layers is a special case of structural damping. It can be assumed that the real and the imaginary part of E_v - the storage and the loss modulus - are frequency dependent. This hypothesis is often experimentally verified for E_{vr}. The frequency dependence of E_{vi} might be taken into account but without adding anything to the main results of the optimization process which does not depend on this parameter ; this is the reason why it is not considered hereafter.

2.2 Modal Response

Frequencies and mode shapes of the undamped associated structures can be considered as a good and simple modal basis to be used for predicting the dynamic behavior of the corresponding damped structure [4].

ω_i and $\emptyset_i$ $i = 1, n$ are the *undamped* frequencies and corresponding mode shapes obtained from the matrix equation :

$$(-|M|\omega^2 + |K|)\{u\} = \{0\} \quad (4)$$

Performing the usual transformation $\{u\} = |F|\{q\}$, and premultiplying by $|F|^T$, several equations (5) are obtained for free vibrations ; they are

$$|\tilde{m}|\{\ddot{q}\} + |c|\{\dot{q}\} + |\tilde{k}|)\{q\} = \{j(t)\}$$
$$|\tilde{m}|\{\ddot{q}\}| + (a|\tilde{m}| + b|\tilde{k}|)\{\dot{q}\} + |\tilde{k}|)\{q\} = \{\varphi(t)\} \quad (5)$$
$$(-\omega^2|\tilde{m}| + (|\tilde{k}_r| + j|k_i|)\{q\} = \{\varphi\}$$

Because of the orthogonality of the modes, $|\tilde{m}|$, $|\tilde{k}|$, $|\tilde{k}_r|$ are diagonal matrices, but not $|c|$ or $|k_i|$. Generally for beams and plates the frequencies are well separated, so the full damping matrices can be considered as diagonal dominant. In these conditions the modal systems (5) are the sum of n uncoupled equations.

2.3 The objective function

It follows from the preceding that in a modal response, a good approximation of the structural loss factor may be easily calculated from each equation of (5) as shown in [6] , so the objective function to be maximized has the general form :

$$\eta = \frac{E_d}{E_S} = \frac{|\Phi|^T|\xi||\Phi|}{|\Phi|^T|K||\Phi|} \quad (6)$$

in which E_S is noted as the elastic strain energy, $|\xi|$ the different forms of the damping matrix, E_d the dissipation energy. Due to the above hypothesis, the damping can be written

$$\eta = \frac{\sum\limits_{k = \text{nb of retained modes}} \alpha_k <\emptyset_k> \chi_k \{\emptyset_k\}}{\sum\limits_{k = \text{nb of retained modes}} \alpha_k k_k} \quad (7)$$

a_k is a weighting modal factor given by the user.

For example, in the case of a damping produced by viscoelastic coatings (7) has the form

$$\eta = \frac{\sum\limits_{k = \text{nb of retained modes}} \alpha_k <\emptyset_k> |k_i| \{\emptyset_k\}}{\sum\limits_{k = \text{nb of retained modes}} \alpha_k k_k} \quad (8)$$

It is pointed out here that the use of the undamped modes simplify readily the calculation of the objective function. The denominator of (7) is invariant during all the optimization process and the numerator depends only on the damping parts.

3 Examples of Optimization

3.1 Optimal locations of dampers

Often in structural engineering the problem of finding what could be the best locations of a given (and few) number of damping devices has to be solved. This optimization problem is generally unconstrained except for the definition of forbidden areas because of technological reasons.

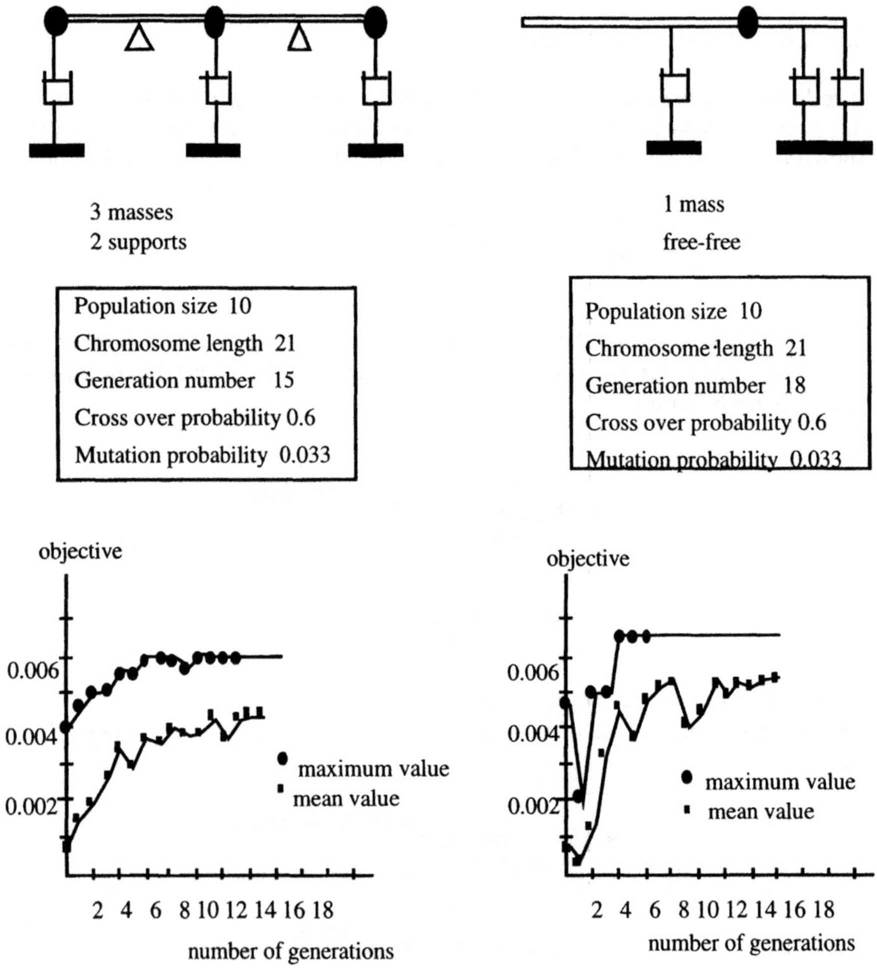

Figures 3.1 : Optimal location dampers

This kind of problem is easily solved by a GA as proved in the following simple and demonstrative example. The damping of the free vibrations of a uniform section beam loaded with concentrated masses is aimed at. 3 identical springs + dash-pots can be used.

The beam is divided in 10 C1 beam finite elements. The beam material damping is neglected. The total length is divided in 3 equal parts. Each position x_i of the dampers is described by a binary number, which means that 3 equal length chains of digits are necessary in order to describe the whole structure (figure 3.1). The 3 first vibration modes are taken into account, each with the weight a_i ($a_1 > a_2 > a_3$)

The results are given on figures 3.1 for the optimal damping of the first 3 modes with 2 different boudary conditions. The first result is physically obvious but was performed in order to verify a symmetrical optimal solution. Clearly the solution procedure can be easily extended to plate problems with any supplementary difficulties.

As the dampers position vary the mode shapes, the modes are calculated at each step of the optimization process.

3.2 Optimal location of viscoelastic constrained layers

Another and efficient technic to damp beams, plates and shells is the use of viscoelastic constrained layers which are glued on the surface of the structure. The optimization problem which can be solved by a GA is the determination of the sizes and the locations of these specific dampers. This problem was studied by Hajela [3] for beams. It is revisited here with a specific application to the damping of skis and plates.

3.2.1 Damping of skis

The goal here is not to optimize the damping of all the resonance frequencies; that would produce a too "soft" ski unacceptable by the users who need a "reactive" structure. On other hand it is known, from previous works, that some resonances are too tight for the stability, especially during a turn. Therefore the objective is to maximize the damping for these specific frequencies.

The technic to obtain this result agrees with severe technological specifications as it is expensive and difficult to modify the ski stiffness distribution. The easiest way is to add damping which can be done by using constraint viscoelastic layers, named here VAS (Vibration Absorbing System) everywhere on the ski surface excepted on the shoe part.

So the optimization problem is as follows : given a variable flexural stiffness distribution, how to distribute a limited length of viscoelastic covering to minimize some specific modes. Figure 3.2 shows how the damping of the first four flexural modes vary when each (but only one) finite element of the mesh is covered with viscoelastic material.

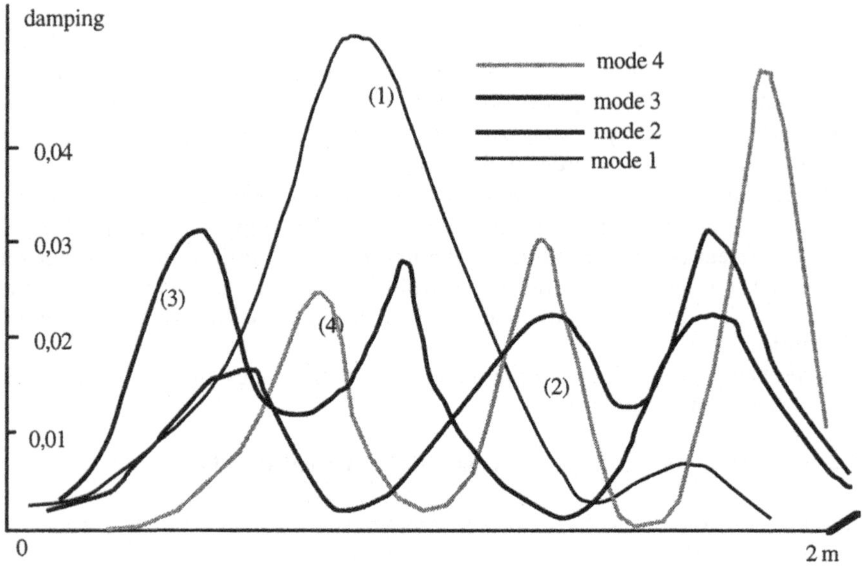

Figure 3.2: damping repartition when one damped finite element is moved along the ski: 4 first flexural modes (ski free-free)

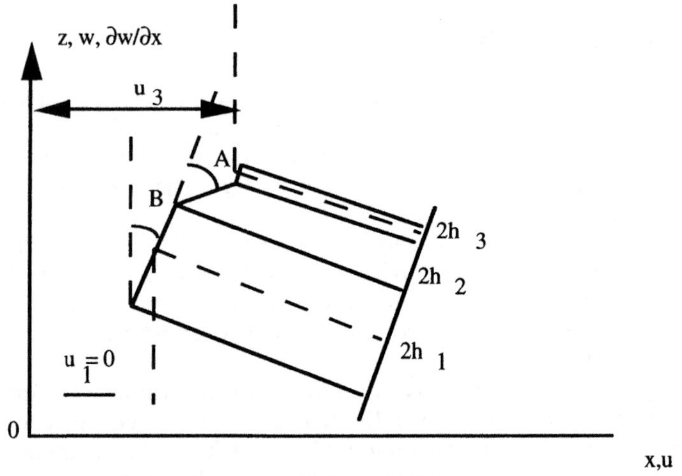

Figure 3.3: damped finite element; selected variables

From figure 3.2 it is obvious that the final result strongly depends on the position of the damped element and that the optimal structure is quite umpredictable especially if several modes are considered.

As in the previous example the GA is once more an efficient way of giving an answer.

The ski is discretized in compatible damped or undamped beam finite elements [6] (Figure 3.3).

Mode 1 : damping = 0.120

mode2 = 0.07 mode3 = 0.06 mode 4 = 0.056

Mode 2 : damping = 0.130

mode1 = 0.06 mode3 = 0.06 mode 4 = 0.15

Mode 3 : damping = 0.106

mode1 = 0.01 mode2 = 0.06 mode 4 = 0.01

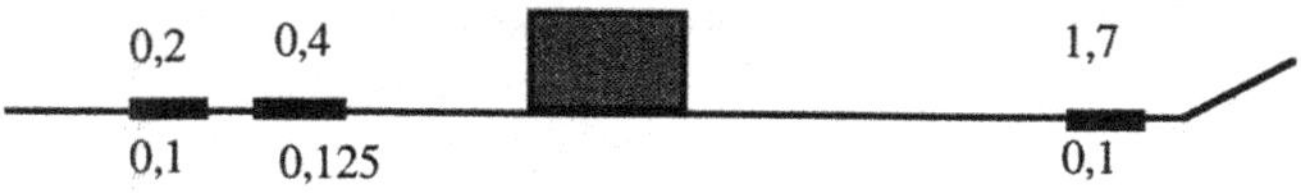

Mode 4 : damping = 0.15

mode1 = 0,05 mode2=0,09 mode3=0,05

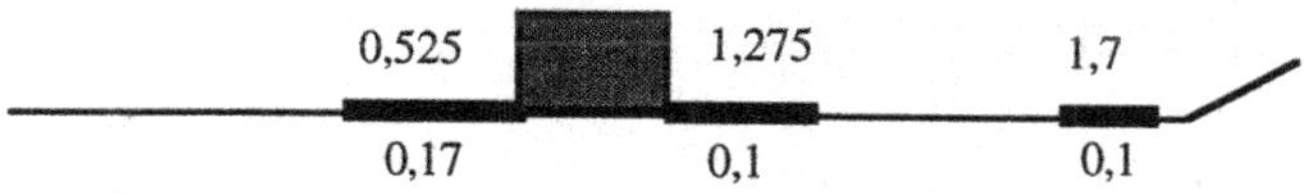

numbers above the ski : abscissa of the constrained layers

numbers below the ski : lengths of the layers

Figure 3.4: optimal solutions for each mode; ski free-free loaded by a mass (72kg)

Undamped (resp. damped) parts are coded by 1(resp. 0) and form the chromosome. Position constraints are easily taken into account by freezing corresponding parts of the chromosome (or the mesh) (schema type), for example the shoe part. Some partial results are reported in figure 3.4 above.

The ski is free-free and loaded by a mass which represents the skier. The damping of the ski itself has not been considered as the question is focused on the search of the most efficient locations of the dampers. Some of the results reported here have been obtained and patented by trial and error experimental methods.

3.2.2 Damping of plates

The same kind of optimizing process was applied to plates. Specific 3-layer finite elements were built. They are a 2 dimensional extension of the preceding beam elements [7,8].

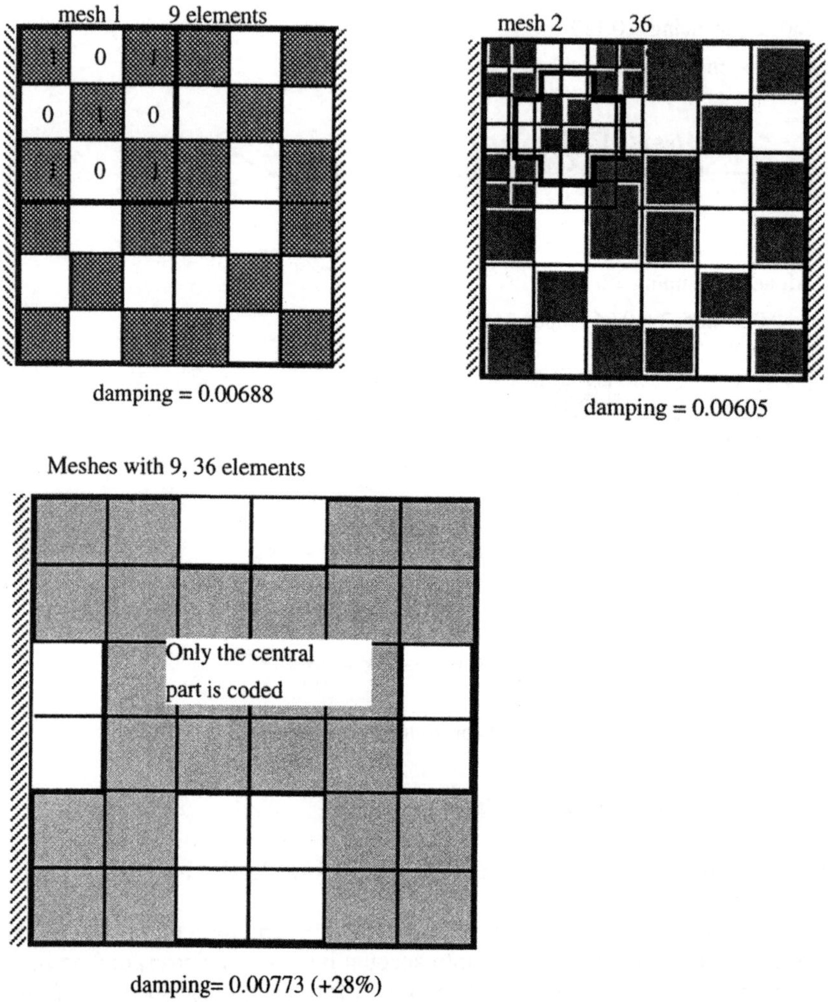

Figure 3.5 : Optimal first mode damping of a clamped-clamped square plate

In the optimization steps, first the main damped parts are determined using a coarse 9x9 mesh, and the convergence is reached in 10 generations ; then the damped area can be specified by refining the mesh (16x16). Finally, in order to obtain a more precise result without increasing substantially the chromosome lengths, a supplementary step may be performed by dividing the chromosomes in subchromosomes, as reported on figure 3.5.

In this last example all the finite elements might be damped. In several applications only a few numbers have this possibility. Then, it is possible to change the meaning of the coding. The division of a chromosome in subchromosomes may be associated to the number of the finite element which is damped.

In this example, as in the preceding, several optimal solutions are possible. The use of "sharing functions" is found effective for determining these different solutions. In the following, only six finite elements can be covered and four local solutions are researched (and obtained).

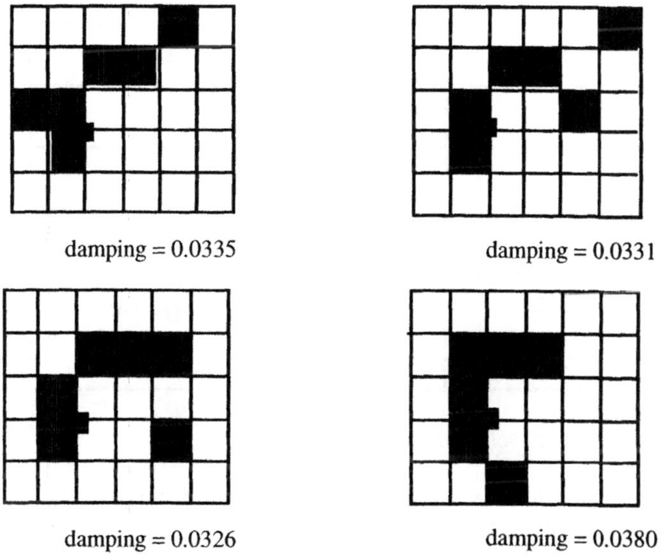

Figure 3.6 : Plate free-free : multiple optimal solutions
for the damping of the mode 2

3.3 Optimisation of a multilayered beam

In this application several optimisation problems are presented and solved by a GA.

3.3.1 Optimisation of a 3 layers beam loss factor

Although the following example is rather simple it is chosen in order to demonstrate the efficiency of the GA used. A beam 2m(L)x20mm(width)x

15mm(height) composed of 5mm thick layers has to be optimised. For each of the first 3 bending modes a stacking sequence of layers, consisting respectively of different materials (table 3.1) is examined to determine which sequence gives the best damping.

Young's modulus *Pa*	Loss factor	Density kg/m3	Index number
0.170 E11	0.003	1900	1 Fiberglass
0.170 E10	0.0241	1062	2 Polyethy.
0.600 E06	0.105	1250	3 Rubber
0.208 E10	0.0143	991	4 Resin ABS
0.934 E10	0.001	1500	5 VK 573
0.732 E09	0.0261	600	6 Polyureth
0.183 E11	0.0141	1400	7 Phenol
0.200 E12	0.0	7800	8 Steel
0.171 E11	0.001	1500	9 PH + VK
0.700 E11	0.001	2800	10 Aluminium

Table 3.1 : List and properties of selected materials

As the convergence may lead to a nearly optimal solution all the calculations done by a GA are checked by an enumerative method.

Properties of the GA used. A *m*-coding is used instead of a binary one to identify the permissible materials. By doing so the problem of invalid coding numbers with any material correspondence is avoided (for example for translating 10 materials in a binary coding, 4 bits are necessary and 6 codes have no meaning). In the same way Le Riche and Haftka have used a ternary coding (0-1-2) and found it efficient.

As the stacking sequences of composites are assumed to be symmetrical, an additional operator called inversion is implemented to improve the convergence of the GA. It is defined as 01234 equivalent to 43210

Finite element model. To calculate accurately the loss factor of a multilayered beam special finite elements were designed which take into account the parabolic shear stress distribution accross the thickness of each layer. More information on this theory is given in [6]. The parameters defining the GA are given in table 3.2.

Pop size:	$n = 50$
Number of materials	$m = 10$
Probability of partially matched crossover	$p = 0.6$
Probability of mutation	$p_c = 0.1$
Probability of inversion	$p_i = 0.5$
Sharing radius	$s = 0.75$

Table 3.2 : Parameters of the GA

As the best solution is already known by enumeration it is interesting to see if the GA could reach it more rapidly. After 20 generations, the probabilities of finding the optimal stacking are the following

Simple GA :	46.7%
with sharing:	46.7%
with inversion :	73.3%
with sharing and inv.	86.7%

Table 3.3 : Probability of finding the optimal stacking within 20 generations.

By using the sharing function several near-optimal solutions are found and reported in the table 3.4.

It is interesting to note that the rubber is always in the middle of the beam which is in agreement with the presumed shear stress distribution.

Multicriterion of the 3 first three modes. A good compromise for taking into account simoustaneously the first 3 modes is to weight by the loss factor the strain energy associated with each of them.

The single overall-fitness selected is

$$F = \frac{\sum\limits_{k = \text{nb of retained modes}} \eta_k E_k}{\sum\limits_{k = \text{nb of retained modes}} E_k} \qquad (9)$$

E_k strain energy of the mode k. The maximisation of F leads to the results reported in table 3.5.

	stacking seq.	dampness
Mode 1	1 3 1	0.1219
	7 3 1	0.1192
	9 3 1	0.1189
	7 3 7	0.1·168
Mode 2	5 3 5	0.1106
	4 3 4	0.0960
	5 3 1	0.0914
	4 3 2	0.0904
Mode 3	4 3 4	0.1124
	4 3 2	0.1124
	2 3 2	0.1111

Table 3.4 : Optimal and near optimal stacking sequences.

Stacking sequence	fitness
4 3 4	0.0902
4 3 2	0.0876
5 3 5	0.0862
2 3 2	0.0856
1 3 1	0.0654

Table 3.5 : Optimal stacking sequences of damping over the first three modes.

3.3.2 Constraints and penalty function

If the optimisation problem is constrained, as no derivatives are available, the constraints are taken into account by using a penalty function p.

Several classes of penalty functions were tried ; here are some results obtained with a multiplicative penalty term ; the overall fitness F is calculated by F = p x damping.

For a composite beam it is important that its weight be as low as possible while still retaining a particular rigidity. The natural frequencies provide a good compromise between stiffness and weight. The penalty function has been chosen to ensure a minimum stiffness ; it has the general form

$$p = 1 - \left| \frac{f\ design}{f} - 1 \right|^a \qquad (10)$$

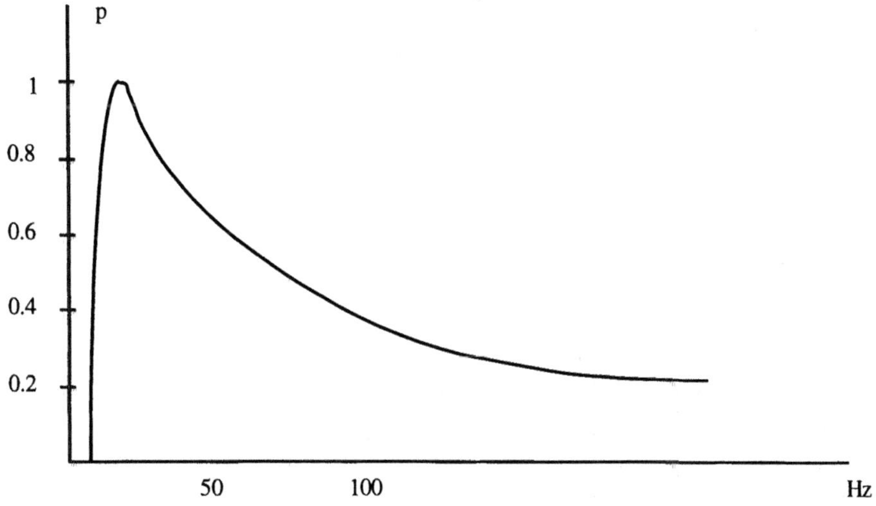

Figure 3.7 : The penalty function p.

Application. For a medium density of 1000kg/m^3 and a medium Young's modulus E = 10^{10} Pa, the first natural frequency of the corresponding beam is closed to 12Hz. If this value is chosen as the reference and if a = 4, then the variations of the function p are shown in figure 3.7. The lower limit is $0.5f_{design} \approx 6.$ Hz. The advantage of this method is to penalise only the modes whose natural frequencies are too small. The results obtained with this penalty function are reported in table 3.6.

The penalisation is effective for the first frequency which is 7.6 Hz but rather invalid for the others for which p is nearly 1.

mode	stacking sequence	dampness factor	natural frequency	overall fitness
1	7 3 7	0.1168	7.6Hz	0.1013
2	5 3 5	0.1106	11.9Hz	0.1106
3	4 3 4	0.1124	15.7Hz	0.1121

Table 3.6 : Stacking sequence for the first 3 modes of the constrained beam.

Composite beam. In table 3.7 the results found by the optimisation of a beam (2mx20mmx15mm) consisting of 3 5mm thick layers are shown concerning the unconstrained and constrained first mode. In addition the mass and the modal stiffness are given.

mode 1	unconstr.	constr.
stacking sequence	1 3 1 7 3 1	7 3 7 7 3 9
dampness factor	0.1219 0.1192	0.1168 0.1164
natural frequency	6.6 7.1	7.2 7.5
mass Kg	3.03 2.73	2.43 2.49
modal stiffness	5211 5433	4973 5529
overall fitness	0.1219 0.1192	0.1013 0.0986

Table 3.7 : The 2 best sequences for the constrained beam.

4 Conclusion

All these examples demonstrated that the GA is an efficient method able to optimise the factor of dampness in several situations. As only input a special value called the fitness is needed without any derivative calculations. The optimisation of other structural elements like composite plates or shells can be easily performed. With this object works are presently done to link the use of GA to a general finite element program. The main advantages of the GA are the possibilities of exploring several regions of the design space at the same time and of using a codification of the physical space instead of natural coordinates which is the only way of solving the problem easily. The main drawback is the number of objective function evaluation which can be computer costly if the structure is divided in a large number of finite elements. For example it seems difficult to extend this method if a new modal basis has to be calculated too many times.

References

[1] Marcelin J.L, Dornberger R., Trompette P., Optimisation of Composite Beam Structures Using a Genetic Algorithm. To be published in *Structural Optimization.*, 1995

[2] Hajela P. and Lin C.Y. Genetic search strategy in multicriterion optimal design. *Structural Optimization* 6, 151-159, 1992

[3] Hajela P. and Lin C.Y. Optimal design of visco-elastically damped beam structures", *Appl Mech Rev* **44**(11) pp. 96-106 1991.

[4] Johnson and Kienholz. Finite Element Prediction of Damping in Structures with Constrained Viscoelastic Layers. *AIAA J.*, vol 20, n°9, 1284,1290. 1982.

[5] Le Riche, Haftka R. Optimization of Laminate Stacking Sequence for Buckling Load Maximization by Genetic .Algorithm. *A.I.A.A. Journal* ,Vol 31, n° 5, 951, 956. 1993

[6] Trompette P., *DYNASKI*, Ref. manual pp 1-28, 1983.

[7] Trompette P., Marcelin J.L and Schmeding C., "Optimal damping of beams and plates by genetic algorithms", *IUTAM Symposium Discrete Structural Optimization,* Poland, August 31-Sept 3,1993.

[8] Marcelin J.-L.Trompette P., Amortissement optimal de plaques et coques contraintes par algorithmes génétiques. *Colloque international CNES/ESA " Structures des véhicules spatiaux"* 21-24 juin 1994 PARIS.

Neural Networks For Embedded Control Designs Applied To The "Smarter Car"

Mark J. Embrechts,[1] Mark J. Luetzelschwab,[1] Frank Dicesare,[2] and Matthew C. Mattina[2]

[1] Department of Nuclear Engineering & Engineering Physics
Rensselaer Polytechnic Institute, Troy, NY 12180.

[2] Department of Electrical, Computer, and Systems Engineering
Rensselaer Polytechnic Institute, Troy, NY 12180.

Abstract. Artificial neural nets have shown promise for pattern recognition tasks and control applications. As embedded microprocessor systems become ubiquitous in the commercial world, it seems natural to develop ways of embedding artificial neural nets into microprocessor based systems. The flexibility and low cost of microprocessors make them ideal for consumer products. This paper illustrates how neural nets that were software trained can be implemented in an off-the-shelf Motorola 68HC11 embedded microprocessor system. The "Smarter Car" is an autonomous battery powered laboratory car that follows a line painted on the floor. The development of the "Smarter Car," combining fuzzy logic control with artificial neural nets, is demonstrated. Even though just one specific example is illustrated in detail, this approach is generic and can be readily extended for the implementation of similar systems such as the control of robot arms and more general industrial neural net applications.

Keywords. Artificial neural nets, embedded control, microprocessors, fuzzy logic, smarter car, backpropagation.

1. Introduction

Artificial neural net (ANN) applications are relatively easy to develop in a software environment. Once the software has demonstrated an acceptable neural net performance, it is often desirable to develop a consumer product in hardware. As embedded microprocessor systems become ubiquitous in the commercial world, it seems natural to develop ways of embedding neural nets into microprocessor based systems. Fuzzy logic controllers have also become more prevalent. Better methods for combining fuzzy logic control and neural net recognition are emerging on a daily basis.

Because the training and testing of ANNs is often a computationally intensive process, ANN development is usually delegated to high-powered computers. However, once a network has been trained, the process of testing a single pattern can be done with modest computer requirements. To produce a system that is commercially viable, it would be unreasonable to expect each consumer to purchase a relatively expensive computer to evaluate neural net patterns. However, lower cost embedded microprocessors have been in the market for years. By first training a neural network on a powerful development system and then passing the trained network to a cheap, low-powered microprocessor, cost-effective neural network systems could easily enter the consumer market. The added flexibility of removable ROMs could elicit user-upgradable systems: if a better training set-up is developed after a product is delivered, the consumer would only need to upgrade a single ROM chip rather than the entire system.

This paper describes the implementation of neural nets and fuzzy control into embedded microcontrollers. The paper addresses specific pitfalls and modifications that are necessary for such an implementation. A simple system, the "Smarter Car," that demonstrates how computer hardware and software can work together to control a real device will be explained in detail (figure 1). This application can be readily modified for other neural net applications such as the control of a robot arm and the design of structural and mechanical systems.

The objective of this "Smarter Car" is to follow a line painted on the floor by using a simple camera to sense how the road curves and shifts as shown in figure 2. The "Smarter Car" describes the simulation of an embedded system that incorporates neural nets and fuzzy logic in a unique fashion. Neural nets are used to recognize 4x5 patterns sensed by a crude camera mounted on the car. The outputs of the neural net are then used as inputs for the fuzzy set controller.

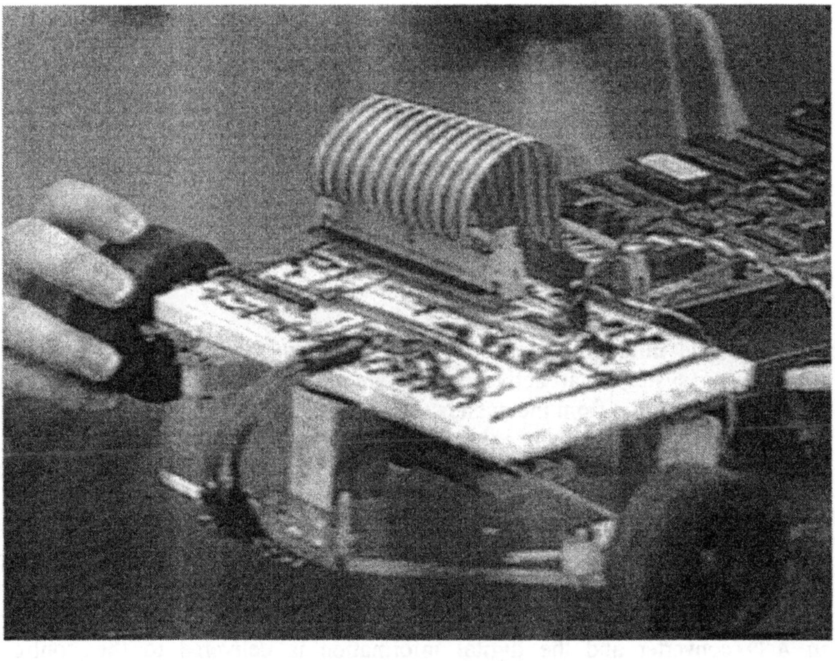

Figure 1: Detail of the "Smart Car."

Figure 2: The "Smart Car" following a curved raceway.

This paper is divided in four sections: after this introduction generalities about microprocesor based neural nets are discussed. The third section describes the development of the "Smarter Car", comments on its performance, and gives suggestions for a more refined system. The final section summarizes the project.

2. Microprocessor Based Neural Nets

2.1 Embedded Control

An embedded microprocessor control system can be loosely defined as an autonomous system that activates software driven decisions based on input parameters measured from the outside world. For example, an embedded controller in a clothes dryer might change the amount of heat being delivered to clothing in the dryer as the humidity of the dryer bin changes. No user input is necessary during operation, though user input is often gathered by embedded controllers before they begin operation.

Four basic facets of embedded control are: sensing, command, control, and display. For the example of the clothes dryer, the sensing might be a hygrometer that has an analog output. This output is converted into a digital signal through an A/D converter and the digital information is delivered to the controller.

Command inputs usually come from the user; in this case it might be an on/off switch or a safety switch that shuts the dryer off when the door opens. The display could be a panel of lights at the top of the dryer that lets the user know what the dryer is doing, i.e. how much time is left. The control is activated by the embedded controller. In the case of a clothes dryer, a microcontroller might decide how fast the bin spins or how much current should flow to the heating element. In general, an embedded microprocessor control system receives commands from a user, senses the environment, displays information, and controls a system.

Microprocessors that are used for embedded systems vary greatly from application to application. Factors such as cost, speed, availability and memory restrictions usually limit the list of plausible solutions for particular applications. There is no need for a RISC chip to be placed inside a simple stoplight controller; however many of the smaller controllers wouldn't be fast enough for image recognition.

2.2 Embedding Neural Nets

Currently, to implement a neural net in a custom application, one may have to buy expensive software and tie up a desktop computer to control a relatively simple system. Another option is to design hardware based neural nets: they are fast and reliable, but also expensive and rather rigid. The turn-around time for modifications can be in the order of days rather than hours for embedded neural nets. Furthermore, commercial software and hardware packages often require royalties if a trained network is to be used in a delivered system. This motivates the implementation of trained neural nets on embedded microprocessor systems. An embedded ANN is a flexible and inexpensive alternative for the control of engineering systems. No royalties are required as long as the executable algorithm is public domain and modifications can be made easily and timely.

An embedded neural net is a flexible and inexpensive alternative for the control of relatively simple systems. In this approach the neural nets are developed off-line in software (the training and testing mode), and then embedded in a microprocessor to be executed on-line in the user's mode.

In this project full control was maintained over all the steps of the implementation: from the training of neural nets, to adjusting the weights to enhance performance, to embedding the software and the weight file into a microprocessor. The application for this project utilized an in-house standard backpropagation code (MetaNeural™) for training, and a stripped down user's version for downloading to the microprocessor. The neural net is trained off-line. The weight file a is transferred to a customized subroutine in the microprocessor. This approach allows us a way to make quick changes to our system without major hardware redesign.

The flexibility of the microprocessor system motivated us to produce tools for all the facets of the entire process. The software is written in C and involves the creation of graphical interfaces for test patterns, training neural nets, building fuzzy rule bases and fuzzy set definitions, creating fuzzy control look-up tables, and incorporating weight files and look-up tables into a single program for a Motorola 68HC11 microprocessor.

It is prudent to mention some of the pitfalls related to embedded neural nets. The smaller chips can be too slow to process larger neural net applications in real time. Some neural nets might have weight files that exceed the memory limits of a microprocessor. Real-time training is usually not plausible due to the computational demands for on-line training. Because most microcontrollers have only a few input channels, multiplexing of inputs is required for most applications. Multiplexing involves serializing one or more parallel input channels and sending this serial information signal to the microprocessor in a time-coordinated fashion. Although multiplexing is a standard technology that can be implemented with off-the shelf components, it results in a more elaborate set-up.

There are a few differences between software simulations and embedded neural nets. A major consideration with microprocessor control is the usual lack of memory available. This motivates to train off-line (in software) and to develop the microprocessor application for a user's mode only. Patterns are delivered to the controller program one at a time. No feedback on the validity of the output is measured. There is therefore no on-line learning once the system has been implemented. By training in software, and executing the net via a microprocessor controlled user's mode version of this software, great care has to be taken to ensure that both versions are compatible (e.g., exactly the same transfer functions are to be used in both versions).

The first step in the development is no different from most traditional neural net based systems. The patterns are generated and given to the neural net for supervised training (in software). Necessary actions such as weight pruning and temperature adjustments can be made to efficiently train the network.

Once the net was trained successfully, the weight files must be downloaded so that they can be accessed by the microprocessor system. Unlike a PC or workstation based system, the Motorola 68HC11 microprocessor has no ability to read and write to any storage device. Thus, the weight data must be compiled directly into the code. A utility called "Executor Prep" was created to take any weight file produced from MetaNeural™ and create a subroutine that could be compiled and linked to the neural net engine.

An interface between the real system and the neural net executor on chip is necessary. Usually this consists of a subroutine that gathers data from the system. It then preprocesses these data to convert them into a pattern that is compatible with the patterns used for training the ANN. This pattern is then passed to the microcontroller based user's version of the neural net engine (via "Executor Prep"). This engine uses the pre-loaded weight file subroutine and creates an output for the specific pattern. The output of the neural net engine is then passed to a function that decides a control action (e.g., steering to the left or right).

All the coding must be integrated and compiled into a single downloadable file. This file is sent to the microcontroller and the neural net is ready for action. The C code for the main function of the program is similar to the one listed in figure 3. The most important restrictions relate to insuring that the preprocessing of the pattern inputs is consistent with the patterns used for training and that exactly the same transfer function is used for the ANN software during training and the user's mode on chip.

```
main() {
/* main loop for program that demonstrates */
/* embedded neural nets                    */
/* initialization */
  float myPattern[NUM_PATS];
  while (1) {
     /* infinite loop                                  */
     /* the example reads two A/D ports and put        */
     /* numbers on [0,1] into myPattern                */
     myPattern[0]  =  ReadADPort(0);
     myPattern[1]  =  ReadADPort(1);
     /* execute the network */
     executeNN(myPattern, outputNN, wt, xlen,
        temp);
    /* Get the output of the neural network and  */
    /* send it out to the LED                     */
    ShowOutput(outputNN[NUM_LAYERS-1][0]);
  }  /* end infinite loop */
}  /* end program */
```

Figure 3: C code for the main function illustrating the infinite input loop.

Once the net has been trained, the process of incorporating it into the embedded system begins. An overview of this process is shown in figure 4. We used the Introl C compiler (available for Macintosh, PC, and some UNIX based systems) to cross-compile our code for the Motorola M68HC11 microprocessor evaluation board (EVB). A simple terminal transfers the program to the memory of the evaluation board. The neural net engine is modified version of the error-calculating subroutine that was pulled out of MetaNeural™. It was slightly modified to accommodate limited memory provisions. The code is freely available upon request from the authors.

3. The "Smarter Car"

3.1 Description

Neural Nets and Fuzzy Logic can be combined in different ways, depending on the specific application. Neural nets have been used to develop valid fuzzy rules and membership functions for complex systems [1]. Fuzzy logic has also been used to improve neural network learning [2]. The literature has widely demonstrated the effectiveness of fuzzy logic and/or neural nets as controllers [3-4]. For the "Smarter Car" the neural net and the fuzzy logic engine are separate entities integrated as one controller: the neural network gathers inputs for the fuzzy controller.

The "Smarter Car" evolved from the "Smart Car" that students are required to build in a junior-level general engineering course at Rensselaer. The goal is to create a battery operated microprocessor controlled car that autonomously follows a line painted on the floor. The standard system has two light sensors mounted on the front of the car; the input from the sensors is fed into the microprocessor.

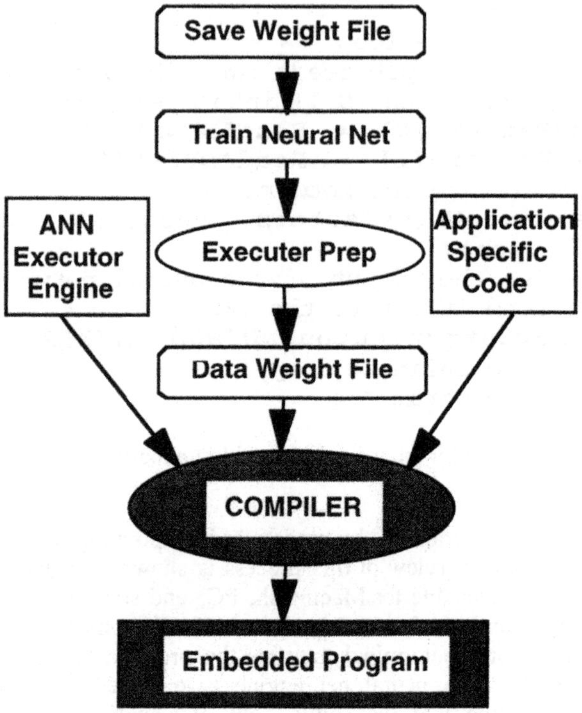

Figure 4: Schematic procedure for embedding a neural net.

The decision of how far to turn the steering wheel is governed by the two inputs. Though this system works, it has its obvious limitations. For this project we decided to add a crude camera. This allows the car to drive itself in a fashion similar to a human controlled car and provides the possibility to "loosely" follow the line. Humans drive along roads in their lanes, but realize that the lines are not barriers. Often, we cut corners or drive off-center depending on our speed and how the road ahead varies. The basics of the system were designed to emulate how humans drive their automobiles. To straddle a line, the driver must first recognize how the line changes in curvature and how far the car is from straddling the line. Once the relevant parameters are recognized, the driver then decides how to alter the direction by a steering wheel action. This process can be mirrored by neural net recognition and fuzzy logic control.

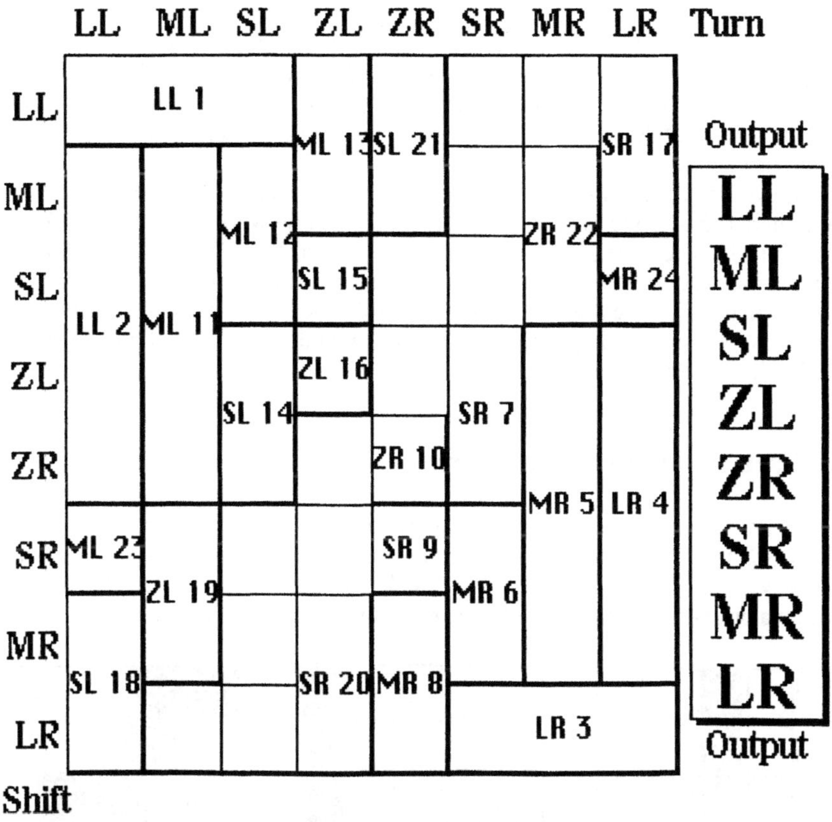

Figure 5: Fuzzy rule space, deciding a steering action based on TURN and SHIFT.

The problem of getting the spatial characteristics of the line into the controller is a daunting task. A crude camera with 4X5 photosensors is currently being developed for this task. For the time being, the track and camera input are software simulated. Assuming that a simple spatial input device can be built, a neural network was trained to take this input and provide outputs for the fuzzy logic controller inputs. The inputs to the controller were based on two parameters: SHIFT and TURN. Rules such as "IF SHIFT is SmallLeft AND TURN is MediumRight THEN SmallTurnRight" were created on the rule space shown in figure 5.

Fuzzy logic is an extension of conventional "crisp" sets. Where elements in crisp sets are either a member or not a member, fuzzy sets allow for varying degrees of membership in sets. This fuzziness provides the ability for computers to deal with imprecise numbers that are prevalent in the real world. A good introduction to fuzzy logic can be found in the works by Kosko [4-5]. The rules for figure 5 were developed based on common sense, experience, and trial and error. More sophisticated approaches for developing fuzzy rules are certainly encouraged, but are not necessary for our application.

The inputs for the fuzzy controller were generated from neural nets that were trained to extract SHIFT and TURN from patterns similar to the ones illustrated in figure 6. The network structure consists of 20 input nodes, 9 hidden nodes, and 1 output node. Note that for easier training and better performance separate networks were trained for each of the two parameters. Around 100 patterns were used to train each network.

The system is demonstrated in figure 7. In a continuous loop, the input from the camera (4X5) is fed to the neural networks. Each network (one for SHIFT and one for TURN) delivers a variable to the fuzzy controller. The controller fuzzifies its inputs and executes the rules stored in memory. The output sets are defuzzified and a scalar value is sent to the steering control. The car is assumed to move at a constant speed (in pixels per steering iteration), and a steering action is taken at regular time intervals.

3.2 User Definable Parameters

The user can set a few of the variables associated with automobile control. Unless otherwise mentioned, these parameters remain constant throughout each test run. Future versions of the "Smarter Car" will allow for dynamic variations. Here follows a brief discussion of the effects of changing the run parameters.

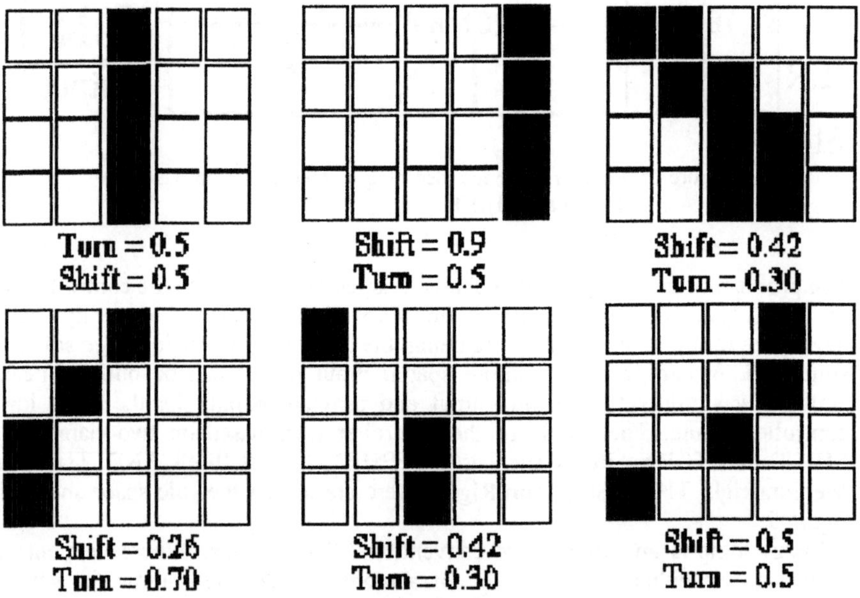

Figure 6: Typical 4x5 camera patterns for the "Smarter Car".
The ANN senses the TURN and SHIFT from these patterns.

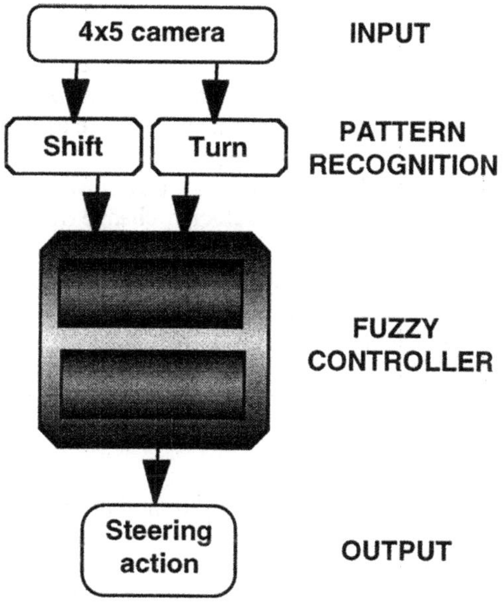

Figure 7: System for deciding steering action in the "Smarter Car."

The first parameter is the car speed (in pixels per iteration). If the speed is set to a value that is too low, the car actually will turn too hard for sharp turns and will stay on the inside of a circular path. If the speed is too high, the car will not be able to make sharp turns, and will lose sight of the path. Unless the car happens to cross the path again by chance, the car will continue until it reaches the end of the space and crashes.

The distance that the car "looks ahead" can be changed by the user. This value has the inverse effect of changing the car speed. If this value is too short the car will not anticipate the upcoming curves, and will not be able to make sharp turns at higher speeds.

The starting location of the car and the angle at which it begins its journey are variable. This allows the user to have the car approach the path at different angles, much like choosing the curvature of an entrance ramp. If the car is aimed in a direction perpendicular to the specified path, the car will most likely not make the required turn. Under low speeds, the car had a tendency to go right at such intersections. A refinement of the algorithms will remove this idiosynchrasy.

The maximum turning angle can be specified to avoid unrealistic turning motions. This parameter is equivalent to the maximum angular velocity of a steering wheel. The output of the fuzzy set controller is a value on [-20, 20]. This value is normalized onto the range of the maximum turning angles [-AngleMax, AngleMax]. If this value is set too large, the car can actually begin to oscillate around the path rather awkwardly. If it is set too low, the car can not navigate the sharper turns.

The field of view of the camera is set relative to the anticipated thickness of the lines. If the field of view is set too small, a thick line will actually black out the camera's view at some angles. If the field is set too large, the line will be ignored when the simulator averages the pixels into the 4X5 matrix. The neural networks could handle breaks in the lines and lines that were twice as thick as expected.

A toggle between variable and constant speed allows the user to specify if they want the car to speed up on the straight-aways and slow down on the curves. A crude controller specifies the rate of velocity change for each situation. The simulations suggested the need for a better controller for dynamic speed adjustment: the car often picked up too much speed in the easier turns and then couldn't slow down fast enough for the sharper curves. The trials documented in this paper are for constant speed.

3.3 Simulator Overview

The path (called the raceway) is simulated by dragging the mouse around an area on the computer screen. Once the simulator is started, the driving space is created into memory by "drawing" the lines saved in the chosen raceway onto a large matrix of 1's and 0's. Noise is added to the lines through a rather imprecise line drawing algorithm. After this space is created, the car is ready to be tested. The user is prompted for some of the variable parameters such as speed, starting location, and time.

The car begins its trip by first looking forward the specified number of pixels and averages the image to the correct size. If no image is present, the nets were trained to give no response, and the car moves forward the amount of pixels set by the speed. It should also be noted that the space that is observed by the car is rotated by the angle at which the car is turned. Some pixels are dropped because of the discrete nature of the data and the field of view is not changed between iterations.

Once the car locates the path, the neural nets recognize the sharpness of the turn and the severity of the offset at regular time intervals. These values are passed to the fuzzy set controller, which then decides on the steering action.

3.4 Results

The fuzzy controller acts on the Shift and Curve outputs from the neural nets and a steering action is taken. The path of the car is tracked and stored into a separate file and can be overlaid on top of the desired path. Sample results are given in figure 8. This figure illustrates several trajectories and trial runs (at different velocities, kept constant during each run).

The controller had moderate success with following the path defined by the user. Notice how it loses the path on sharp curves at the higher velocities. Also notice that it is able to pick up the path from different starting angles. Anomalies such as the rather large point in one of the raceways (due to the automated noise generation procedure used) did not seem to affect the performance of the "Smarter Car": it generally followed the trend of the defined path.

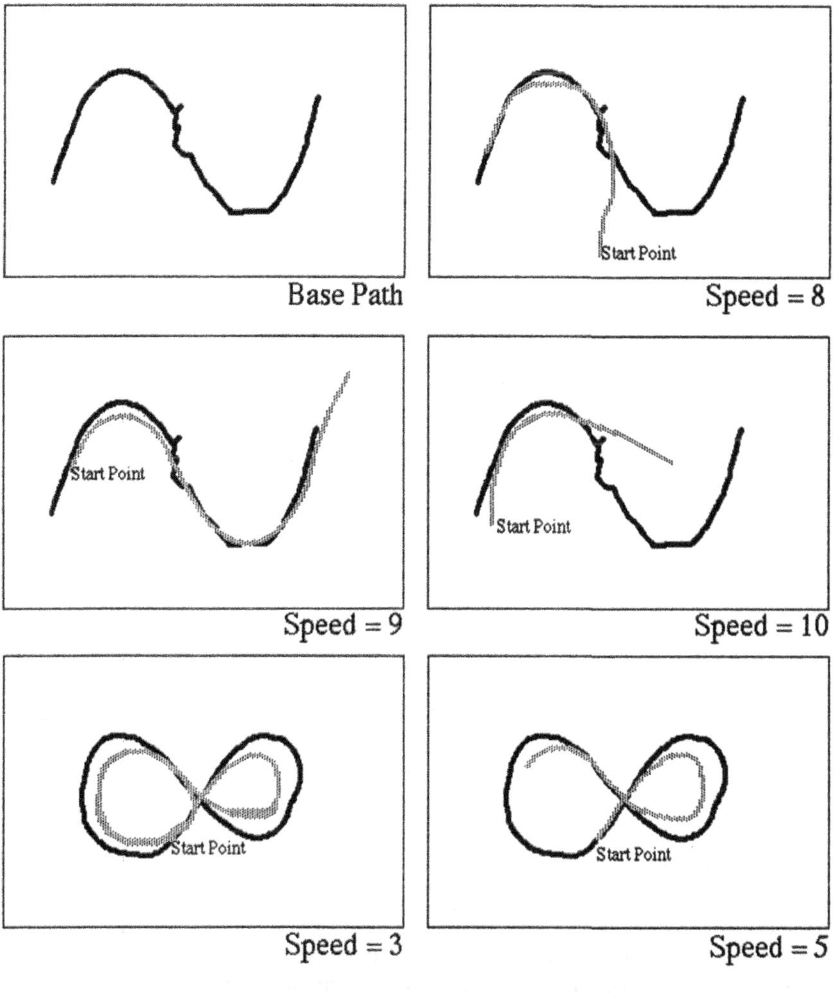

Figure 8: Raceway outlay and neural net results for several simulations at various velocities.

3.5 Problems with the Car Controller

The speed controller that we implemented was too crude to be effective. A simple fuzzy set controller can be added to adjust the car speed and to change the distance that the car looks ahead to reflect the speed. This additional controller can also be used to let the angle at which the car is looking ahead differ from the angle of travel; this would be equivalent to giving a driver a neck to turn his/her head.

Because of the way that one of the neural net trained, the car has a slight tendency to drift to the right. This misaligned "wheel" becomes apparent when the car needs to make a decision on a path that is perpendicular to its traveling direction; it will always go to the right. Also, when there is no path visible to the controller, the car will travel in a wide circle.

3.6 Suggestions for Improvement

In this version it was decided to use a look-up table for the rather dense rule set representing the fuzzy controller. This was appropriate for a number of reasons. Because the set was so dense, the amount of rules that would need to be processed would slow down the fuzzy decision making process considerably on slower controllers. For sparser sets, we looked into sparse-array linked lists that saved quite a bit of memory, but were not easily transferred between the simulator and the real controller as the linked lists require defined memory locations. Because the memory addressing of the simulator and the real controller were drastically different, this was not considered except for simulation purposes. If the rule set was smaller, a real time rule execution would have provided satisfactory results for comparable memory requirements. If additional fuzzy controllers were to be added (e. g., for speed control), the memory needed for the multiple rule bases and the fuzzy rule executor would be less than what would be necessary for multiple look-up tables.

Additional networks could be trained to recognize colors and the raceway could be given a third dimension by changing the color of contours in the plane. If hills were added, a fuzzy controller could be added to act as a cruise control in order to keep the speed constant. The car could also follow lines of different colors or learn to stay between two lines of different colors (i.e., between a white line and a yellow line).

4. Conclusions

While training artificial neural networks requires powerful processors, once trained, they can be embedded in relatively inexpensive microprocessors. Because of their low cost and high availability, microcontrollers employing neural networks and fuzzy logic promise cheap innovative consumer products and R&D implementations.

Both fuzzy logic and neural nets can control systems on their own. However, the hybrid approach demonstrated in this paper delivers a more flexible system, permitting rapid design changes and effective implementations. The processes required for embedding neural nets, fuzzy logic control, and a hybrid control scheme were demonstrated in this paper.

Currently an effective imaging system is being implemented which will allow the spatial information necessary for an effective control of the "Smarter Car". Other parameters such as dynamic speed control will be added to the fuzzy controller.

A point worth considering for this application is the ability for neural networks to model fuzzy system outputs. In the cases with more inputs, a neural network might be trained to emulate the fuzzy controller outputs. If this were possible, the speed of a fuzzy logic controller with run-time rule execution and a trained neural net should be compared.

Acknowledgement

The authors thank Linda Lem for digitizing the images of the "Smart Car."

References

1. C. S. Chin-Teng Lin and George Lee, "Neural-Network-Based Fuzzy Logic Control and Decision Systems," IEEE Transactions on Computers, Vol. **40**, No 12, 1320-1336 (Dec. 1991).
2. Y. Hayashi, J. J. Buckley, and E. Czogala, "Fuzzy Neural Network with Fuzzy Signals and Weights," International Journal of Intelligent Systems, Vol. **8**, 27-537 (1993).
3. Y. H. Pao, S. M. Phillips, and D. J. Sobajic, "Neural Net Computing and the Intelligent Control of Systems," Intelligent Control, Vol. **56**, No. 2, 263-289 (1992).
4. Bart Kosko, Neural Networks and Fuzzy Systems, Prentice Hall (1991).
5. Bart Kosko, Fuzzy Thinking, Prentice Hall (1993).

Conceptual Design Using Emergent Computing Techniques

Donald E. Grierson
Civil Engineering, University of Waterloo
Waterloo, Ontario, N2L 3G1, Canada

Abstract. The paper concerns the development of a computer-based capability for conducting conceptual design using emergent (self-organizing) computing papradigms that arrive at a solution through simulated evolution and artificial learning. Specifically, a genetic algorithm is coupled with a neural network to create a computational model for conceptual design that relies directly on the experience and judgement of the designer to arrive at alternate best-concept solutions. The proposed model is yet in the early stages of development and a number of its shortcomings are discussed in the concluding section of the paper.

Keywords. conceptual design, emergent computing, genetic algorithm, neural network

1 Introduction

Design can be classified as being either *creative, innovative* or *routine*. Creative design encompasses all things that designers do for which it is not possible to apriori specify a solution mechanism. In fact, it has been argued that any successful attempt to describe the solution mechanism for a creative design activity will lead immediately to redefining the activity as being noncreative, [1]. Even if possible, a computational mechanism for creative design must deal with the fact that the problem is generally poorly defined at any stage of the solution process. That is, both the attributes of the design and their possible values are not predefined at any point of a creative design activity. Alternatively, an innovative design activity involves predefined design attributes for which possible values are not known beforehand. Finally, for routine design the attributes and possible values are fully specified in advance, and the solution mechanism reduces to searching a predefined design domain.

The process of design can be viewed as involving the three stages of *conceptual, preliminary* and *detailed* design. The conceptual design stage is the least structured part of the process and involves applying experience and judgement in a qualitative manner to arrive a number of alternative best-concept scenarios. The preliminary design stage is somewhat less open-ended and involves refining the best-concept scenarios from the conceptual design stage in an approximate quantitative manner to arrive at one or more preliminary solutions worthy of further detailed consideration. Finally, the detailed design stage is the most structured and involves finalizing the details of the worthy solutions from the preliminary design stage in a rigorous quantitative manner to arrive at the final detailed solution.

The detailed design stage generally has a complete algorithmic format that is readily computerized using a conventional procedural computing paradigm. Moreover, even though they may be based on heuristics, many activities of the preliminary design stage are also algorithmic and readily computerized. On the other hand, the conceptual design stage generally has no algorithmic basis and is therefore difficult to computerize using procedural computing techniques that require the path to the solution to be laid out in advance.

During the conceptual design stage, the designer's experience and judgement are applied in the form of heuristics that are understood either consciously or subconsciously. Some heuristics are simple time-saving devices, other heuristics are based on fundamental theoretical and experimental knowledge, while still other heuristics derive from considerations concerning such things as style, appeal and aesthetics, [2]. For a new or difficult problem area, the precise way to apply these various heuristics is generally not known beforehand. Rather, the designer is guided through the conceptualizing process by a sequence of judgements derived from heuristics that are applied on an as-needed basis, i.e., the designer progressively learns a way to apply heuristics such as to arrive at a solution to the problem. This suggests, then, that conceptual design may be conducted on a computer by employing emergent(self-organizing) computing techniques to develop a computational model that has the capabilty to learn how to solve a problem as it proceeds. The remainder of this paper illustrates the development of such a computational model for the conceptual design of a bridge structure.

Discussion is confined to the 'routine' conceptual design of a bridge structure, in the sense that all design attributes and their ranges of possible values are specified beforehand. For example, the bridge *Superstructure* is identified as a design attribute that can be assigned any one of the eight possible alternative configurations shown in Figure 1, [3]. As discussed in the following, five other design attributes and their possible values are similarily predefined for the bridge conceptual design problem.

2 Conceptual Design

Intentions are generally quite vague at the outset of a conceptual design activity. They only begin to evolve and sharpen as judgements and choices are made by the designer based on analogies drawn between specific problem information and stored information in memory. The solution process advances best if, in addition to lists of design attributes and possible values,the problem data includes meaning-related and goal-related information so that judgements and choices can be made accounting for semantic overlapping between problem-specific and memory-resident information, [4]. However, the provision of such semantic information is beyond the scope of the present study. Instead, as shown in Table 1, the problem data is herein restricted to providing only the information required to identify the problem attributes and the range of possible values that each attribute may assume in conceptual solutions to a bridge design problem, [5].

Table 1: Input Data for Conceptual Bridge Design

ATTRIBUTES	POSSIBLE ATTRIBUTE VALUES
Superstructure	Arch,Truss,Cable-stay,Suspension,Box-girder,etc.
Foundation	Pile,Pedestal,Mat,Caisson,Excavate,etc.
Length	50m, 75m, 100m, 150m, 200m, etc.
Clearance	5m, 10m, 20m, 30m, 50m, etc.
Spans	1, 2, 3, 4, 5, etc.
Traffic Lanes	2, 3, 4, 6, 8, etc.

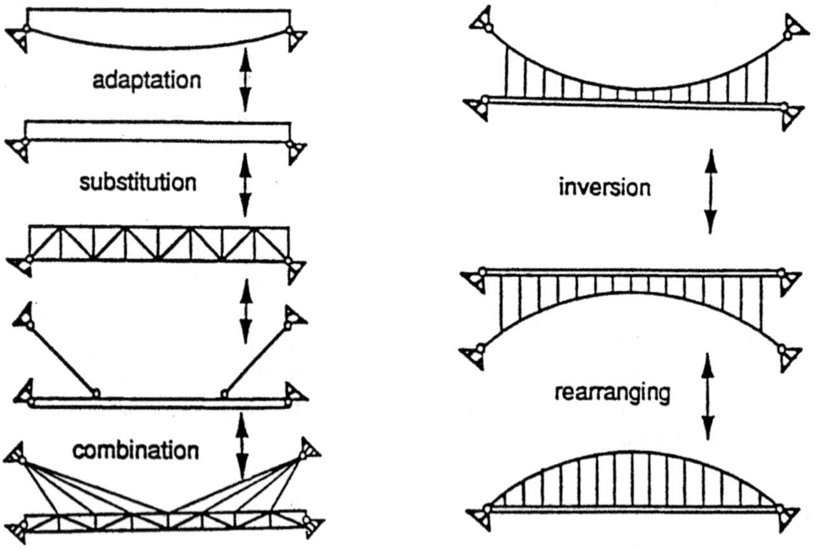

Figure 1: Possible Configurations of Bridge Superstructure

The information in Table 1 is intended to be problem-specific only insofar that it refers to a bridge structure, and not to any particular bridge design problem . In fact, the designer is encouraged to specify ranges of attribute values that are quite wide so as to allow for many conceptual design scenarios, no matter how unreasonable any one of them may be for the specific design problem at hand. At the same time, it is expected that the ranges of attribute values are such that they also embrace reasonable conceptual solutions to the specific design problem.

The expected outcome from the computing system is one or more conceptual design solutions worthy of further investigation (i.e., in a subsequent preliminary design stage). It is computationally difficult to evaluate the merits of generated design outcomes. As such, this study adopts a semi-automated design approach where computational techniques are employed to generate design outcomes while their evaluation is delegated to the designer. A conceptual design scenario is defined by a set of generated attribute values and the corresponding user-specified likelihoods of the attribute values occurring individually and/or together in that particular combination. The attribute likelihoods (e.g., poor, fair, good) reflect the designer's judgement concerning the merit of the conceptual solution in relation to the specific design problem under consideration. For example, two conceptual scenarios for the design of a bridge structure may be as given in Table 2.

Table 2: Conceptual Bridge Design Scenarios

SCENARIO	ATTRIBUTE	VALUE	LIKELIHOOD
	Superstructure	Truss	Poor
	Foundation	Caisson	Good
1	Length	100m	Poor
	Clearance	30m	Fair
	Spans	4	Poor
	Traffic Lanes	2	Poor
	Superstructure	Cable-stay	Fair
	Foundation	Caisson	Good
2	Length	200m	Fair
	Clearance	10m	Poor
	Spans	2	Fair
	Traffic Lanes	4	Good

The attribute-likelihood ratings given in Table 2 are based on the designer's past experience and current knowledge in relation to the specific design problem under consideration. A 'poor' likelihood rating for an attribute value in a particular design scenario reflects the designer's judgement that the value is inappropriate by itself and/or in combination with the other attribute values in that scenario. A 'fair' likelihood rating probably reflects the fact that an attribute value basically meets the requirements of the problem at hand but that the designer is not entirely sure as to its merit by itself and/or in combination with the other attribute values in the design scenario. Finally, a 'good' likelihood rating reflects the designer's judgement that an attribute value well meets the problem requirements and is quite suitable in combination with one or more of the other attribute values in the design scenario. Evidently, then, the second scenario in Table 2 is felt by the designer to be a better conceptual solution for the design problem at hand than the first scenario. Even then, the designer yet feels that the second conceptual design solution could be further improved upon by virtue of the fact that some of its attribute values have a poor or, at most, a fair likelihood rating. In fact, the final output from the computing system for conceptual design is expected to be a design scenario(s) for which the attribute values are all rated by the designer as having a good likelihood of occurring together in a best-concept solution(s) to the design problem under consideration.

3 Computational Model

The computing system is to have the capability to evolve to a best-concept solution(s) to the design problem by learning a search strategy guided by the judgement of the designer. Computer and computational scientists have been working for some time now on computing paradigms that are 'emergent' or 'self-organizing' systems in the sense that, rather than specifying the solution process in the form of an algorithm, the precise mechanisms required to solve a problem are learned by the system as it progresses. One representative of such a computing paradigm are genetic algorithms, which use 'guided random search' techniques based on the genetic evolution principle of 'natural selection' to progressively improve a problem solution by recombining the most desirable features of existing solutions until convergence occurs to a 'maximum fittness' solution(s). Another representative of a self-organizing computing paradigm are neural networks, which use a large number of simple processing elements based on the structure and biology of the human brain to 'learn' about a problem domain. The operations of both genetic algorithms and neural networks are parallel in nature, thereby allowing for the possibility to carry out a massive number of computations efficiently. This capability is of importance to the development of a computational model for conceptual design because of the need to search through a large number of possible solutions in the problem domain before arriving at best-concept solutions. As we know, computers are getting better and better in doing these massive computational exercises largely because of the availability of parallel computing hardware and software.

Described in the following is an emergent or self-organizing computing system that couples a genetic algorithm with a neural network to achieve a computational model for conceptual design that uses directed random search guided by artificial learning to generate best-concept design solutions. To aid in the description of the computing model it is convenient to first recast the bridge design problem using more concise notation. To this end, the six attributes listed in Table 1 are identified by the variables Xi (i=1,2,...,6), while their possible values are identified by the base-10 indices 0,1,2,etc.. Assuming that each attribute has, say, eight possible values, Table 1 can then be reformatted as in Table 3.

Table 3: Concise Input Data for Conceptual Bridge Design

BASE-10 INDEX	X1 Superstructure	X2 Foundation	X3 Length	X4 Clearance	X5 Spans	X6 Lanes
0	Arch	Pile	50m	5m	1	2
1	Truss	Pedestal	75m	10m	2	3
2	Cable-stay	Mat	100m	20m	3	4
3	Suspension	Caisson	150m	30m	4	6
4	Box-girder	Excavate	200m	50m	5	7
5	...	...	...	...	...	...
6	...	...	...	...	...	...
7	...	...	...	...	...	...

Therefore, for example, the second conceptual design solution in Table 2 can be concisely expressed as X1=2, X2=3, X3=4, X4=1, X5=1 and X6=2.

The flowchart in Figure 2 illustrates the operation of the evolutive-cognitive computing model for conceptual design. First, the design problem is defined using the information in Table 3 concerning attributes and their ranges of possible values. A neural network is created for which the input nodes refer to a set of attribute values, the output nodes refer to the corresponding attribute likelihoods, and the connection weights are initially randomly set. Then, a set of attribute values is randomly selected to create an initial conceptual design solution and the neural network is employed to predict the corresponding attribute likelihoods. For example, the bridge design problem has six attributes and, therefore, the associated neural network will have six input nodes corresponding to their values and six output nodes corresponding to their likelihoods. To illustrate further, assume that the attribute values for the second design scenario in Table 2 are randomly selected as input and that the corresponding likelihoods given in Table 2 are predicted by the neural network, as shown in Figure 3.

```
                    Define the problem
                      * Attributes
                      * Ranges of attribute values
                                 |
                                 V
                                 |
                    Create a Neural Network (NN)
                    * Input = design attributes
                      * Output = attribute likelihoods
                                 |
                                 V
                                 |
                    Train NN  to learn designer-specified
                    attribute likelihoods for randomly
                    selected sets of attribute values
                                 |
                                 V
                                 |
                    Apply a Genetic Algorithm (GA) to
   +--->---         find best-concept design scenario(s)
   |                  * Assess design fitness using
   |                    currently-trained NN
   |                             |
   ^                             V
   |                             |
   |                Query designer if best-concept          yes
   |                design scenario(s) has been found ?  --->--- STOP
   |                             |
   ^                             V  no
   |                             |
   |                Solicit designer to modify attribute likelihoods
   |                of current best-concept design scenario(s)
   |                             |
   ^                             V
   |                             |
   |                Train NN to learn designer-specified
   |                attribute likelihoods for
   +---<---           * current best-concept design scenario(s)
                      * plus all previous input-output training pairs
```

Figure 2: Flowchart of Computational Model for Conceptual Design

	ATTRIBUTE VALUES		LIKELIHOODS (predicted)
(Superstructure =Cable-stay)	X1=2 **--->O<**		**>O--->** 0 =Fair
(Foundation =Caisson)	X2=3 **--->O<**	**Hidden**	**>O--->** +1 =Good
(Length=200m)	X3=4 **--->O<**		**>O--->** 0 =Fair
(Clearance=10m)	X4=1 **--->O<**	**Layer(s)**	**>O--->** -1 =Poor
(Spans=2)	X5=1 **--->O<**		**>O--->** 0 =Fair
(Traffic Lanes=4)	X6=2 **--->O<**		**>O--->** +1 =Good

Figure 3: Attribute Value --> Likelihood Neural Network

Note that the output of the neural network in Figure 3 is normalized over the range -1 to +1, where -1, 0 and +1 define poor, fair and good likelihood, respectively.

Having the likelihoods predicted by the neural network for a randomly selected set of attribute values, as in Figure 3, the designer is then prompted to critique the conceptual design scenario in the sense of agreeing or disagreeing with the likelihoods predicted for the attributes values, either in whole or in part. In the case of disagreement, the designer is allowed to modify the attribute likelihoods as experience and judgement may dictate, and the neural network is then trained to learn the new design scenario (i.e., input-output training pair) defined by the current attribute values and the corresponding designer-modified likelihoods. For example, suppose for the bridge design scenario in Figure 3 the designer feels that *Superstructure=Cable-stay* has a 'good' rather than a 'fair' likelihood of being part of a best-concept solution, but that *Spans=2* has a 'poor' rather than a 'fair' likelihood . The attribute likelihoods are modified accordingly as shown in Figure 4, and the neural network is then trained to predict the designer-modified likelihoods as output for the current set of attribute values given as input. The foregoing training of the neural network is carried out for a specified number of randomly selected sets of attibute values so as to embody in the network connection weights some initial measure of the designer's judgements and choices concerning the problem domain.

	ATTRIBUTE VALUES		LIKELIHOODS (predicted)	(modified)
(Superstructure =Cable-stay)	X1=2 **--->O<**	**>O--->**	0 =Fair	+1 =Good
(Foundation =Caisson)	X2=3 **--->O<**	**>O--->**	+1 =Good	+1 =Good
		Hidden		
(Length=200m)	X3=4 **--->O<**	**>O--->**	0 =Fair	0 =Fair
		Layer(s)		
(Clearance=10m)	X4=1 **--->O<**	**>O--->**	-1 =Poor	-1 =Poor
(Spans=2)	X5=1 **--->O<**	**>O--->**	0 =Fair	-1 =Poor
(Traffic Lanes=4)	X6=2 **--->O<**	**>O--->**	+1 =Good	+1 =Good

Figure 4: Attribute Value --> Likelihood Neural Network (Including Designer-modified Likelihoods)

Having the currently-trained neural network, a population of conceptual design solutions is randomly created in a form suitable for application of a genetic algorithm .For this study, each such solution is defined by a set of attribute values expressed as a base-2 binary string. For example, as each of the attributes for the bridge design problem has eight possible values, this suggests that each attribute may be represented by a base-2 binary string having three bits. To illustrate, suppose that the value of the *Foundation* attribute is represented by the 3-bit binary string 110, which decodes to the base-10 value X2= 1*2^0+1*2^1+0*2^2= 3, which identifies the *Foundation=Caisson* from Table 3. As there are six attributes for the bridge design problem, a complete conceptual design solution is represented by a 6x3= 18-bit binary string. For example, the conceptual design solution shown as input in Figure 4 is represented by the 18-bit binary string 010 110 001 100 100 010, which decodes to X1=2, X2=3, X3=4, X4=1, X5=1 and X6=2.

Each conceptual design solution represented in the randomly generated population of base-2 binary strings is decoded to its equivalent set of base-10 attribute values, as illustrated in the foregoing, and given as input to the currently-trained neural network, which in turn predicts the corresponding attribute likelihoods. Then, the 'fitness' of the design solution is established as the summation of the individual likelihoods of the attributes (this is a reasonable measure of design fitness since the output from the neural network is based on a numerical scale for which better likelihoods are assigned larger values that poorer likelihoods). For example, for the first design scenario in Table 2 the fitness = -1+1-1+0-1-1 = -3, while for the second design scenario the fitness = 0+1+0-1+0+1 = +1.

Upon establishing the fitness of every design in the population, as described in the foregoing, the operators of the genetic algorithm (i.e., reproduction, crossover, mutation, etc.) are then applied to create the next generation of conceptual design solutions, and the currently-trained neural network is applied again to establish the corresponding attribute likelihoods and, hence, fitness for each design. This evolution process results in the creation of successive generations of conceptual design solutions having progressively higher average population fitness, until convergence occurs to a set(s) of attribute values and likelihoods corresponding to a maximum-fitness conceptual design scenario(s).

The designer is then prompted to critique the current maximum-fitness conceptual design scenario(s) and modify the attribute likelihoods as felt necessary. Then, while accounting for all previously considered scenarios (i.e., input-output training pairs) of attribute values and likelihoods, the neural network is retrained to also learn the additional design scenario(s) defined by the current maximum-fitness attribute values and the corresponding designer-modified likelihoods. The genetic algorithm is then reapplied, now using the retrained neural network containing the expanded measure of the designer's judgements and choices as the basis to evaluate design fitness during the genetic search to establish an improved maximum-fitness conceptual design scenario(s).

As shown in Figure 2, the foregoing evolutive-cognitive process is repeated until a maximum-fitness conceptual design scenario(s) is found that is deemed acceptable by the designer without modification, at which point a best-concept design solution(s) has been established.

4 Concluding Remarks

The described emergent or self-organizing computational model for conceptual design learns qualitative likelihood features about a problem solution in keeping with the experience of the designer, and evolves to a best-concept design scenario(s) by progressively recombining positive features and eliminating negative features. The model accounts for all judgements and choices made by the designer over the course of the conceptualizing process and, as such, provides a consistent basis for the determination of best-concept design solutions. Moreover, the simplicity of the model suggests enormous potential for application. Since it only requires the names of design attributes and their ranges of possible values as input, and is otherwise not problem specific, the model can be applied to find best-concept design solutions in essentially any problem domain (e.g., engineering, medicine, etc.).

At the same time, for many design problems it is often a difficult task to identify the attributes and their possible values. There is a need to develop 'exploration' techniques that can be applied to establish the design space for such problems, [1]. As well, the simplicity of the computational model also raises serious concerns about its feasibility and practicality for doing conceptual design. By limiting the problem input data to a list of design attributes and corresponding possible values, any analog reasoning involving problem-specific and memory-resident information can only be carried out to the extent of making judgements based on the superficial similarity of attribute names and values. Moreover, the described computational model for conceptual design involves a neural network for which the designer essentially creates every input-output training example. This is not at all practical as the training of such networks typically involves many training examples. There is evidently a need to modify the computational model of the designer's memory so that it is able to retrieve best-concept design solutions with much less required intervention on the part of the designer. Finally, there is no quantifyable evidence that the computational model allows for innovativeness and ingenuity when finding best-concept design scenarios. To achieve this, it is necessary to develop a human-computer interface environment that maintains the fluidity of the creative design process while harnessing together the 'thinking' power of the human and the 'computation' power of the computer through a communication mechanism that allows for their creative interaction, [6].

Research in a variety of areas is being conducted to investigate how the human brain works in general and, more specifically, how it may be modelled on the computer. Cognitive science researchers are working on modelling the human brain from a variety of different perspectives, including the operation of the human memory system. The latter research is of prime interest to the development of a computational model for conceptual design because it deals with the mechanisms that designers use to express opinions/preferences about entities by drawing analogies with stored information in memory. Brain scanners are now able to identify the processes of the mind, down to the last 1,000th of a second, [7]. They operate by highlighting the hunger of brain cells for blood when they are in action, and by tracking the rapidly changing magnetic field that results from the electrical activity generated within the conscious brain. In essence, the scanners are able to capture a thought as it develops on the fly. To date, they have been used to capture but simple thought processes, such as putting a name to an object when shown it's picture. It seems reasonable to assume, though, that this technology will further develop over the next number of years, and that it could be of great future benefit to the development of a computer-based system for conceptual design.In a different area, researchers are currently developing virtual-reality tools to make people and computers more compatible. It also seems reasonable to think that virtual reality technology will be of future benefit in developing human-computer interfaces for conceptual design systems. In fact, it is not too far-fetched to imagine a human-computer interface environment, involving multimedia text, picture, audio, touch, smell and antimated information, where the human communicates directly with the computer through a process of thought to teach it about the subject domain and to then direct it to do the computations necessary to search the domain for the very best solution to the problem at hand.

Acknowledgements

The research reported in this paper is financially supported by the Natural Science and Engineering Research Council of Canada.

References

1. Gero, J. S. and Maher, M. L., (Editors): Modeling Creativity and Knowledge-Based Creative Design". Lawrence Erlbaum Associates, Publishers, New Jersey, USA, 1993

2. Miles, J. C. and Moore, C. J.: Some Thoughts on Conceptual Design: Heuristics and their Utility in Knowledge-Based Systems. *Structural Engineering Review*, Vol. 5, No. 2, pp. 87-95,1993

3. Roorda, J.: Computer Aided Conceptualization. Report of Centre for Engineering Educational Technology, Univ. of Surrey, UK, 1993

4. Thagard, P., Holyoak, K. J., Nelson, G. and Gochfeld, D.: Analog Retrieval by Constraint Satisfaction. *Artificial Intelligence*, Vol. 46, pp. 259-310, 1990

5. Grierson, D. E.: Conceptual Design using Evolutive-Cognitive Computing Techniques. Proceedings of the ASCE Congress on Computing in Civil Engineering, Washington, DC, USA, June 20-22, 1994

6. Grierson, D. E.: A computational Model for Conceptual Design?. Proceedings of Workshop on the Future Directions of CAE, Bridging the Generations, Carnegie-Mellon University, Pittsburgh, PA, USA, June 17-19, 1994

7. The Globe and Mail, Toronto, Canada, April 23, 1994

Role of Emergent Computing Techniques in Multidisciplinary Rotor Blade Design

P. Hajela and J. Lee

Mechanical Engineering, Aeronautical Engineering and Mechanics
Rensselaer Polytechnic Institute, Troy, NY 12180-3590

Abstract. Design synthesis in multidisciplinary systems is typically characterized by a large number of design variables and constraints; additionally, there are complex interactions between the participating disciplines which must be both identified and then suitably represented in the design process. In a number of applications, the design space itself may be multimodal, thereby introducing the need for a global search strategy; the latter may contribute to an excessive demand on computational resource requirements. This paper discusses some of the above problems in the context of a multidisciplinary design of a helicopter blade, where the underlying analysis is inherently nonlinear, computationally demanding, and the design space is nonconvex. The use of newly emergent computing techniques such as genetic algorithms and neural networks is discussed in the context of this problem. The role of neural networks as a function approximation strategy is discussed; the multilayer backpropagation network (BP network) and a variant of the counterpropagation network (CP network) are examined. The paper also shows how a trained BP network can be used to identify couplings in a multidisciplinary system. Such an identification provides a rational approach for determining the topology of problem decomposition, and a sequence of smaller, more tractable subproblems can be generated.

Keywords. genetic algorithms, neural networks, multidisciplinary design

1. Introduction

There has been considerable recent interest in applying formal optimization methods to the design of multidisciplinary systems [1,2]. Inspite of an increased awareness of the potential of such methods, their application in real-life multidisciplinary systems continues to be a challenge. These challenges stem from the fact that such systems are characterized by the presence of a large number of design variables and constraints, and analysis from various contributing disciplines which are not completely independent. In several cases, the dimensionality of the design space may increase to a degree that obtaining reliable solutions to the optimization problem is rendered questionable when using traditional mathematical programming methods. Furthermore, there is a diminished capacity to evaluate the acceptability of solutions in higher dimensionality design spaces.

Decomposition based methods have been proposed as a solution to large-scale coupled problems [3,4], wherein the original problem is decomposed into a set of smaller, more tractable subproblems. The coupling between the subproblems may either be hierarchical, where it is possible to identify distinct tree-like patterns of influence (Fig.1), or, no obvious hierarchy may exist and there may be multiple one or two-way couplings among disciplines (Fig. 2). The use of decomposition techniques to create a number of smaller subproblems which represent the full complexity of the original problem may allow for parallel processing, and contribute to a better understanding of the problem domain. In general, there are two principal difficulties that have been identified in using decomposition based design techniques. First, in all but the simplest of problems, it is very difficult to establish the topology for decomposition - i.e., assign a subset of design variables and constraints to a particular subsystem. Even when a decomposition topology has been established and smaller subproblems have been created, there is a need to ensure that the coupling between these subproblems is properly taken into the consideration as the design progresses.

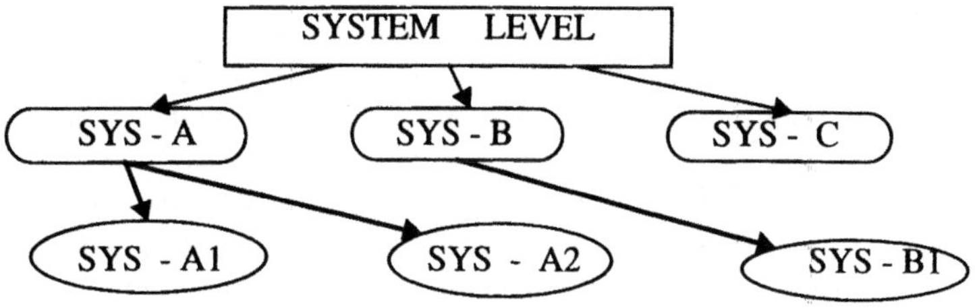

Fig.1 Hierarchical system with top-down couplings

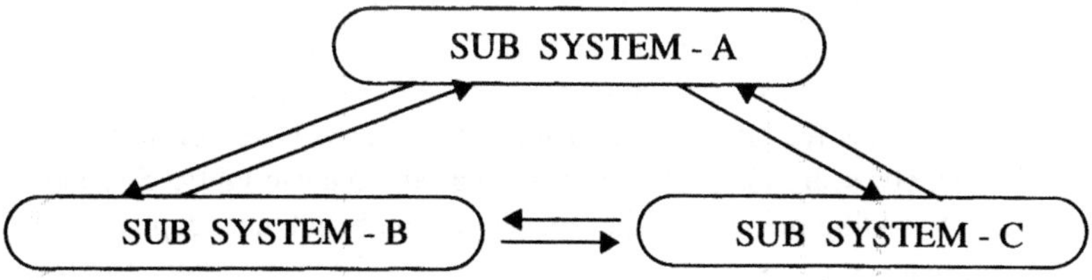

Fig.2 Nonhierarchical system with two-way couplings

A major concern in all optimal design studies related to complex multidisciplinary systems in aerospace design, is the computational cost that is associated with the function evaluations required during the iterative optimization process. A significant emphasis is placed, therefore, on investigating techniques for creating function approximations with the intent of minimizing the overall computational effort required in the design process. In this context, it is important to recognize that *wall-clock time* is of as much concern in preliminary design as the total CPU effort that is expended, and methods which may require a larger upfront investment of CPU costs but provide robust and fast approximations during the actual design process, cannot be dismissed.

Search algorithms based on principles of mathematical programming have an undeniable role in multidisciplinary design. The more efficient of these class of methods is gradient-based, and the search progresses from one design point to the next in a direction determined by the gradients of the objective and constraint functions. Computational expediency in using such methods has been obtained through the use of first-order Taylor series approximations of the functions. However, there are a number of problems where the applicability of the gradient-based methods is in doubt. In particular, in problems where the design space is nonconvex or disjoint [5], such methods are known to yield suboptimal designs. Problems where the design variables are of discrete or integer type are also more difficult to accommodate using mathematical programming techniques.

Nontraditional search methods such as genetic algorithms (GAs) have been useful in such applications. These methods represent a highly exploitative stochastic search that is global in nature. Unlike the gradient-based methods, the search progresses from one set of multiple points in the design space to another, incorporating information from all points to establish the direction of move. While such an approach offers an increased probability of locating the global optimum, the computational costs are high due to the necessity of multiple design point evaluations. A Taylor series type approximation scheme would also be unacceptable in this approach as multiple approximations would have to be constructed. Furthermore, the integrity of these approximations would be in question, as the method does not impose explicit limits on design changes during an iteration. Global approximation techniques that use only function information in constructing the approximations are a candidate of choice for such an approach. The present paper discusses the application of neural networks in this context.

The effectiveness of genetic search methods in multidisciplinary design will hinge upon their successful extension to large scale problems. As shown in Fig. 3, there are two distinct strategies for such an extension. The first is based on improving the efficiency of genetic search in problems of increased dimensionality. Strategies that are equivalent to adding memory to the stochastic search and that adaptively identify the most promising regions of the design space have been recently examined [5]. The second approach relies on decomposition strategies to create smaller dimensionality subproblems in which genetic search is directly applicable; the issue here is to investigate methods by which the interaction between the temporarily decoupled subproblems can be represented as the design is advanced. In both of these approaches, the use of neural networks as a function approximation tool is important from the standpoint of computational efficiency.

In subsequent sections of this paper, a more detailed discussion of the applicable elements of genetic search and neural networks is presented. The design of a helicopter rotor blade is discussed and formulated as a typical multidisciplinary design problem for which specialized solution strategies are required. Both direct and decomposition based implementations of genetic search are discussed in the context of this problem. The role of neural networks as a global function approximation tool, for establishing the topology of problem decomposition, and for including the interactions among the temporarily decoupled subproblems, is also discussed.

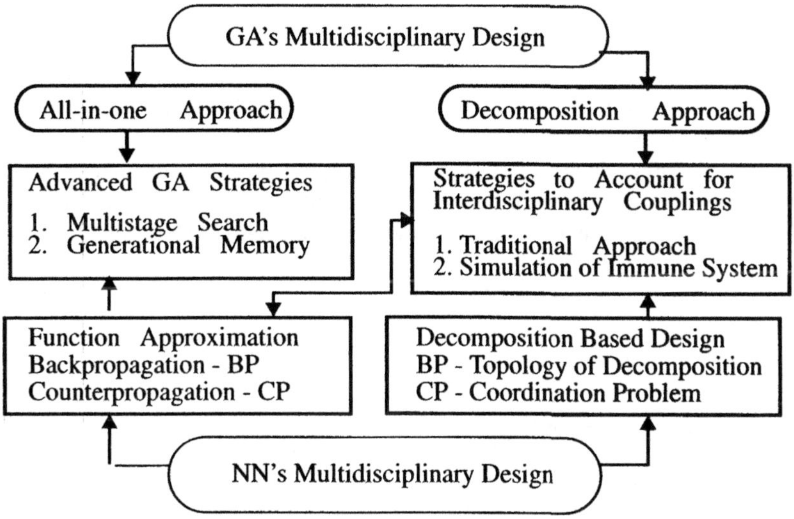

Fig.3 GA's in large scale multidisciplinary design

2. Neural Networks in Multidisciplinary Design

Neurobiological methods of computation are distinctly different from traditional numerical computing and the more recently emergent strategies of symbolic processing and computation. The basic computational processor in neural networks is an artificial neuron; a network comprises of a number of interconnected neurons, and the characteristics of the network are determined by the nature of the interconnections. One commonly used architecture is that of a multilayer feedforward network shown in Fig. 4. At the most basic level, the network can be viewed as providing a mapping between input and output vectors of interest. There is an input layer of neurons which serves as a fan-out layer for the input signal, a number of *hidden* layers of neurons, and a layer of output neurons where the output vector is available. The h-hidden layer network shown in Fig. 4 is fully connected, in that a neuron in each layer is connected to all neurons in adjacent layer. The strength of the interconnection between the i-th and j-th neuron of the k-th and l-th layer, respectively, is represented by interconnection weight w_{ij}^{kl}. In terms of these interconnection weights, the input to the i-th neuron of layer l can be written as follows:

$$Z_j = \sum_i w_{ij}^{kl} x_i^k$$

This weighted sum of inputs is processed through an activation function F(Z) to generate an output signal Y_j^l which is then communicated to other neurons in downstream layers in a similar manner. Several forms of activation function F(Z) may be chosen; a sigmoid function of the following form has been widely adopted:

$$F(Z) = \frac{1}{1 + exp\left(-Z - B_j^l\right)}$$

Here, B_j^l is a bias constant which regulates the output of the neuron. During network training, a number of input-output pairs (training patterns) are presented to the network. The interconnection weights and bias constants are adjusted so that for all the training patterns, the error between the output predicted by the network and the actual output is minimized. Such a network is considered to be trained, and can be used to generate an approximate output to an input vector that belongs to the domain of training but was not part of the training set.

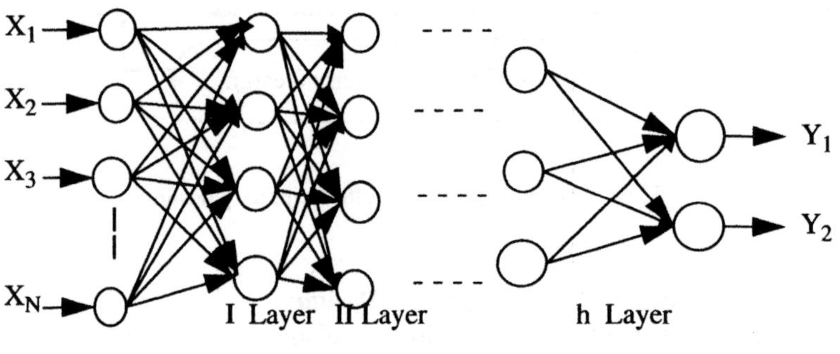

Fig.4 A BP network with h-hidden layers

A number of issues have been raised about the usefulness of this approximation procedure, including questions about the number of input-output samples that must be included in the training set, the choice of network architecture, and if the computational effort involved in network training offsets any savings that may result from the use of the network based approximations during optimization. The approximation procedure is very similar to constructing a polynomial based response surface, where the sigmoid function provides a very rich nonlinear function for surface manipulation without the need for having to specify the order of the polynomials. While the issue of training set size and network architecture have been studied in recent work [6,7], other network architectures such as the functional link network [8] and the counterpropagation (CP) network [9] have been developed to explicitly overcome the issue of network training time. The CP network can be used to generate both forward and inverse mappings between input and output vectors, and has interesting applications in multidisciplinary design problems.

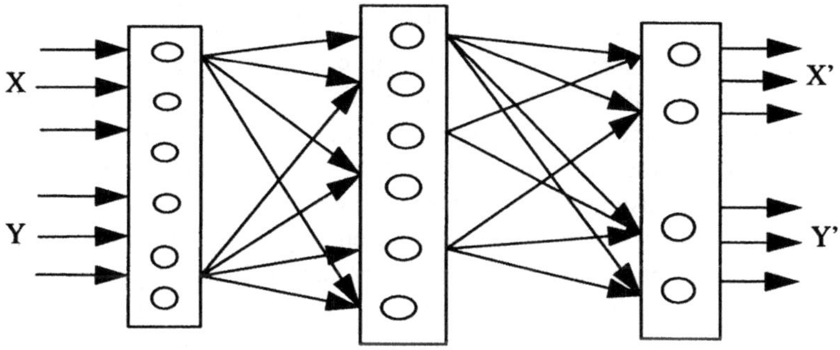

Fig.5 Schematic representation of a CP network

This network is a combination of two layers of neurons referred to as the Kohonen and Grossberg layers, respectively (Fig. 5). The network is designed to mimic the functioning of the brain in a closer sense in that the data presented to the network is first classified as belonging to a specific category by the Kohonen layer of neurons, and the Grossberg layer of neurons translates this data into the desired output. In this sense, the network behaves as a look-up table capable of generalization. The weights leading to any neuron in the first layer are simply the agregate center of all patterns that belong to the category (cluster) represented by that neuron. The interconnection weights between that neuron and any output layer neuron are components of the output vector corresponding to that cluster. In the improved CP network described in [9], the generalization operation proceeds by determining the extent to which a new pattern belongs to any or all clusters, and then uses a nonlinear averaging scheme to determine the output; note that the output is simply a weighted average of output weights corresponding to all participating Kohonen neurons.

Among the many advantages of this network are its ability to train rapidly and since classification is based on a nearest neighbor principle, the network is also able to work with incomplete input patterns. In practical terms, the ability to work with incomplete patterns implies that if some components of the input vector are missing, the network estimates the missing components in producing the generalized output response. This feature has been shown to be important in decomposition based design, where a different subset of design variables is used in temporarily decoupled subsystems.

2.1 Identification of Decomposition Topology

In using decomposition principles in multidisciplinary design optimization, an issue that has to be considered at the very outset is the search for a rational approach to decompose the problem. This paper shows how neural networks can be used to identify dependencies among design variables and design objectives, and which can be used as a guideline for problem decomposition.

Consider the strengths of the interconnections between the neurons of the various layers of a trained multilayer feedforward network. This matrix of numbers can be viewed as a hologram representing the mapping between the desired input and output quantities, and offers little in terms of a physical interpretation of the problem domain. The flow of information between the input and output layers, however, is a strong function of these interconnection weights, and an analysis of the weights provides meaningful insight into the problem domain.

Consider a network with a single hidden layer of neurons, where the interconnection weights between the input layer and the hidden layer and between the hidden and output layers are denoted as w_{ij} and u_{jk}, respectively. If the magnitude of interconnection weights is used to determine flow-paths of information, then the fraction of signal received at the j-th hidden layer neuron which can be attributed to the i-th input neuron is $|w_{ij}|/S_j$, where S_j is the sum of the magnitude of weights between all input layer neurons and the j-th neuron of the hidden layer. If the output signal of interest is that at node k of the output layer, then the signal from the j-th hidden layer node is further multiplied by $|u_{jk}|$ and a sum of all such signals through the different neurons J in the hidden layer can be written as follows:

$$\eta_{ik} = \sum_{j=1}^{J} \frac{|w_{ij}||u_{jk}|}{S_j}$$

If the sum of all η_{ik} taken over all input layer neurons is denoted by Θ_k, then the fraction of an output weight that is attributable to the i-th input neuron can be expressed as the elements t_{ik} of a transition matrix T as follows:

$$t_{ik} = \frac{\eta_{ik}}{\Theta_k}$$

These can be further normalized in the following manner:

$$\hat{t}_{ik} = \frac{t_{ik}}{\sum_i t_{ik}}$$

The elements of this matrix show the contribution of the i-th input to the k-th output. Note that the above operation is equivalent to a matrix multiplication of normalized or partitioned interconnection weight matrices of successive layers, taken in order; the extension, therefore, to multiple hidden layer networks is straightforward. Inspection of a particular column of this matrix (e.g. the k-th column), the influence of each input component on the k-th output component can be qualitatively assessed.

Since the sum of all elements in a column is unity, this influence is represented by a fractional quantity, and allows for easy identification of dominant input parameters, and hence a topology for decomposing the design problem (this approach is referred to as ABS in subsequent sections of the paper). Note further that a change in the transition matrix resulting from a change in size of the training set can be used to determine if an adequate number of training patterns has been considered.

In the jargon of optimization methods, this matrix yields an aggregate sensitivity that is valid over a design range that is defined by the domain over which the neural network was trained. If the design region is large, a number of networks can be developed, each one trained for a cluster of the design region with similar characteristics. An alternate approach for computing this aggregate sensitivity matrix is described in [10]; this approach simply considers the matrix product of the interconnection matrices, an approach that incorporates the effect of the sign of interconnection weights in the analysis. As shown in [10], this approach (referred to in this paper as ALT) is useful in identifying both the magnitude and direction of input-output interactions.

2.2 Decomposition in Non-Hierarchical Systems

The multidisciplinary design problem of interest here is nonhierarchical, and the Concurrent Subspace Optimization (CSSO) approach [11] is a candidate solution strategy. This section of the paper describes how neural networks can be used to further enhance the CSSO solution approach. The essence of this method is to decompose the large dimensionality global design variable vector into a number of smaller design variable groups (subspaces), each of which is used to optimize the system level objective function and simultaneously account for constraint satisfaction in all subspaces, albeit to different degrees. In a traditional implementation, the mathematical statement of the problem can be defined as follows:

$$minimize \rightarrow F(X_k)$$

$$subject\ to$$

$$C^p \le C_o^p\left[s^p\left(1 - r_k^p\right) + (1 - s^p)\, t_k^p\right] \qquad p = 1, NSS$$

$$X_k^L \le X_k \le X_k^U$$

where, NSS is the number of subspaces; C^p is a measure of all constraints of subspace p; superscript p and subscript k denote the influence of the p-th subspace on the k-th subspace. The coefficient r_k^p represents the responsibility assigned to the k-th subspace for reducing the constraint violation of the p-th subspace. The coefficient t_k^p represents the trade-off associated with each subspace that allows for the violation of a constraint in the p-th subspace, provided a corresponding compensation is sought in the k-th subspace. The switch parameter s^p simply enables or disables the responsibility or trade-off coefficients. The constrained optimum resulting from these subspace optimizations is a function of the coefficients r and t, and a second level problem referred to as a co-ordination level problem must be solved. The principal intent of the coordination problem is to account for the differences in optimal objective function values that result from working with a limited set of design variables in each subspace. If proper caution is not exercised in the choice of the coefficients r and t, it is quite possible to obtain a new design point (concatenation of subspace optima) that in addition to being nonoptimal, will not even be feasible.

The goal of a recently initiated study is to explore the extent to which neural networks would enhance this approach. The use of a transition matrix based identification of the decomposition topology is certainly applicable; in addition, the use of a neural network based approximation of constraint functions for use in subspace optimization has shown promise. Neural networks are also being implemented in this problem to provide estimates of the responsibility and trade-off coefficients to use in the approach. In previous work, these coefficients were selected in a somewhat ad-hoc manner. Finally, and perhaps most importantly, neural networks are also useful in providing alternative methods to replace the coordination problem phase of this strategy. One such approach is discussed in this paper in the context of decomposition in a genetic search based (using function information only) design.

3. Genetic Algorithms in Multidisciplinary Design

The most well-defined role for genetic algorithms in the multidisciplinary design problem continues to be their application as an optimization tool. In this approach, a particular design is represented by a chromosome-like string - a simple approach for doing this would be to take a binary number representation of each design variable, and attach all representations of variables head-to-tail to construct one string like structure that defines a design. A number of such design representations can be created to generate a population of designs which are then subjected to a number of genetic evolution-type operators to evolve into new generations. The new generations are better adapted to an environment which could be defined in terms of the objective function value and/or the level of constraint satisfaction.

Even though the effectiveness of genetic algorithms in optimization problem has been demonstrated, its application to multidisciplinary design problems is not quite straightforward. In such problems, when the number of design variables is large, the chromosome-like string representations of designs get quite long, and the number of design alternatives in the search space that must be examined also increases dramatically. In order to sample an adequate number of design alternatives to meaningfully evolve a current population, the size of the population must also be increased, thereby resulting in increased computational requirements. In order to apply genetic algorithms in large scale multidisciplinary design, special strategies within GA must be established, or alternatively, the application of GA's in decomposition based design should be examined further (Fig. 3). The former approach has resulted in the proposition of two specialized strategies, the directed crossover and the multistage search [5], and these are applied to the multidisciplinary design problem in this work.

The directed crossover strategy is based on the idea that random crossover in genetic search is disruptive once the process has started to converge. Additionally random crossover in regions where bit string values have converged does not result in an exploration of the design space. The directed crossover operation develops a generational memory pertaining to the performance of various segments of the design string representation in producing favorable changes in the design objective. It then assigns a higher probability of performing crossover in those select regions of the string. The multistage search evolved from the need to increase the population size as the string lengths increased to accomodate larger number of variables represented with higher and higher precision. The approach here is to initiate the search with coarse representation of design space (and hence smaller string lengths), and then searching with an increased precision in only the most promising regions identified in the previous stage. In essence, both strategies draw upon a recognized fact that searching with string representations that have a shorter defining length will result in better convergence characteristics [12].

The second approach adapts genetic search into a decomposition based solution strategy. As described in an earlier section, the transition matrix of a trained feedforward neural network is first used to determine the topology for decomposition. Each decomposed subproblem consists of a smaller number of design variables that can be handled more effectively by genetic search. To account for the interactions that exist among the temporarily decoupled systems, two distinct strategies are being explored in the present work, and these are summarized as follows.

3.1 Neural Network Approximations

In this approach, a single CP network would be trained to construct the mapping between the global design variable vector and the system objective and constraint values. The decomposed optimization problems will be processed in parallel; each would have access to the CP network and seek estimates of subspace constraints from the network as the subspace design variables are updated by the optimizer. The pattern completion capability of the CP network would be used in this task as exact values of other design variables (being changed in other subspace optimizations) would not be available. As shown in the section of numerical results, the resolution towards a common design can be improved if design variable changes from each subspace are periodically updated in all other subspaces.

3.2 Immune Network Based Resolution of Interactions

In biological immune systems, foreign cells and molecules, denoted as antigens, are recognized and eliminated by type-specific antibodies. The antigen-antibody pair may be conceptualized as a lock and key combination, and an appropriate antibody (key) must be made to fit a specific antigen (lock). The number of antigens is quite large, and the immune system must respond to these foreign elements by generating antibodies from a limited number of genetic resources. The immune system does so very effectively by using combinatorial mechanisms. The genetic information for an antibody is encoded by separate gene segments, each of which is drawn from a library of gene segments. In order for the combinatorial mechanism to function, the contents of the libraries must act cooperatively and partition the antigen recognition task evenly among the segments. This cooperative partitioning requires some organization, and an evolutionary approach can be used to develop this characteristic. The evolutionary approach is somewhat similar to that used in GA's; the principal difference is in the choice of the fitness function. In immune systems, the fitness of an individual would be determined by its ability to recognize specific antigens. The individual would be exposed to a set of antigens and the expressed antibodies of that individual would be matched against the antigens to determine the similarities. More similarity would imply higher fitness for the individual. The dynamics of the immune system has been modeled extensively in theoretical biology [13] and have been embraced in machine learning applications. The application of the immune network to the present problem can be understood as follows.

The stringlike chromosome structure representing the design contains a definition of all design variables. After partitioning of the design variable vector for each subsystem has been performed, genetic evolution is carried out in each subsystem in parallel, with the fitness function described in terms of the system level objective function (note that only that subsection of the chromosome string which corresponds to the design variables for that particular subsystem is changed). This process can be carried out for a fixed number of generations, and then a predetermined number of fit strings from each subsystem are introduced as antigens to all other subsystems. A second stage of genetic search is then performed in each subsystem where the fitness function is defined in terms of each individual's ability to manufacture antibodies that match the newly introduced antigens. This stage of the genetic search is seen as a correction step that introduces compatibility between the different subpopulations, so that the process eventually converges on the desired design. Note, that as for the first stage, this search can also be performed in parallel. This approach is currently under implementation, and results of this work will be reported in a separate publication.

4. Multidisciplinary Rotor Blade Design - Problem Statement

The multidisciplinary design of a rotor blade requires an integration of the disciplines of acoustics, aerodynamics, dynamics, and structural analysis within the optimization framework. This simplified example illustrates some of the inherent complexities in such design problems, and where emergent computing techniques are particularly applicable.

The use of composites in rotorcraft blade design provide opportunities for enhanced aerodynamic, structural, and dynamic performance. The use of composites makes it practical to fabricate non-rectangular blades with variations in twist distribution and airfoil sections along the blade span, thereby contributing to increased flexibility in aerodynamic design. Satisfactory aerodynamic design requires that the required horsepower for all flight conditions not exceed the available horsepower, that the rotor disk must retain lift performance to avoid blade stall, and that the vehicle remain in trim. Important factors in structural design include material strength considerations for both static and dynamic load conditions. A combination of flapwise, inplane, torsion, and centrifugal forces typically comprise the static loading. Another important consideration that encompasses both structural and aerodynamic design, is the autorotation capability. The autorotation requirement pertains to maintaining the mass moment of inertia of the rotor in the rotational plane at an acceptable level. This is a function of the vehicle gross weight, rotor aerodynamic performance, and the rotor system mass moment of inertia. Finally, dynamic design considerations of the rotor blade pertain to the vibratory response of the blade under the applied loads; this design limits the dynamic excitation of the fuselage by reducing the forces and moments transmitted to the fuselage.

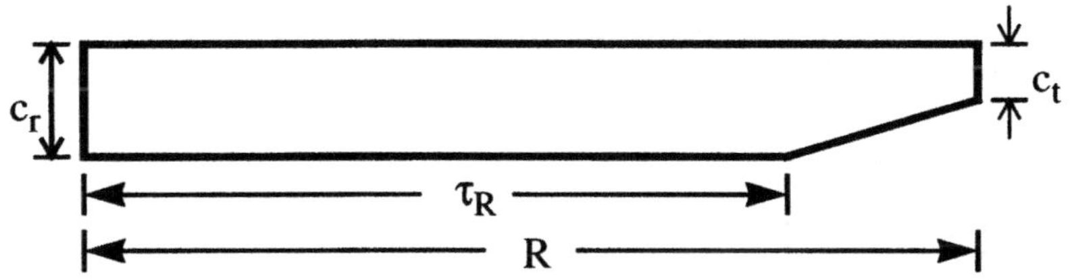

Fig.6a Rotor blade planform geometry

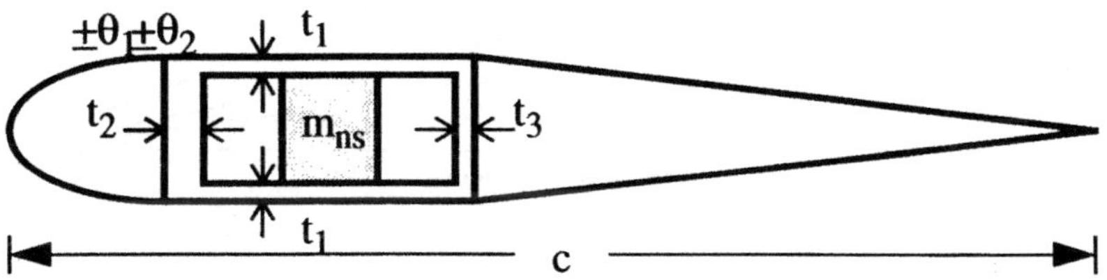

Fig.6b Cross section of blade airfoil

A finite element in time and space formulation was used to model the dynamics of the blade [14]. This formulation is based on a multibody representation of flexible structures undergoing large displacements and finite rotations, and requires that the equations of motion be explicitly integrated in time. An unsteady aerodynamic model is used to obtain the induced flow and to calculate the aerodynamic forces and moments in hover and forward flight. In addition to the geometric nonlinearities that are inherent in this problem, the loading on the blade varies as it moves around the azimuth - on the advancing side the flow velocity over the blade is additive to its tangential speed; on the retreating side, these velocities substract. Consequently, in order to maintain force and moment equilibrium, the pitch of the blade is continuously changed as it rotates around the azimuth, and the hub loads are a function of the blade rotational frequency. There is a transient period during which equilibrium of the vehicle is established (trim), and this is generally obtained in 8-10 revolutions of the blade. This process of obtaining trim condition prior to load calculation introduces additional computational costs. The nature of the solution process used here is such that designs that correspond to aeroelastically unstable configurations will simply yield a divergent response in time. As described in a later section, this requires that the neural network based approximation procedure be treated somewhat differently.

The objective of the design problem was to design the blade geometry and internal structure to minimize a weighted sum of hub shear force and bending moments for a hingeless rotor blade in forward flight; aerodynamic, performance and structural design requirements are considered as constraints, and dynamic requirements constitute a multicriterion objective function. The premise behind the approach is that a minimization of the hub loads and moments translates into lower vibrations that are transmitted to the fuselage structure.

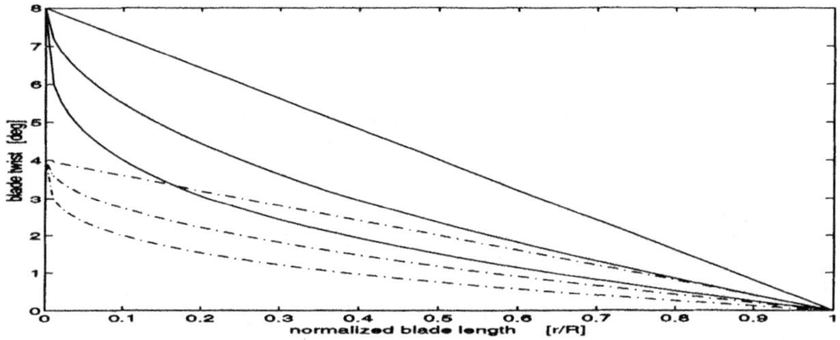

Fig.6c Spanwise variation in blade twist

The planform geometry of the blade is shown in Fig. 6a. The blade is tapered in both the chord and the depth, and all loads are assumed to be carried by the structural box shown in Fig. 6b. The design variables for this problem are also shown on the figures and summarized in Table 1. The blade planform geometry is defined by the chord ratio λ_c and the point of inception of taper along the span denoted by τ_R.The blade twist varies from θ_t at the root to zero at the tip; this variation maybe linear or nonlinear (Fig. 6c), and is controlled by the twist shape parameter δ. Both θ_t and δ were design variables in the problem. Two nonstructural masses were positioned along the span of the blade, and the magnitude of these masses (m_1 and m_2) as well as their locations along the span (d_{m1} and d_{m2}) were considered as design variables. The rotational speed of the rotor Ω was also selected during the design process. The remaining variables, although descriptive of the internal structure of the blade, have a strong influence on both aerodynamic and dynamic performance. These were the horizontal flange thickness t_1, and the left and right vertical sections of the box beam denoted by t_2 and t_3, respectively. For the Graphite/Epoxy rotor blade, the horizontal flanges are symmetric ± 45 deg laminates; this layup is also present on the outer half of both vertical sections of the box beam. The inner half of the vertical walls are divided into two segments, with a layup of $\pm\theta_1$ and $\pm\theta_2$ deg, respectively. This accounts for a total of 14 design variables for the problem. Note that it is relatively easy to increase the problem dimensionality by simply varying the thickness and orientations of plies in discrete segments along the span.

The mathematical statement of the optimization problem can be written as follows:

$$minimize \rightarrow F = c_1 F_z + c_2 M_y + c_3 M_z$$

$$subject\ to$$

$$g_1 \equiv \frac{HP_h}{HP_a} - 1 \le 0 \qquad g_2 \equiv \frac{HP_f}{HP_a} - 1 \le 0$$

$$g_3 \equiv 1 - \frac{\frac{C_T}{\sigma}}{\left(\frac{C_T}{\sigma}\right)_L} \le 0 \qquad g_4 \equiv \frac{\frac{C_T}{\sigma}}{\left(\frac{C_T}{\sigma}\right)_U} - 1 \le 0$$

$$g_5 \equiv 1 - \frac{FM}{(FM)_L} \le 0 \qquad g_6 \equiv \frac{FM}{(FM)_U} - 1 \le 0$$

$$g_7 \equiv 1 - \frac{AI}{(AI)_L} \le 0 \qquad g_8 \equiv \frac{W}{(W)_U} - 1 \le 0$$

$$g_9 \equiv \bar{R} - 1 \qquad g_{10} \equiv \frac{\sigma_{buck}}{\sigma_{all}} - 1 \le 0$$

In the above, F_z, M_y and M_z are the maximum peak-to-peak values of the scaled shear force, flap bending moment, and lead-lag bending moment, respectively; HP_h and HP_f denote the horsepower required during hover and forward flight, respectively, and HP_a is the available horsepower; C_T and σ are the rotor thrust coefficient and solidity, respectively, and the subscripts L and U denote the lower and upper bounds of this ratio. These bounds are necessary to limit the lift performance of the rotor disk to avoid blade stall. The parameter FM is a figure-of-merit for the helicopter performance out of the ground effect, and subscipt L is used to denote its lower bound; the blade must have sufficient autorotation (AI) capacity to safely descend in an out-of-power configuration, and this is represented in constraint g_7; W is the sum of the structural and nonstructural weights and W_U denotes its upper bound; $\bar{R}$ is used to denote the structural failure criterion, and a Tsai-Wu measure was used for the composite structure in this problem; σ_{all} is the allowable stress for the material and σ_{buck} are the static stresses due to buckling. The constraint set represents response quantities that are traditionally associated with structures, dynamics, and aerodynamic performance of the helicopter. For one given set of design variables, the analysis time required for the evaluation of the objective and constraint functions is substantial (about 18 CPU minutes on a SPARC station) and clearly not amenable to integration with a genetic algorithm based optimization routine.

Table 1: Range of Design Variables Used in Neural Network Training

design variable	symbol	min	max	precision
mass 1 location	m_1	1	5	1
mass 2 location	m_2	6	10	1
tuning mass 1 [kg]	d_{m1}	0.0	5.0	0.1
tuning mass 2 [kg]	d_{m2}	0.0	5.0	0.1
horizontal flange thickness ratio	t_1	0.25	0.35	0.01
left vetical flange thickness ratio	t_2	0.20	0.30	0.01
right vertical flange thickness ratio	t_3	0.20	0.30	0.01
blade twist [deg]	θ_t	4.0	8.0	0.1
twist shape parameter	δ	0.3	1.0	0.1
taper inception point	τ_R	0.50	0.80	0.01
chord ratio	λ_c	1.0	2.0	0.1
rotational speed [rad/sec]	Ω	24.0	34.0	1.0
ply angle of inner vertical flange [deg]	θ_1	30	90	1
ply angle of outer vertical flange [deg]	θ_2	90	90	1

5. Results and Discussions

This section presents the results from a number of numerical experiments designed to assess the applicability of genetic algorithm and neural network strategies discussed in earlier sections, to the multidisciplinary design problem. The rotor blade design problem comprises the test-bed for these numerical experiments.

5.1 Neural Networks in Function Approximation

The first set of experiments was designed to validate the effectiveness of both the BP and CP neural networks in generating the mapping between the design variables and the response quantities of interest. Some of these mappings are quite nonlinear, and require careful consideration of the choice of network architecture and of the number of training patterns. The range of design variable variations used in this training are shown in Table 1. A number of training patterns was generated, in the range of design variable variation, and this included both stable and unstable designs. A BP network with 14 input layer neurons, 10 hidden layer neurons, and 5 output layer neurons (a 14-10-5 network), corresponding to 5 output quantities, viz., vertical hub shear, flap moment, lag moment, rotor thrust, and the failure criterion index $\overline{R}$, was established and trained with 550 training patterns. This training presented problems in that it was difficult to reduce the training error to below 3%. The training data was then sorted to separate the designs that yielded stable and unstable responses; it was considered expedient at this stage to establish 5 networks, each mapping the design variables into one output only (14-10-1 networks for all but the lag moment, where a 14-10-8-1 network was used). The training of each of these networks could be done in parallel. The training using these separated patterns procceded well, converging to an error of less than 1%, with the exception of the lag moment (error was 2.6%). Fig. 7 shows the time response of the hub shear and moments for a typical design; the pattern of variation for the lag moment is quite erratic, and provides an explanation for the discrepancy in training. Table 2 shows the results of testing these networks for generalization performance using 8 sets of design variables that were not part of the original training set, and the results are extremely encouraging. Similar generalization performance was obtained from networks trained with the unstable data. The difficulty in training the networks for the combination of stable and unstable data can be attributed to either a) a completely different input-output relationship in one or more components or b) excessive data for the number of weights and bias constants in the network that could be varied to fit the data. The weight analysis procedure to obtain the transition matrix sheds some light on this question and is discussed later in this paper.

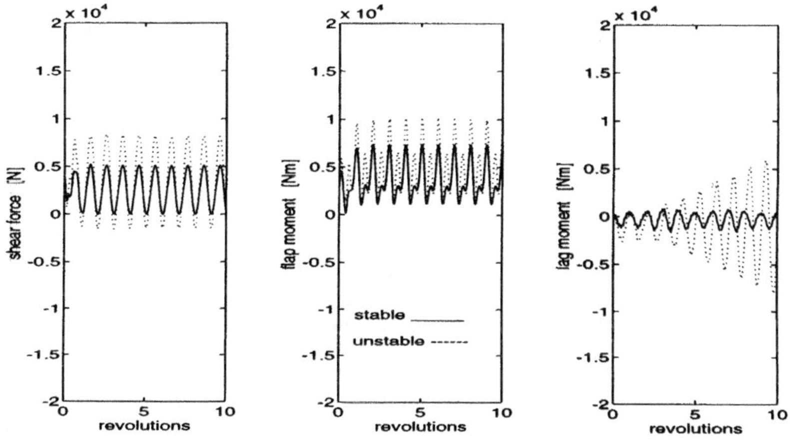

Fig.7 Time response of vibratory force and moments at hub

Similar experiments were also performed using the full CP network, that generates an identity mapping of the type [X,Y] -> [X',Y']. This network architecture requires a much larger number of input patterns, and the quality of generalization depends upon the number of Kohonen layer neurons that are permitted (indirectly a measure of maximum cluster radius). If the cluster radius were set to zero, the number of Kohonen neurons would be equal to the number of training patterns, and the network would simply "memorize" all training data. The results of numerical experiments designed to test this network are shown in Table 3.

Table 2: Generalization Performance of BP Network

shear force (N)			flap moment (N m)			lag moment (Nm)			thrust (N)			failure criterion		
NN	actual	%	NN	actual	%	NN	actual	%	NN	actual	%	NN	actual	%
13695	13546	1.10	11403	11711	2.08	5278	5266	0.24	778	784	0.76	0.261	0.261	0.01
8666	8649	0.19	6295	6891	0.50	4059	3994	1.65	517	506	2.17	0.243	0.241	0.83
6864	6790	1.09	6700	6745	0.67	4036	3880	4.72	369	374	1.33	0.247	0.249	0.84
9160	9373	2.27	8409	8406	0.04	3917	3772	3.84	498	502	0.79	0.313	0.308	1.62
14952	15028	0.51	12905	12984	0.61	4082	4086	0.10	857	873	1.83	0.250	0.250	0.02
17002	17001	0.03	12768	12736	0.25	5974	5899	1.27	926	921	0.54	0.319	0.316	0.94
7284	7339	0.75	6934	6957	0.33	11606	11559	0.40	435	427	1.87	0.238	0.244	2.45
6754	6591	2.47	7542	7569	0.35	8217	7980	2.97	348	352	1.13	0.306	0.307	0.65

Table 3: Generalization Performance of CP Network

shear force (N)			flap moment (N m)			lag moment (N m)			thrust (N)			failure criterion		
NN	actual	%	NN	actual	%	NN	actual	%	NN	actual	%	NN	actual	%
8942	8687	2.93	7316	7504	2.53	3663	3705	1.13	472	483	2.28	0.259	0.258	0.39
14963	14272	4.84	10056	10264	2.02	8273	7892	4.82	766	763	3.93	0.312	0.312	0.01
12384	12579	1.55	10568	10458	1.05	5894	5758	2.36	763	754	1.19	0.235	0.237	0.84
8926	8919	0.08	7045	7058	0.18	5685	5713	0.49	509	502	1.39	0.264	0.266	0.75
10253	10156	0.96	13093	13185	0.69	11796	11183	5.48	933	907	2.86	0.329	0.333	1.20
8113	8046	0.84	8936	8841	1.01	12529	11932	5.00	376	371	1.35	0.347	0.349	0.57
9735	9653	0.85	9618	9685	0.69	22940	23695	3.18	467	454	2.86	0.394	0.390	1.02
14885	14100	5.56	10856	10547	2.93	36633	31888	8.10	650	642	1.25	0.510	0.507	0.59

A BP network was also trained to differentiate between stable and unstable patterns. In the present work, stable patterns were denoted by a 0 and unstable patterns by a 1. The purpose of this network was to act as a filter during optimization to discard the inclusion of unstable patterns in the genetic population. A testing of this BP network showed that it was able to distinguish between stable and unstable patterns correctly 95.4% of the time.

5.2 Identification of Input-Output Couplings

The weights of the trained BP neural networks were analyzed to establish the topology for decomposition of the problem. Both the ABS and ALT strategies were implemented for networks trained with stable and unstable patterns. These transition matrices are shown in Tables 4 and 5; a qualitative description of the strength of couplings between input and output quantities taken from [15] is shown as Table 6. Note that both the ABS and the ALT strategies show in quantitative terms the dominance of specific variables on the 5 outputs considered, and there is qualitative agreement with the assertions made in Table 6. The ALT approach offers additional insights into the direction of change which is unavailable in the ABS approach. These matrices were also obtained for data that was trained with unstable samples only (also shown in Tables 4 and 5). The ABS strategy yielded a transition matrix that was not qualitatively different from the one obtained for stable patterns. This may point to the conclusion that there is no significant difference in the input-output relationship between stable and unstable patterns, and that the difficulty in training encountered with a combined set of training patterns is attributable to the network size. If one looks at the transition matrix using the ALT approach, however, there are some differences in signs between dependencies of stable and unstable patterns, and these may very well contribute to training difficulties.

5.3 Genetic Algorithm Based Optimization

A number of optimal design experiments were conducted in which the trained neural networks were used as a function approximation tool in conjunction with the genetic algorithm based search process. In using the finite element in time and space formulation adopted in this study, the analysis of unstable designs results in a somewhat abrupt termination of the program. In using the exact analysis coupled to an optimizer, special attention would have to be placed in recognizing unstable designs proposed by the optimizer for analysis. Although this is not a concern in an approach where neural network based approximations are used, there is a possibility of generating designs that may actually be unstable. In the first set of experiments, a neural network trained with stable patterns only was used in the optimization. The results of the genetic search based optimization using the plain GA, and using GA with specialized strategies such as directed crossover and multistage search are presented in Table 7. It is easy to see from these results that the use of specialized strategies results in a considerable decrease in functional evaluations that are required to obtain the optimum. The designs themselves are improved through the use of these strategies. The savings in CPU effort realized in this approach are worthy of note. On a SPARC 10 machine, the average cost of a GA based optimization using exact analysis is about 750 hrs; this compares with 14.5 hrs using a neural network based approximation scheme that includes the time required for generating the training data, network training, and actual use in the optimization process. Note that 5 neural networks were used to map design variables into each of the 5 output components of interest. If a single network were used, the time used for neural network based approximations would be further reduced to about 3 hrs.

Since the optimizer does not recognize the difference between stable and unstable designs, it is quite conceivable that unstable designs generated by the optimizer were evaluated (generalized) using the neural network trained with stable patterns only, and incorrectly assumed to be viable candidate designs. To circumvent this problem, the BP network trained to recognize stable and unstable patterns was used as a filter between the optimizer and the neural network based approximation scheme; designs recognized by the BP network as being unstable were assigned arbitrarily low values of fitness and were consequently eliminated by the optimizer. Alternatively, the filter could be placed in a closed loop, where the optimizer would be required to generate the requisite number of stable patterns for analysis. The results from this experiment are summarized in Table 8; note the differences in these designs with those presented in Table 7.

5.4 Interactions in Decoupled Subspaces

The CP network described earlier was also studied for its effectiveness in providing approximations that would allow for simultaneous optimization to proceed in temporarily decoupled subspaces. As a test of this approach, the 14 design variables were split into three groups, X_1, X_2 and X_3. From the data shown in Tables 4-5, design variables d_8 and d_9 have the strongest influence on the thrust requirement, and are designated by the vector X_1; similarly, X_2 includes the components $d_1, d_2, d_3, d_4, d_{10}, d_{11}$, and d_{12} of the design variable vector that impact the structural failure criterion. The vector X_3 represents the remaining components of the design variable vector that have a very weak influence on both the thrust requirements and the stress failure criterion.

In a typical decomposition based design, the vectors X_1 and X_2 would be the design variable vectors in different subsystems, and would be independently varied to optimize the problem objective and satisfy a subset of design constraints; the components of X_3 would remain fixed during this procedd in the subsystem level. It would be important, however, to ensure that an independent design variable move in one subsystem does not offset the constraint gains attained in another subsystem; hence knowledge of all design and constraint changes must be transparent during the design process.

As a test of the CP network in providing this information, a nominal design variable vector was chosen, and components corresponding to X_1 and X_2 were perturbed about this nominal value. The CP network was used to produce an estimate of constraints corresponding to both the thrust requirement and the stress failure. This result is equivalent to the case where design variable moves are completely transparent across the subsystems; such transparency is impractical to expect in a decomposition based design. For an appropriately trained CP network,

Table 4: Transition Matrix (ABS)

	shear force		flap moment		lag moment		thrust		failure criterion	
	stable	unstable	stable	unstable	stable	unstable	stable	unstable	stable	unstable
d_1	0.03963	0.05510	0.02519	0.03157	0.02277	0.02655	0.01596	0.02846	0.07730	0.06421
d_2	0.06413	0.06977	0.03252	0.07892	0.08537	0.09471	0.06045	0.04780	0.06694	0.06386
d_3	0.03936	0.06443	0.04822	0.08629	0.03049	0.05269	0.02060	0.06797	0.08644	0.10749
d_4	0.05282	0.01849	0.03964	0.05136	0.06738	0.07140	0.03811	0.06759	0.11038	0.07909
d_5	0.04526	0.03749	0.04998	0.04397	0.04362	0.04782	0.06372	0.05035	0.02800	0.07108
d_6	0.02087	0.03935	0.03894	0.08324	0.04450	0.05115	0.02381	0.03967	0.04531	0.05909
d_7	0.02638	0.09225	0.01763	0.02360	0.04764	0.07581	0.01445	0.06355	0.01514	0.04886
d_8	0.21750	0.13423	0.22217	0.13084	0.11375	0.06294	0.26282	0.14025	0.03108	0.04383
d_9	0.10604	0.08424	0.18861	0.13366	0.09472	0.06937	0.10714	0.09136	0.01723	0.04900
d_{10}	0.10622	0.06723	0.09299	0.05654	0.09421	0.08567	0.08276	0.05604	0.15239	0.11301
d_{11}	0.13746	0.11016	0.07280	0.10711	0.09575	0.08502	0.12044	0.11388	0.17763	0.09099
d_{12}	0.11844	0.15019	0.09308	0.11493	0.06381	0.10510	0.09699	0.15849	0.14064	0.13522
d_{13}	0.00677	0.04511	0.05996	0.03461	0.09827	0.09606	0.05119	0.04810	0.01591	0.04367
d_{14}	0.01909	0.03193	0.01828	0.02332	0.09771	0.07569	0.04156	0.02646	0.03562	0.03054

Table 5: Transition Matrix (ALT)

	shear force		flap moment		lag moment		thrust		failure criterion	
	stable	unstable	stable	unstable	stable	unstable	stable	unstable	stable	unstable
d_1	0.02074	0.03663	- 0.00667	- 0.00880	- 0.17626	0.06606	0.00957	0.05035	0.26567	0.08070
d_2	0.01966	- 0.02833	0.01917	- 0.16332	0.74172	0.42942	- 0.02640	- 0.00298	0.24132	0.10599
d_3	0.03711	0.13995	- 0.01747	0.08013	- 0.27764	0.25895	- 0.00027	0.03257	0.12938	0.04444
d_4	0.17700	0.07239	0.09541	- 0.00472	1.00000	0.56254	0.06866	- 0.00501	0.63550	0.37660
d_5	0.10597	0.13043	0.16833	0.21025	0.03724	0.00710	0.05728	0.16790	- 0.08011	- 0.12244
d_6	0.06228	- 0.01499	0.00187	0.08546	- 0.37850	- 0.07705	0.02465	- 0.04619	0.15416	0.00583
d_7	- 0.03873	0.03275	- 0.00175	- 0.01133	0.02929	0.04537	- 0.02913	- 0.00721	- 0.05039	0.01807
d_8	1.00000	1.00000	1.00000	1.00000	0.43362	- 0.11590	1.00000	1.00000	0.11463	0.01481
d_9	- 0.59041	- 0.58762	- 0.67066	- 0.77645	- 0.36576	- 0.54168	- 0.50425	- 0.60111	- 0.05210	0.00199
d_{10}	0.22830	0.07621	0.18173	0.12070	- 0.11079	- 0.06268	0.15303	0.12496	0.48158	0.07720
d_{11}	- 0.34442	- 0.64685	- 0.16574	- 0.42171	- 0.68886	- 0.40604	- 0.19693	- 0.38843	- 0.93228	- 0.64906
d_{12}	0.52357	0.99539	0.34654	0.75995	0.74762	1.00000	0.38188	0.78055	1.00000	1.00000
d_{13}	0.00312	- 0.06252	- 0.07220	- 0.05765	0.19221	0.63768	0.02557	- 0.03348	0.01729	- 0.07959
d_{14}	0.00949	- 0.01208	0.03209	0.01364	- 0.00116	0.42581	- 0.03692	- 0.01530	0.07082	- 0.00213

Table 6: Interactions Among Disciplines [15]

variable	Aerodynamics	Dynamics	Structures
planform	S	S	S
twist	S	S	W
tip speed	S	S	S
stiffness	S	S	S
mass distribution	W	S	S

the quality of this approximation is expected to be good. Presenting the perturbed components of either X_1 or X_2 while keeping the other at its nominal value to the CP network produces approximations to the constraints that are representative of a case where no knowledge of design move in another subsystem is available, and may result in considerable error in approximation (extent of error determined by the strength of coupling). In most practical implementations, updates of design variable vectors corresponding to other subsystems can be provided as the design progresses. As an example, if in addition to new X_1, an approximation of X_2 is provided to the CP network, the quality of constraint approximations is considerably improved. The results of this experimentation are summarized in Table 9; optimization results using this approach will be presented in another publication.

6. Closing Remarks

The paper discusses the applicability of emergent computing techniques in the optimal design of multidisciplinary systems. In particular, the application of neural networks and genetic search are examined in the context of the multidisciplinary design of a helicopter rotor blade. The advantages of this approach are significant for problems that have complex, multiple interactions, have design spaces that may be nonconvex or even disjoint, and for which the computational costs for function evaluations are high. Although genetic search offers an increased probability of locating a global optimum in these generically difficult problems, it does so at the cost of significant increase in the number of function evaluations. Neural networks are shown to provide a robust global function approximation strategy that can yield savings in computational resource during GA based search. They also provide a rational approach to temporarily decouple multidisciplinary design problems, and make available a mechanism by which the interactions between these decoupled systems are taken into consideration.

Acknowledgements

Support for this work received under ARO Contract DAAH-04-93-G-0003 to the Rensselaer Rotorcraft Technology Center is gratefully acknowledged.

Table 7: GA Experiment with Advanced Strategies

	d_1	d_2	d_3	d_4	d_5	d_6	d_7	d_8	d_9	d_{10}	d_{11}	d_{12}	d_{13}	d_{14}	OBJ	NGEN
plain	3	8	0.7	0.0	0.25	0.24	0.26	6.8	0.6	0.80	2.0	24	30	70	7222	126
DC	2	8	0.4	0.0	0.25	0.22	0.26	7.2	0.7	0.79	2.0	24	31	90	7196	90
MS	2	8	0.4	0.0	0.25	0.22	0.26	7.1	0.7	0.79	2.0	24	30	80	7099	84
DC+MS	1	8	0.4	0.0	0.25	0.22	0.26	7.0	0.7	0.79	2.0	24	30	86	7050	62

Table 8: GA Experiment with Advanced Strategies (Unstable Design Filtered)

	d_1	d_2	d_3	d_4	d_5	d_6	d_7	d_8	d_9	d_{10}	d_{11}	d_{12}	d_{13}	d_{14}	OBJ	NGEN
plain	3	8	0.2	0.0	0.25	0.21	0.22	7.4	0.8	0.79	2.0	24	46	88	7137	117
DC	1	8	0.4	0.0	0.25	0.20	0.24	7.4	0.8	0.79	2.0	24	30	82	7059	86
MS	3	8	0.1	0.0	0.25	0.20	0.26	7.4	0.8	0.79	2.0	24	31	59	7109	82
DC+MS	2	8	0.1	0.0	0.25	0.22	0.26	7.2	0.8	0.79	2.0	24	30	84	6965	73

Table 9: CP Experiment for Decomposition Based Design

Test ID	change of X_1 and X_2		X_1 changed X_2 unchanged		X_1 changed X_2 estimated		X_2 changed X_1 unchanged		X_2 changed X_1 estimated	
	thrust	$\overline{R}$	thrust	$\overline{R}$	thrust	$\overline{R}$	thrust	$\overline{R}$	thrust	$\overline{R}$
	411	0.230	411	0.230	411	0.230	411	0.230	411	0.230
1	709	0.296	566	0.231	686	0.302	513	0.292	634	0.296
2	401	0.244	389	0.229	472	0.282	419	0.243	459	0.244
3	860	0.343	631	0.231	763	0.296	592	0.337	705	0.336
4	772	0.377	481	0.228	657	0.319	651	0.382	651	0.382
5	407	0.265	355	0.229	440	0.290	497	0.263	480	0.263
ave error (%)	-	-	22.64	7.17	10.22	11.21	24.96	1.11	16.24	1.52

References

1. J. Sobieszczanski-Sobieski, "Multidisciplinary Design Optimization: An Emerging New Engineering Discipline", World Congress on Optimal Design of Structural Systems, Rio de Janeiro, Brazil, August 2-6, 1993.

2. R.H. Tolson and J. Sobieszczanski-Sobieski, "Multidisciplinary Analysis and Synthesis: Needs and Opportunities", AIAA Paper No. 85-0584, 1985.

3. J. Sobieszczanski-Sobieski, "A Linear Decomposition Method for Large Optimization Problems - Blueprint for Development", NASA TM 83248, 1982.

4. J. Sobieszczanski-Sobieski, "Optimization by Decomposition: A Step from Hierarchic to Non-Hierarchic Systems", 2nd NASA/Air Force Symposium on Recent Advances in Multidisciplinary Analysis and Optimization, NASA CP-3031, Sep. 1988.

5. C.-Y. Lin and P. Hajela, "Genetic Search Strategies in Large Scale Optimization", proceedings of the 34th AIAA/ASME/ASCE/AHS/ASC SDM Conference, La Jolla, California, 1993.

6. P. Hajela and Z. Szewczyk, "On the Use of Neural Network Interconnection Weights in Multidisciplinary Design," AIAA Paper 92-4808, proceedings of the 4th AIAA/NASA/Air Force Symposium on Multidisciplinary Analysis and Optimization, Cleveland, Ohio, September 1992.

7. P. Hajela and L. Berke, "Neurobiological Computational Models in Structural Analysis and Design", Computers and Structures, Vol. 41, No. 4, pp. 657-667, 1991.

8. Y.-H. Pao and R.D. Beer, "The Functional Link Net: A Unifying Network Architecture Incorporating Higher Order Effects", International Neural Networks Society, First Annual Meeting, Boston, Massachusetts, September 1988.

9. Z. Szewczyk and P. Hajela, "Neurocomputing Strategies in Decomposition Based Structural Design", proceedings of the 34th AIAA/ASME/ASCE/AHS/ASC SDM Conference, La Jolla, California, 1993

10. S. Goel and P. Hajela, "Identification of Couplings in Complex Engineering Design

Problems Using Neural Networks", proceedings of the 5th AIAA/NASA/Air Force Symposium on Multidisciplinary Analysis and Optimization, Panama City, Florida, September 7-9, 1994

11. C. L. Bloebaum and P. Hajela, "Heuristic Decomposition for Non-Hierarchic Systems", proceedings of the 32nd AIAA/ASME/ASCE/ASC SDM meeting, Baltimore, MD, April 1991.

12. J. H. Holland, "Genetic Algorithms and Adaptation", Technical Report No. 34, Univ. of Michigan, Cognitive Science Dept., 1981.

13. P. H. Richter, "A Network Theory of the Immune System", Eur. J. Immunol., 5, pp. 350-354, 1975.

14. O. A. Bauchau and N.K. Kang, "A Multibody Formulation for Helicopter Structural Dynamic Analysis", Journal of American Helicopter Society, Vol. 38, No. 2, pp. 3-14, April 1993.

15. H. Adelman and W. A. Mantay (eds), Integrated Multidisciplinary Optimization of Rotorcraft: A Plan for Development, NASA TM 101617, May 1989.

Salience Analysis of Mapping Samples for Multilayer Feedforward Networks

Hauhua Lee[1] and Prabhat Hajela[2]

[1]Engineering Mechanics Laboratory, GE Corporate R&D Center, Schenectady, N.Y. 12301, USA
[2]Department of Mechanical Engineering, Aeronautical Engineering & Mechanics
Rensselaer Polytechnic Institute, Troy, N.Y. 12180, USA

Abstract. Salience analysis is to identify non–salient (insignificant) training samples and sample components. By removing non–salient information, one can speed up off–line training processes considerably. The salience information is also helpful in decomposing a high–dimensional mapping into a number of sub–mappings so as to predict certain outputs separately and more accurately. Moreover, with the sample salience information, the cross–validation training algorithm can be made more reliable, e.g., via using salient samples as training set, and non–salient samples as testing set.

Keywords. salience analysis, MFN, mapping nonlinearity, stubborn samples, trainability, and learnability.

1. Introduction

Novel schemes for analyzing the salience of training data are presented. The salience information refers to the sensitivity a sample or a component of sample possesses with respect to the overall mapping nonlinearity that the entire set of mapping samples exhibit. The salience information offers insights into the understanding of the underlying mapping function, and is useful in many ways. For instance, by removing non–salient samples, one can speed up off–line training processes. Furthermore, one can improve the cross–validation training algorithm by using salient samples as training set and non–salient samples as validation set. In addition, the component salience can serve as a valid criterion for decoupling a high–dimensional mapping into sub–mappings that are more manageable and easier to train.

The approach of salient analysis is fundamentally different from the conventional statistics schemes that study the similar topics. Statistics approaches are essentially based on mapping models that are linear or mildly non–linear, and are inappropriate for highly nonlinear mappings. In dealing with nonlinear mappings, the present research suggests that in the first place mapping nonlinearity must be quantified, and then the salience of samples and components can be evaluated based on their sensitivities to the overall mapping nonlinearity.

The salience analysis scheme presented here is developed from the authors' previously published works, the hyper-reshaping interpretation of multilayer feed-forward networks(MFN)[1], as well as the mapping nonlinearity indicators, distribution angle and distribution gradient[2]. The present paper recapitulates these works first, followed by a presentation of the salience analysis for both training samples and sample components.

2. The Hyper-Reshaping Interpretation of MFN

The MFN is essentially a successive hyperspace transformation mechanism, which transforms an m-dimensional input space R^m into an n-dimensional output space R^n via one or several intermediate spaces, and the dimension of each intermediate space corresponds to the number of nodes in the respective hidden layer. For example, let $m \times h \times n$ denote a 3 layer MFN that has m, h, and n nodes in the input, hidden, and output layers respectively, such a network indeed conducts the successive hyperspace transformation: $R^m \rightarrow R^h \rightarrow R^n$.

The continuous space transformation is depicted by a set of discrete mapping samples. Suppose the mapping function f: $R^m \rightarrow R^n$ is depicted by p samples, ($\mathbf{I}_i$, $\mathbf{O}_i$), $i=1,...,p$, $\mathbf{O}_i=f(\mathbf{I}_i)$, $\mathbf{I}_i \in R^m$, $\mathbf{O}_i \in R^n$, and is trained by an $m \times h \times n$ MFN. One can represent the input vectors and output vectors as two separate matrices, $[\mathrm{I}]_{p \times m}$ and $[\mathrm{O}]_{p \times n}$, and similarly the point vectors in the intermediate space R^h can be represented as $[\mathrm{H}]_{p \times h}$. In this manner, the successive space transformation $R^m \rightarrow R^h \rightarrow R^n$ is then represented as the successive matrix transformation $[\mathrm{I}]_{p \times m} \rightarrow [\mathrm{H}]_{p \times h} \rightarrow [\mathrm{O}]_{p \times n}$. In any space, one can think of the entire set of p points as an imaginary hyper-dimensional pliable object, referred to as a *training object* (and the associated matrix is referred to as the state matrix of the training object). Then the successive matrix transformation process can be conceptually realized as the successive *hyper-reshaping* of the imaginary *training object*.

The *hyper-reshaping* operation can be further decomposed into a *hyper-stretching* and a *hyper-squeezing*. Mathematically, the *hyper-stretching* corresponds to the multiplication of source state matrix and a weight matrix, and the *hyper-squeezing* corresponds to the processing through an activation function. Hence, it is through the alternate *hyper-stretchings* and *hyper-squeezings*, the training object is transformed from the input state to the desired output state.

The *hyper-reshaping* interpretation is valid for multidimensional (n-to-m) mappings, as well as for both binary and continuous mappings. This interpretation provided a new perspective to the understanding of the dynamics of MFN, and clarified many persistent questions pertaining to MFN. For example, it is now obvious that the training of an MFN can be realized as the process to find out an appropriate way that can successively reshape the training object from its input state to its desired output state; the approximation capability of MFN indeed results from reshaping the continuous space that encloses the training object. In particular, as elaborated in [2], the generalization capability with MFN is due to the *hyperspace folding*, achieved through the repetitive *hyper–stretching* and *hyper–squeezing*. The role of activation function is also clearer: the slope of the activation function can be compared to the resistance property of the imaginary training object, and the precise form of the activation function should not be critical to the performance of MFN.

2.1. Illustrative Example

Figure 1 shows a continuous 1–to–1 mapping function with 15 samples, labeled from *A* to *O*. The mapping relationship is shown as sinusoidal curves: the solid line represents the desired mapping, and the dashed line represents the predicted (by the MFN) mapping. The minor deviations between desired outputs and predicted outputs are considered negligible. The mapping relationship is indeed unrelated to the present discussion of hyper–reshaping, and it is obvious that visualizing the mapping relationship for higher–dimensional mapping is difficult. Figure 1 also shows the training object at input state as well as at both desired and predicted output states.

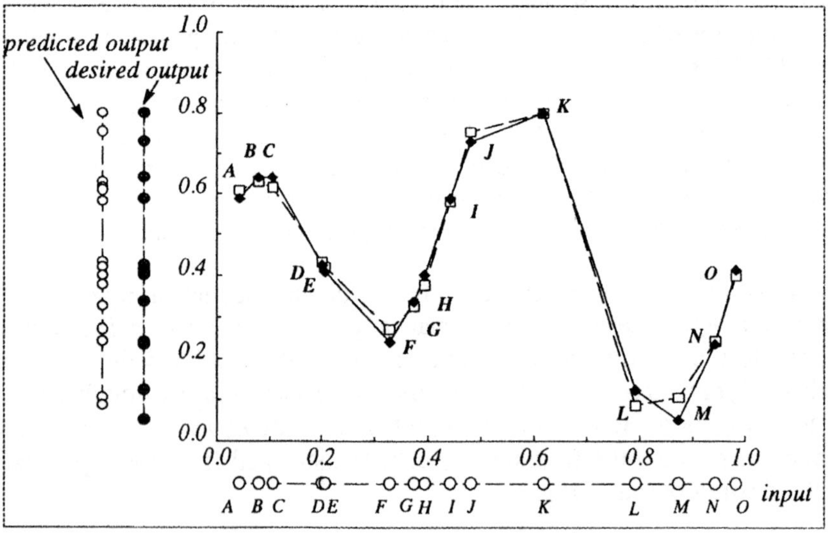

Figure 1. An illustrative 1–to–1 mapping example.

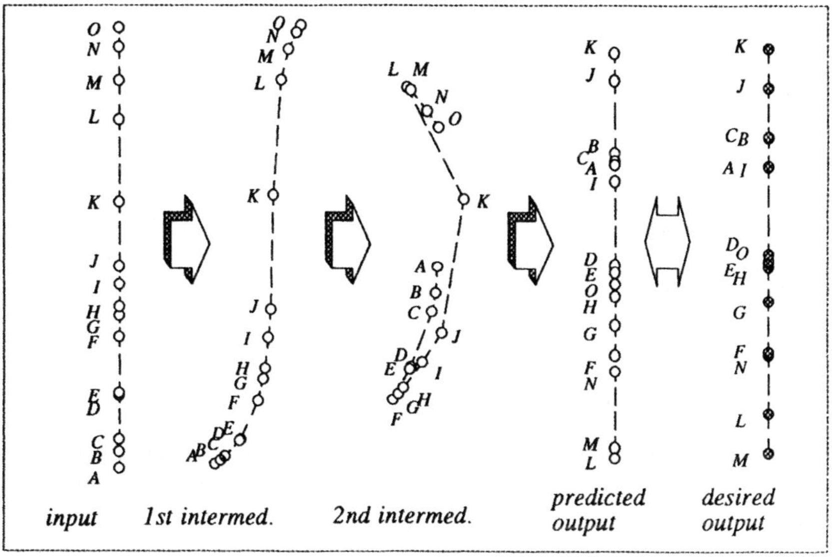

Figure 2. An illustrative example of successive *hyper–reshaping*.

This mapping function is trained by a 1x2x2x1 MFN. The use of 1x2x2x1 configuration was intentional – in this manner, the intermediate states are 2D and hence can be visualized easily. Figure 2 shows the training object at each state associated with the 1x2x2x1 MFN: the input state, the first intermediate state, the second intermediate state, the predicted output state, and the desired output state. (The two intermediate states are 2D, and the other three states are 1D.) One should observe how the training object is *reshaped* from the input state to the output state – during the reshaping process, the points which are close to (or far away from) each other at output state tend to get closer to (or further away from) each other. In other words, during the successive hyper–reshaping process, the *appearance* of training object evolves toward that of the desired output state.

3. Mapping Nonlinearity Indicators

The *hyper–reshaping* interpretation of MFN allows one to quantify mapping nonlinearity based on the spatial characteristics of training samples. Such approach is different from the conventional statistical approach (such as [3]) where the samples are assumed to follow some type of statistical distributions, which does not hold for many real world applications. In [2], the distribution angle α and distribution gradient β have been introduced as measures of mapping nonlinearity of training samples – the larger the α and β, the more difficult is the training, and also a poorer prediction performance is likely. The mapping nonlinearity represented by α and β have been validated with *empirical trainability* in [4, 5] with two model problems being the mapping between turbine performance and blades throat design, as well as the mapping between geometrical properties and inhomogeneous planar shapes.

In this paper, mapping nonlinearity and *mapping similarity* are used as antonyms. Mapping similarity refers to the similarity between the *appearance* of the input and output states of a training object. Recall that the two states are often with different dimensions, hence to quantify their similarity, one needs to have a dimension–independent representation of the state in the first place. The *object distribution vector*, $\mathbf{V}_{\text{dist}}$, was introduced for this purpose. The mapping similarity/ nonlinearity were then quantified based on the two associated $\mathbf{V}_{\text{dist}}$'s in terms of α and β.

3.1. Object Distribution Vector $\mathbf{V}_{\text{dist}}$

The $\mathbf{V}_{\text{dist}}$ characterizes the spatial distribution of the point vectors on a training object. A $\mathbf{V}_{\text{dist}}$ is obtained by sequentially enumerating all distances between every two point vectors on the associated training object. Hence, let $\Phi_{n,m}$ denote the state of a training object which consists of n point vectors in an m–dimensional space, i.e.,

$$\Phi_{n,m} \equiv \{\mathbf{p}_i \mid i=1,..,n,\ \mathbf{p}_i \in R^m\}, \tag{1}$$

then

$$\mathbf{V}_{\text{dist}}(\Phi_{n,m}) \equiv \{d_{ij} \mid d_{ij} = \overline{\mathbf{p}_i\mathbf{p}_j};\ \mathbf{p}_i,\ \mathbf{p}_j \in \Phi_{n,m};\ i > j;\ i, j=1,..,n\}. \tag{2}$$

Clearly, the number of components in $\mathbf{V}_{\text{dist}}$ equals $n(n-1)/2$, with n being the number of samples. Therefore, the $\mathbf{V}_{\text{dist}}$ representation transforms an m–dimensional object with n point vectors into a point vector with dimension $n(n-1)/2$, regardless of the original dimension m. Consequently, two states of the same training object but with different dimensions are then represented as two point vectors with the same dimension.

3.2. Distribution Angle α and Distribution Gradient β

Suppose $\Phi_{n,s}$ and $\Phi_{n,t}$ represent the source and target state of a training object, $\mathbf{V}_s$ and $\mathbf{V}_t$ are the corresponding $\mathbf{V}_{\text{dist}}$, and $d^{(s)}{}_{ij}$ and $d^{(t)}{}_{ij}$ denote a component in $\mathbf{V}_s$ and $\mathbf{V}_t$ respectively, i.e.,

$$\Phi_{n,s} \equiv \{\mathbf{s}_i \mid i=1,..,n,\ \mathbf{s}_i \in R^s\}, \tag{3}$$

$$\Phi_{n,t} \equiv \{\mathbf{t}_i \mid i=1,..,n,\ \mathbf{t}_i = F(\mathbf{s}_i),\ \mathbf{t}_i \in R^t\},$$
$$\text{where } F \text{ is the mapping function,} \tag{4}$$

$$\mathbf{V}_s \equiv \mathbf{V}_{\text{dist}}(\Phi_{n,s}) = \{d^{(s)}{}_{ij} \mid d^{(s)}{}_{ij} = \overline{\mathbf{s}_i\mathbf{s}_j};\ \mathbf{s}_i,\ \mathbf{s}_j \in \Phi_{n,s};\ d^{(s)}{}_{ij} > 0\}, \tag{5}$$

$$\mathbf{V}_t \equiv \mathbf{V}_{\text{dist}}(\Phi_{n,t}) = \{d^{(t)}{}_{ij} \mid d^{(t)}{}_{ij} = \overline{\mathbf{t}_i\mathbf{t}_j};\ \mathbf{t}_i,\ \mathbf{t}_j \in \Phi_{n,t};\ d^{(t)}{}_{ij} \geq 0\}. \tag{6}$$

Then the distribution angle α is defined as the angle between $\mathbf{V}_s$ and $\mathbf{V}_t$:

$$\cos\alpha \equiv \frac{\mathbf{V}_s \bullet \mathbf{V}_t}{|\mathbf{V}_s|\,|\mathbf{V}_t|}\ ; \tag{7}$$

As an example to validate α, Table 1 listed all α's between any two V_{dist}'s of the five states in the previous 1x2x2x1 MFN model; the α's are listed in degrees. The smaller the α, the more similar the two states. One sees that in the first row, data increases from left to right, which is consistent with the observation that the input state is most similar to the first intermediate state, and becomes more and more dissimilar to the states thereafter. In fact, the data in the same row or column all follow the same trend (decreasing from top to bottom in a column). In addition, note that the differences in magnitudes of α also appear appropriate in indicating the degree of similarity, comparing with one's visual estimations.

	1st Inter.	2nd Inter.	Predicted Out.	Desired Out
Input	8.9	25.8	41.2	41.6
1st Inter.	–	20.4	41.0	41.4
2nd Inter.	–	–	38.3	38.6
Predicted Out.	–	–	–	6.2

Table 1. Distribution angles for the 1x2x2x1 MFN.

The distribution gradient, β_{ij}, is another type of nonlinearity indicators, and is defined with respect to any two samples:

$$\tan \beta_{ij} \equiv \frac{d^{(t)}_{ij}}{d^{(s)}_{ij}}, \text{ where i, j are sample labels.} \tag{8}$$

The *gradient* refers to the rate of similarity change (i.e., the *distance change*) from the source space to the target space for two specific samples. The *maximum distribution gradient* β_{max}, and the *average distribution gradient* β_{avg}, are defined as:

$$\tan \beta_{max} \equiv \textit{maximum of all } \frac{d^{(t)}_{ij}}{d^{(s)}_{ij}}, \tag{9}$$

$$\tan \beta_{avg} \equiv \textit{average of all } \frac{d^{(t)}_{ij}}{d^{(s)}_{ij}}. \tag{10}$$

Both α and β characterize different aspects of mapping nonlinearity. When these indicators are small, the training would be easier to be successful. In particular, a large β_{ij} indicates either one of the two associated samples would be difficult to train – i.e., either the i–th or the j–th sample is a so–called stubborn sample.

The object distribution vector V_{dist} has two variations, the *modified* object distribution vector $V^m{}_{dist}$ and the *normalized* object distribution vector $V^n{}_{dist}$. The $V^m{}_{dist}$ is defined by subtracting the minimum component in a V_{dist} from every component (hence the minimum component becomes 0). The $V^n{}_{dist}$ is defined by normalizing a V_{dist} so that its norm equals 1. Accordingly, the nonlinearity indicators defined based on these variations are noted with a corresponding superscript, such as α^m and $\beta^n{}_{ij}$.

4. Salience Analysis

The mapping nonlinearity for a set of training samples would be affected when the mapping samples are "*perturbed*". In this research, the perturbation is done by removing either one sample or one component from the original mapping set; such perturbations lead to the concept of salience analysis. Illustrated in Figure 3 is a pair of input and output matrices each with dimension $p \times s$ and $p \times t$, where p is the number of samples, s and t are the dimension of input and output space. In the two matrices, a sample is represented as a pair of row vectors, and a component is associated with a column vector in either matrices. By removing a pair of row vectors, the change of mapping nonlinearity is defined as the *sample salience*. Similarly, by removing a column vector in any matrix, the change of mapping nonlinearity is defined as the *component salience*. This concept is further illustrated in the following sections.

4.1. Salience of Samples

With a simple one-to-one mapping (i.e., $y=F(x)$), the salient samples can be realized as the extreme points (i.e., the local minimum/maximum points, where the gradients are zero). As shown in Figure 4, the extreme points are marked by circles. Note that the non-extreme points like p and q might be taken as salient points too. In other words, zero-gradient is a necessary condition for a salient point, not a sufficient condition. However, it is important to note that when the output space is multi-dimensional, the extreme points are no longer meaningful. In this regard, the mapping nonlinearity could offer alternative criteria over the gradients.

The sinusoidal mapping example used in Figure 1 is once again used to study the salience of each sample. Figures 5 shows changes in mapping nonlinearity measured by α, β_{avg} and β^n_{avg} (all in degree). The changes are obtained by removing one sample at a time from the set of 15 – i.e., each data is computed using 14

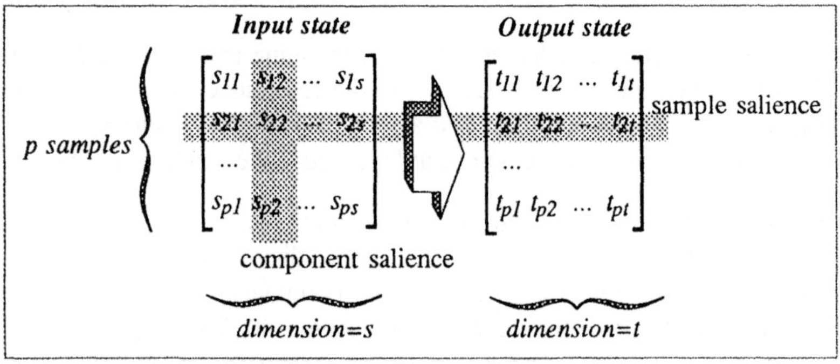

Figure 3. Salience analysis of training samples and components.

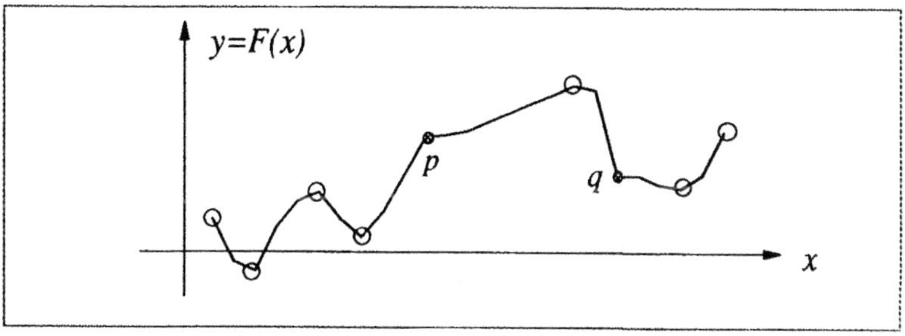

Figure 4. Salient points for a 1–to–1 mapping function.

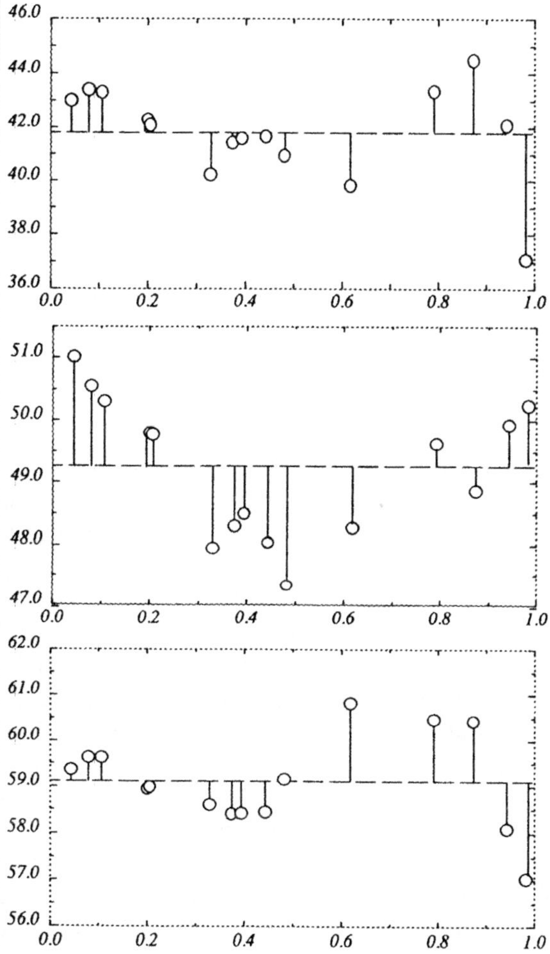

(a) sample salience based on α:

Each α is obtained by removing the corresponding sample from the mapping. The dashed line represents the original α, 41.792°, for the 15 samples.

(b) sample salience based on β_{avg}:

Each β_{avg} is obtained by removing the corresponding sample from the mapping. The dashed line represents the original β_{avg}, 49.25°, for the 15 samples.

(c) sample salience based on β^n_{avg}:

Each β^n_{avg} is obtained by removing the corresponding sample from the mapping. The dashed line represents the original β^n_{avg}, 59.1°, for the 15 samples.

Figure 5. Sample salience analysis for the sinusoidal mapping example.

samples with a target sample being omitted. For example, by removing the sample *O*, α dropped sharply to 37.08°, β_{avg} and β^n_{avg} became 50.25° and 57° – for the original mapping with all 15 samples: $\alpha = 41.79°$, $\beta_{avg} = 49.25°$ and $\beta^n_{avg} = 59.1°$, which are marked by the horizontal dashed lines.

Clearly, Figure 5 (a), (b) and (c) exhibit different salience information. There does not seem to be a clear correlation between these three types of information, except that the overall trend in 5(a) is more similar to 5(c) than 5(b). In Figure 5(a), the points *D*, *E*, *G*, *H*, *I*, and *N* are shown with a very low salience – their associated α's are very close to the original α. This observation is consistent with the fact that these points happen to be at those locations which do not offer extra nonlinearity. In other words, it seems the underlying mapping function can be characterized sufficiently without the non–salient *D*, *E*, *G*, *H*, *I*, and *N*. With the nine most salient samples only, the α becomes 41.48°, and $\beta_{avg} = 46.62°$, β^n_{avg}=54.33°. The fact that the new α is still close to the original one (41.79°) supports the assertion that the nine salient samples are sufficient to characterize the mapping function as a whole, while the new β_{avg} and β^n_{avg} becomes lower (comparing with 49.25° and 59.1°) may imply that the new mapping would tend to have a smaller training error.

Figure 5 does not show β_{max}, as it is only sensitive to the elimination of those samples which are the principal contributor to the β_{max}. In the current example, $\beta_{max} = 77.33°$, and is offered by points *N* and *O*. When either one is dropped, β_{max} is reduced to 75.58°, which is offered by points *L* and *M*. However, with β^n_{max}, the change is more noticeable for all samples.

In Figure 6, efforts were made to further study the change in α when non–salient samples are incrementally removed from the mapping. First, Figure 6(a) shows the samples sorting by their salience, from the least salient *I* to the most salient *O*, where the salience is decided by the magnitude of the variation in α (with respect to the original α). Figure 6(b) then shows the change in α when certain number of non–salient samples are removed from the mapping. For example, when x=0, α equals 41.792 and corresponds to the original mapping; when x=1, the sample *I* is omitted and α is still close to the original value; next, two samples (*I* and *H*) are omitted from the mapping, then three samples (*I*, *H* and *N*) are omitted, and so on. It appears that upto six least salient samples (i.e., *I*, *H*, *N*, *E*, *G* and *D*) one can omit without affecting α noticeably. This result reconfirmed that the non–salient samples so identified are not critical to the overall mapping nonlinearity. In other words, these non–salient samples might be redundant and might be ignored during the training process. When they are omitted, a batch mode training process would speed up considerably, for the computational cost of the objective function is proportional to the number of samples used.

The above assertion is further explored with results shown in Figure 7. Here, the six non–salient samples are omitted and the remaining nine samples are used for

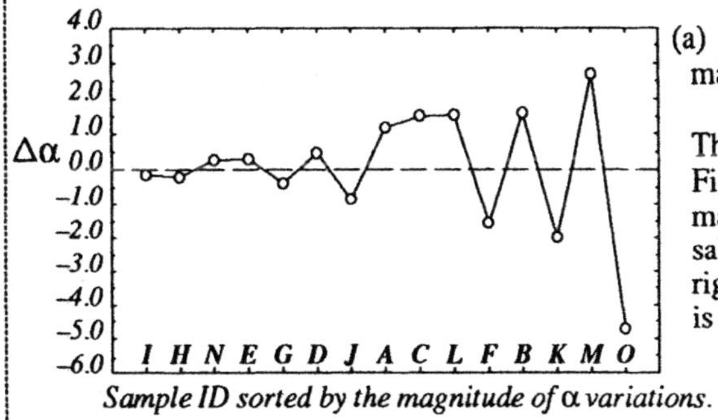

(a) Sorting samples by the magnitude of α variations.

This figure is derived from Figure 5(a) by sorting the magnitude of α variation. The salience increases from left to right, i.e., the left most sample is the least salient one.

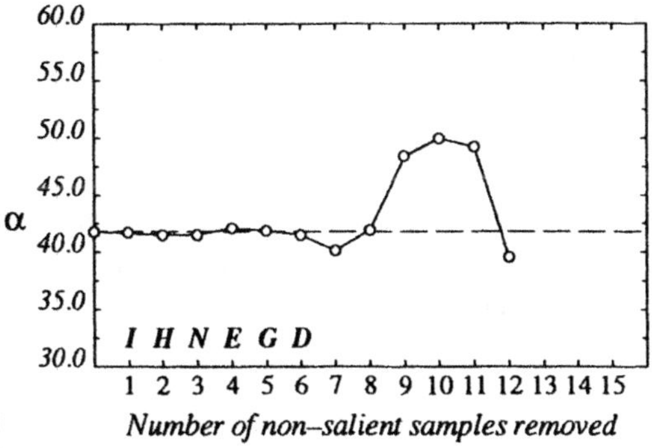

(b) *α vs. the number of non–salient samples removed.*

The x–axis represents the number of the least salient samples removed from the mapping set. For example, the first α is obtained by removing sample *I* only, then removing two samples (*I* and *H*), then three samples (*I*, *H*, and *N*), and so on.

Figure 6. Changes in mapping nonlinearity when removing non–salient samples.

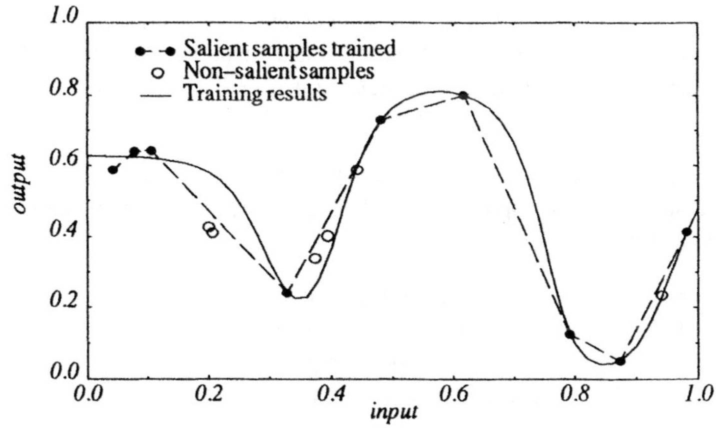

Figure 7. Training with salient samples.

a training, which is terminated when the fitting error converged. The solid line shows the performance of the fit across the entire input space. Obviously, comparing the performance of the fit between the nine training samples, marked by solid circles, and the six non–salient samples, marked by blank circles, the former are fitted better than the latter in general. This is because theoretically there exist an infinite number of hyper–surfaces that can fit all training samples perfectly (in the current example the hyper–surface is shown as a 2D curve), and a perfect–fit of training samples does not necessarily guarantee a perfect–fit of testing samples.

4.2. Salience of Components

Upon modeling a large multi–dimensional (m–to–n) mapping, one may want to decouple the mapping into a few sub–mappings that have smaller dimensions and are hopefully easier to train and more accurate to predict. The decoupling usually requires a priori knowledge and assumptions, without which the decoupling would be difficult and may have to base solely on trial and error. In this regard, the salience analysis of sample components may offer a simple and efficient criteria for decoupling.

A simple 4–to–3 mapping is used to illustrate the component salience analysis. The input and output parameters are denoted as x_1, x_2, x_3, x_4 and y_1, y_2, y_3, respectively. The mapping is represented by the following function,

$$y_1 = \sin 4(x_1 + x_2 + x_3),$$
$$y_2 = -0.7x_1,$$
$$y_3 = 0.25\,(x_1 + x_2 + x_3 + x_4).$$

Intuitively, one sees that among the four inputs, x_1 should offer the most nonlinearity, next x_2 and x_3 should offer the same degree of nonlinearity, and x_4 should offer the least nonlinearity. On the other hand, among the three outputs, only y_1 contains nonlinearity, y_2 and y_3 are indeed linear mappings. A set of 1000 mapping samples, randomly generated in the region $x_i \in [0,1]$, $i=1,2,3,4$, was used to conduct the salient component analysis.

Table 2 lists the mapping nonlinearity for the original mapping (case 0), as well as twelve sets of sub–mappings (cases 1 to 12). The 12 sub–mapping cases are organized as follows:

case 0:	all inputs to all outputs
cases 1, 2, 3, 4 :	omit one input component;
cases 5, 6, 7:	omit one output component;
cases 8, 9, 10:	predicting each output component individually using all inputs; (Case 10 also represents mapping function y_3.)
case 11:	mapping function y_1;
case 12:	mapping function y_2;

The nonlinearity indicators used are α^m, $\tan\beta^n_{max}$ and $\tan\beta^n_{avg}$, which are also shown as bar charts in Figure 8 with the data being processed by α^m–30, *log*

$tan\beta^n_{max}$ and $tan\beta^n_{avg}$–0.4 so as to help in a visual analysis. Comparing the profile of the three bar charts, they have a similar trend at a macro level. Hence discussions here only focus on the α^m case. As opposed to being insensitive in sample salient analysis, the β^n_{max} is very sensitive here and exhibits a large variation in magnitudes.

ID	*Inputs*	*Outputs*	Dist. Angle	Dist. Gradient	
			α^m	$tan\beta^n_{max}$	$tan\beta^n_{avg}$
0	x_1 x_2 x_3 x_4	y_1 y_2 y_3	31.93°	5.2	1.001
1	x_2 x_3 x_4	y_1 y_2 y_3	34.51°	86.8	1.228
2	x_1 x_3 x_4	y_1 y_2 y_3	33.39°	86.7	1.223
3	x_1 x_2 x_4	y_1 y_2 y_3	33.68°	131.6	1.169
4	x_1 x_2 x_3	y_1 y_2 y_3	32.22°	25.0	1.130
5	x_1 x_2 x_3 x_4	y_2 y_3	22.31°	1.2	0.642
6	x_1 x_2 x_3 x_4	y_1 y_3	35.54°	5.4	0.973
7	x_1 x_2 x_3 x_4	y_1 y_2	33.57°	5.4	1.013
8	x_1 x_2 x_3 x_4	y_1	38.15°	5.7	0.990
9	x_1 x_2 x_3 x_4	y_2	30.49°	1.7	0.734
10	x_1 x_2 x_3 x_4	y_3	33.52°	1.0	0.429
11	x_1 x_2 x_3	y_1	38.50°	5.4	1.116
12	x_1	y_2	0.001°	1.4	1.000

Table 2. Component salience analysis for the 4x3 example mapping.

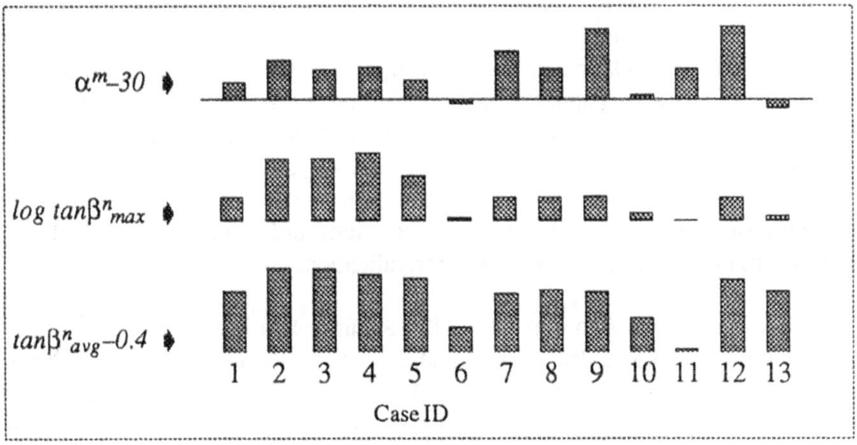

Figure 8. Profile of component salience analysis cases.

Comparing cases 1, 2, 3, 4, and 0, one sees that deleting any single input component resulted in an increase in mapping nonlinearity. The changes in α^m and $tan\beta^n_{avg}$ are consistent with expectation: x_1 is the most influential, followed by

x_2, x_3; x_4 is the least influential. In addition, the sharp increases with $tan\beta^n{}_{max}$ suggest that every input component actually contributes to a better trainability. Comparing cases 5,6,7 and 0, one finds that y_1 is the most salient. Nevertheless, the small α^m and $\beta^n{}_{max}$ in case 5 suggest that predicting y_2 and y_3 together should be an easier task. On the other hand, removing either y_2 or y_3 from the mapping would deteriorate trainability.

Comparing cases 8, 9 and 10, one again confirms that y_1 is the most nonlinear and the most difficult to predict. Since y_2 and y_3 are linear mappings, their mapping nonlinearities appear to be small as expected. Recall that $y_2 = -0.7x_1$. Hence comparing cases 9 and 12, one sees that if the unrelated input components are dropped out, the mapping nonlinearity may be reduced significantly. (Theoretically the α^m in case 12 should be 0; the $0.001°$ is due to round off errors.) However, such a reduction is not guaranteed. For example, both cases 8 and 11 predict y_1. Case 8 contains a redundant input x_4, but actually has a slightly smaller α^m and $tan\beta^n{}_{avg}$ (although $tan\beta^n{}_{max}$ is slightly larger).

4.3. Discussions

4.3.1 Cross–Validation With Sample Salience

The cross–validation is a training algorithm that uses specific termination criteria to avoid over–training (or over–fitting). This algorithm separates samples into a training set and a validation set. During training, both sets are monitored for their training error, but only the training set is used for the error backpropagation. Theoretically, the training errors for both sets should tend to decrease. Once the training error for validation set starts diverging (or increasing), the training process is terminated. This algorithm did not gain its popularity in practice primarily because it lacks a criterion to separate the training and validation samples. It is obvious that should the separation be done randomly, the algorithm would not be reliable. The sample salience information can now serve as a valid criterion for this purpose, e.g., one can decide a salience threshold, with the salient samples used for training and non–salient ones for validation.

The cross–validation algorithm implicitly assumes that training errors for both sets follow a monotonic decreasing trend. Unfortunately, in practice, these errors often do not decrease monotonically, and may even fluctuate considerably. This makes it difficult to determine the right time to stop the training. In this regard, the present research suggests to modify the algorithm as follows: at the beginning, the training set is used until the training error has converged. Then the validation set is taken into account, together with the training set, for error propagation, and the training should be terminated when the validation error reaches its minimum. The idea here is that the salient samples are used to get to the vicinity of optimum weights more quickly (for using fewer samples), and the non–salient samples are then taken into account to fine tune weights to minimize over–fitting. For example, in Figure 7, the fitting curve was trained only by salient samples and apparently have overfitted the non–salient samples. To minimize such an overfit, one could continue training using all samples and terminate training when the overfit is minimized.

4.3.2 Decoupling of Mapping – Caveats

Component salience analysis provides insights toward the nonlinear correlations between inputs and outputs, and one may therefore be able to decouple a large mapping into smaller and less nonlinear sub–mappings. It is important to note that such decoupling is valid only for the output components, not for the raw input components. This is because the salience of an unrelated input component could be high or low: a small component salience shows a lower nonlinear dependency, while a large component salience indeed indicates a higher nonlinear dependency, which does not necessarily imply a non–dependency. In other words, the salience information alone is insufficient to tell whether or not the component is indeed unrelated to outputs – although linear dependency between a subset of inputs/outputs can be easily detected with component salience. However, if the raw inputs are preprocessed appropriately, the preprocessed inputs could be then decomposed. For example, if one applies the Fourier transformation[6] to raw inputs, in the frequency domain, every coefficient is a function of all raw inputs and therefore any subset of such coefficients is still related to all raw inputs and is hence eligible to serve as a source space for a target mapping function.

4.3.3 Some Related Statistics Schemes

Apart from salience analysis, many statistics schemes are also available for exploring input/output mapping relationships, such as regression analysis, ANOVA (analysis of variance), correlation analysis and design of experiments.[7, 8] However, the approach of salience analysis is fundamentally different from these statistics schemes, as is the information the two would provide. Unlike statistics schemes, salience analysis only deals with deterministic mappings, where for each given input, the output is unique rather than following certain statistical distribution. Hence, salience analysis is unrelated to statistics issues such as interval estimation, confidence level, testing hypotheses, and so on.

Salience analysis is valid for any degree of mapping nonlinearity. In contrast, statistics schemes primarily focus on mappings that have low degree of nonlinearity, e.g., simple linear, intrinsically linear (which can be transformed into a linear mapping), and simple nonlinear such as quadratic, logistic, and/or their combinations[7]. In situations where mappings exhibit higher degree of nonlinearity, the step–wise regression model is usually used: it divides an input space into several sub–regions and approximates the sub–regions individually. Obviously, the sub–regional mappings would be less nonlinear – the average mapping nonlinearity of sub–regional mappings is approximately inversely proportional to the number of subregions divided. (This *divide and conquer* concept is similar to the approach of cluster–based neural networks models, such as the counter–propagation nets, or the combination of Kohonen nets and MFN nets.) Hence, it is obvious that in dealing with highly nonlinear mappings, statistics schemes are inadequate. For example, information provided by the correlation analysis is really the degree of *linear* dependency. When there is a strong interaction among input variables (i.e., the mapping nonlinearity is high), such linear correlation information becomes ineffective.

5. Concluding Remarks

The present paper presents the salience analysis schemes for training samples, based on the mapping nonlinearity indicators, *distribution angle* α and *distribution gradient* β, which are originated from the *hyper–reshaping* interpretation of MFN. Essentially, the MFN is a mechanism for successive space transformation; the entire space is hyper–stretched and hyper–squeezed repeatedly according to the weight vectors associated with each hidden (and output) layer. In the course of a successive space transformation, the topological features of the training object evolves toward that of the desired output state. The object distribution vector $\mathbf{V}_{dist}$ is used as a dimension–independent representation of a training object, and is the basis for defining α and β.

The salience analysis is defined as the sensitivity that a sample or a component has with respect to the overall mapping nonlinearity, and can be conducted in two directions: in row vectors that correspond to training samples, or in column vectors that correspond to sample components. One may trim down the size of training data by removing non–salient samples and/or components to speed up the batch–mode training. The cross–validation training algorithm can be improved by using salient samples as training set and non–salient samples as validation set. In addition, with salient components analysis, one may be able to decouple a high–dimensional mapping to a few smaller sub–mappings, which are potentially easier to train and better in performance.

Acknowledgement

The authors would like to thank Professor Mark Embrechts at Rensselaer Polytechnic Institute, for providing the BackProp software, *Metaneural*; GE Corporate Research and Development Center (CRD) for offering the computing facilities. Partial support received under grant NAG 3-1196 from the NASA Lewis Research Center is gratefully acknowledged.

Reference

1. H. Lee and P. Hajela, *On A Unified Geometrical Interpretation of Multilayer Feedforward Networks, Part I: The Interpretation,* proceedings of the World Congress on Neural Networks, Vol. 3, pp. 625–630, San Diego, CA, June 4– 9, 1994. (Revision under review by Neural Networks Journal.)

2. H. Lee and P. Hajela, *On A Unified Geometrical Interpretation of Multilayer Feedforward Networks, Part II: Quantifying Mapping Nonlinearity,* proceedings of the World Congress on Neural Networks, Vol. 3, pp. 631–636, San Diego, CA, June 4– 9, 1994. (Revision under review by Neural Networks Journal.)

3. E. Baum and D. Haussler, *What size net gives valid generalization?*, Neural Computation, vol. 1, pp. 151–160, spring, 1989.

4. H. Lee and P. Hajela, *Prediction of Turbine Performance Using MFN by Reducing Mapping Nonlinearity,* in Artificial Intelligence and Object Oriented Approaches for Structural Engineering, Edited by G.H.V. Topping and M. Papadrakakis, pp. 99 – 105, Civil–Comp Press, Proceedings of the 2nd International Conference on Computational Structures Technology, Athens, Greece, Aug. 30 – Sep. 1, 1994.

5. H. Lee and P. Hajela, *Neural Networks Based Prediction of Geometrical Properties With Reduced Mapping Nonlinearity,* Proceeding of the 5th AIAA/USAF/NASA/ISSMO Symposium on Multidisciplinary Analysis and Optimization, pp. 901–909, (AIAA–94–4347–CP), Panama City, FL. September 7–9, 1994.

6. H. Lee and P. Hajela, *Vector Clustering for Neural Network based Prediction of Geometrical Characteristics,* proceedings of the 3rd. International Conference on the Application of Artificial Intelligence to Civil and Structural Engineering, Edinburgh, UK., Aug. 17–19, 1993.

7. R. Mead, *The Design of Experiments: Statistical Principles for Practical Application,* book, Cambridge University Press, 1990.

8. J. Neter, W. Wasserman and M. H. Kutner, Applied Linear Regression Models, 2nd ed., book, Richard D. Irwin, Inc. 1989.

Fuzzy Neural Networks for Control Design

Grantham K.H. Pang

Department of Electrical and Computer Engineering
University of Waterloo
Waterloo, Ontario
Canada N2L 3G1

Abstract. Neural networks and fuzzy systems have been applied to a wide range of applications in engineering and other disciplines. Rule-based fuzzy control systems offer a high degree of transparency into the system being modelled. Neural networks, on the other hand, have very strong learning capability. Knowledge is stored implicitly in the weights of the network. However, neural networks are only mappings from a multivariate input to output, and they offer no insight to internal dynamics and relationships. In this paper, an architecture of hybrid neuro-fuzzy control system which combines the advantages of both approaches is described. An implementation of fuzzy control systems using neural networks will be described. After the establishment of the neural network, it can be trained using numerical data. Therefore, the method combines fuzzy knowledge representation with the association and learning functions of neural networks. The new algorithm can be used to determine the shapes of membership functions in fuzzy systems.

Keywords. Neural networks, fuzzy systems, control, hybrid systems.

1 Introduction

Neural networks and fuzzy systems have been applied to a wide range of applications in engineering and other disciplines. Rule-based fuzzy control systems are based on fuzzy theory, with expert knowledge represented explicitly using a set of fuzzy if-then rules. They offer a high degree of transparency into the system being modeled by stating the causal relationships of prime process variables. Neural networks, on the other hand, have very strong learning capability. Given some numerical data on how a system should behave, there are many neural network algorithms which try to learn this system behaviour. Knowledge is stored implicitly in the weights of the network. However, neural networks are only mappings from a multivariate input to some outputs, and they offer no insight to internal dynamics and relationships. In this paper, an architecture of hybrid neuro-fuzzy control system which combines the advantages of both approaches is described.

An implementation of fuzzy control systems using neural networks will be described. The technique is based on the functional equivalence of a class of fuzzy systems and the type of neural network adopted in the paper. The combination of both paradigms allows the merging of sophisticated learning algorithm, which is developed in the realm of neural networks, together with the representation of qualitative and cognitive rules in fuzzy systems.

A rule-based fuzzy system can be implemented using a structured neural network. The neural network has three main stages which corresponds to the functionalities of fuzzification, rule inferencing and defuzzification found in fuzzy control systems. It is constructed so that the procedures and membership functions involved can be retrieved from the implementation of the neural network. Hence, the proposed method established a one-to-one mapping between a fuzzy control system and a structured neural network.

After the establishment of the neural network, it can be trained using some numerical data. The weights of the network will be adjusted. The derivation of the learning algorithm is based on a gradient descent algorithm. After the training procedure, the modified membership functions of the fuzzy control systems can be retrieved. Therefore, the method combines fuzzy knowledge representation with the association and learning functions of neural networks. The new algorithm can be used to determine the shapes of membership functions in fuzzy systems.

The first part of the paper contains a survey of some of the hybrid neuro-fuzzy approaches proposed by various researchers. Of great importance will be the issues on architecture, structure selection, training algorithm, transparency and interpretation of trained systems. Next, the implementation of a fuzzy control system using neural networks will be described. This can be achieved by constructing three types of sub-networks : fuzzification network, rule reasoning network and defuzzification network. The three types of sub-networks will be integrated in such a way that the structure and decision-making process of the original fuzzy system can be fully retrieved from its network implementation. The corresponding neural network should have similar performance as the original fuzzy control system.

In the third part of the paper, a learning algorithm to train the three sub-networks will be described. The objective is to train the neural network so that it will refine the membership functions of the fuzzy systems. A set of learning rules similar to the backpropagation learning algorithm can be derived from the gradient descent method. The proposed hybrid neuro-fuzzy methodology has been implemented and simulation results will be used to illustrate the ideas in the paper. An example of controlling an inverted pendulum (a one-link manipulator) using the method will be shown.

2 Fuzzy-Neural Approaches

A methodology called temporal back propagation has been proposed by Jang [1]. The objective is to enhance fuzzy controllers with self-learning capability. The proposed architecture is called ANFIS (Adaptive-Network-based Fuzzy Inference System), which is functionally equivalent to a fuzzy controller. The architecture is used to identify near-optimal membership functions and other parameters of a rule-base for achieving a desired input-output mapping. A special form of gradient descent method similar to the back-propagation algorithm is used for the tuning of the parameters.

The type of fuzzy rules that is represented by ANFIS is :

$$\text{If } x \text{ is } A_1 \text{ and } y \text{ is } B_1, \text{ then } f \text{ is } f_1(x, y)$$

If x is A_2 and y is B_2, then f is $f_2(x,y)$

where A_i and B_i $(i = 1,2)$ are linguistic labels (small, large etc.), and
$f_1 = p_1 * x + q_1 * y + r_1$
$f_2 = p_2 * x + q_2 * y + r_2$
$\bar{w1} = w1/(w1 + w2)$
$\bar{w2} = w2/(w1 + w2)$
The membership functions of A_i and B_i are chosen to be bell-shaped functions like

$$\mu_{A_i}(x) = \frac{1}{1 + \left[\left(\frac{x-c_i}{a_i}\right)^2\right]^{b_i}} \tag{1}$$

where $\{a_i, b_i, c_i\}$ are the parameters that need to be tuned during training. Another set of parameters that will be obtained after training is $\{p_i, q_i, r_i\}$ $(i = 1,2)$. The equivalent neural-network like architecture is represented by Fig. 1. The fuzzy controller is used in a control scheme called temporal back propagation, through which the state variables follow a desired trajectory. The controller and the plant at each time stage is cascaded together to form a trajectory adaptive network to facilitate the learning process. As the training process is based on the back-propagation algorithm, the problem with convergence will be encountered often. In the paper, heuristics on changing the step size are discussed with the aim of speeding up convergence.

Berenji [2] has proposed the Generalized Approximate Reasoning-based Intelligent Control (GARIC) architecture to learn and tune a fuzzy logic controller based on reinforcements from a dynamic system. Learning is achieved by integrating fuzzy inference into a feedforward neural network, which can then adaptively improve performance by using gradient descent methods. The GARIC architecture can learn to adjust the fuzzy membership functions of the linguistic labels used in different control rules. A conjunction operator called softmin was introduced in the paper, which was needed for learning purpose because of the differentiability of the operator.

Other fuzzy neural networks (FNN) have been proposed by Horikawa [3]. The methods have been developed for the modeling of complex nonlinear systems linguistically. The types of neural network structure proposed have connection weights which have particular meanings for constructing fuzzy inference rules and membership functions. This procedure is based on the use of the back-propagation algorithm. Three types of FNN are proposed. In the first type, the consequent part of the fuzzy rule is a constant like :

$$\text{If } x \text{ is } A_1 \text{ and } y \text{ is } B_1 \text{, then } z \text{ is } C$$

In the second type, the consequent part of the fuzzy rule is a first-order linear equation:

$$\text{If } x \text{ is } A_1 \text{ and } y \text{ is } B1 \text{, then } z \text{ is } k0 + k1 * x + k2 * y$$

In the third type, the consequent part of the fuzzy rule has the form :

$$\text{If } x \text{ is } A_1 \text{ and } y \text{ is } B_1 \text{, then } z \text{ is } C_1 \text{ with truth value } t$$

where $C_i (i = 1, 2)$ is a straight line in the universe of discourse of z. The selection of structures of FNN is the first step. Next, using a set of input-output data, the network will be trained and the parameters of the membership functions will be identified. One limitation of the third type is that the fuzzy rules in the consequent part can only have fuzzy variables "small" or "big".

Ishibuchi et al. [4] has also proposed a type of fuzzy neural network for control problem. Fuzzy if-then rules such as

$$\text{If } x_1 \text{ is small and } x_2 \text{ is large, then } y \text{ is small}$$

are generated and adjusted by learning methods using numerical data. Hence, the idea is to utilize fuzzy if-then rules obtained from human experts in the learning of neural networks. That is, both numerical information from measuring instruments and linguistic information from human experts are utilized in the fuzzy neural network. The proposed learning algorithm can be viewed as an extension of the back-propagation algorithm to the case of fuzzy data.

One general disadvantage of the above approaches is that the learning of the fuzzy neural networks are all based on the back-propagation algorithm or its variation. Hence, learning can be very slow and there is no guarantee of convergence in learning. Yamakawa [5] has developed a new neuron model which is combined with fuzzy logic. Because of the simple architecture of this neuron model, the learning speed is exceedingly fast when compared with conventional multilayered feedforward neural network. The structure of the neo fuzzy neuron is shown in Figure 2. The output of this neo fuzzy neuron can be represented by the equation :

$$y = f_1(x_1) + f_2(x_2) + ... + f_m(x_m)$$

The nonlinearity of each synapse $f_i(x_i)$ is determined by fuzzy IF-THEN rules and the synapse output is obtained by fuzzy inference with defuzzification. Each input space x_i is divided into several fuzzy segments which are characterized by membership functions $\mu_{i1}, \mu_{i2}, ...$ Each membership function is triangular and the summation of two neighboring membership functions is always unity and thus these membership functions are complementary. w_{ik} are some constants. With this special structure, the input signal activates only two rules, because the antecedents of all rules are described by complementary membership functions. As a result, the defuzzified value can be computed by a very simple formula which does not even involve any arithmetic division. However, the limitation of this structure is that the membership functions are confined to the triangular shape and they have to be complementary. Also, the content of the fuzzy rule has to be very primitive of the type :

If x_1 is μ_{1i}, then y is w_{1i} $(i = 1, ..., k)$
If x_2 is μ_{2i}, then y is w_{2i} $(i = 1, ..., k)$

Another neural network model for fuzzy logic control has been proposed by Lin and Lee [6] [7]. The network can be constructed from training examples and fuzzy logic rules are developed from the resulting neural network. Hence, a neural network structure is used to develop the fuzzy rule-base. Optimal input/output membership functions are also obtained in the process. One major advantage of their design is that its learning algorithm converges faster than the original back-propagation algorithm.

3 Architecture of a Neuro-Fuzzy System

In this section, an implementation of a fuzzy control system using neural networks will be discussed [8]. Figure 3 shows the structure of the system. The idea is based on [8] and there are four main layers in the implementation with three types of sub-networks. Each sub-network performs a different function. The first one is called fuzzification network, which is used to identify patterns of input variables in terms of membership functions of linguistic terms. The second one is called rule reasoning network, which is used to conduct a conjunction by calculating the strength of firing each of the fuzzy rules. The third sub-network is called defuzzification network which is used to carry out the task of control synthesis by generating a defuzzified control action. The three types of networks are connected sequentially. However, the training and the construction of the networks are performed independently. The objective is to preserve the decision-making procedure in a fuzzy logic control system using a neural network implementation.

3.1 Fuzzification network

The inputs of the system go to the first layer which is consisted of a number of fuzzification networks. The input x_i to each fuzzification network is the same as the input to the fuzzy logic control system. The outputs from the fuzzification network correspond to the degree of membership μ_{ik} where k is the number of linguistic terms corresponding to the x_i. Figure 4 shows some two-layer fuzzification networks for this purpose. Initially, each network should be trained with the specified membership functions of x_i. If two inputs have the identical set of linguistic terms, then the same fuzzification network can be used at the beginning. During the network learning stage described later on, the membership functions of the linguistic terms will be changed subsequently using the training data.

3.2 Rule reasoning network

For each fuzzy rule in the knowledge base of a fuzzy logic control system, there is a rule reasoning network (Fig. 5). The number of inputs to each rule reasoning network depends on the number of components in the precondition part of the rule. This network is used to implement different kinds of conjunction operators like intersection or algebraic product. Again, the network can be easily modified during the learning stage by use of the learning algorithms.

3.3 Defuzzification network

This is the most important stage in the construction of the neuro-fuzzy network. Layers three and four are in this network (see Figure 6). The objective here is to determine the final crisp control according to the firing strengthes of the rule reasoning networks in the fourth layer. This final layer consists of steps to generate fuzzy control, and then convert it back into a crisp value. The main issue concerns with ways to represent membership functions of fuzzy control actions using a network implementation and to perform defuzzification. The defuzzification network consists of two layers. Layer three is introduced to calculate fuzzy controls. The number of inputs to this layer is the same as the number of fuzzy rules r. Let n be the number of discrete control values in the range of control u. Let $U = \{u_1, u_2, ..., u_n\}$ be a set of discrete control values. The initial weights of this layer are calculated from membership functions of the control action as

$$w_{ij} = \mu_{C_j}(u_i) \tag{2}$$

where $i = 1, ..., n$ and $j = 1, ..., r$. Note that C_j is the linguistic label of rule j. Essentially, it means that the weight w_{ij} equals the degree of membership of C_j at $u = u_i$. Layer four in this network performs the task of defuzzification. Here a center-of-area method is used. In this case, the weights of the second layer is given by

$$\beta_i = \frac{u_i}{\gamma} \tag{3}$$

where

$$\gamma = \sum_{i=1}^{n} \mu_i \tag{4}$$

The defuzzified control value u is given by

$$\sum_{i=1}^{n} \beta_i \mu_i \tag{5}$$

Again, the initial weights in layer three of the defuzzification network can be changed later by the learning algorithm and the training data.

However, the learning process will only change the membership functions of control actions in layer three and not the logic operations for control synthesis in layer four.

3.4 A Simple Example

When the individual components of the fuzzification network, rule reasoning network and the defuzzification network have been constructed, they are to be connected together according to the design of the original fuzzy control system. A simple example of a neuro-fuzzy network with two fuzzy rules is shown in Figure 7.

The two fuzzy rules are :

$$\text{If } x_1 \text{ is } A_1 \text{ and } x_2 \text{ is } B_1, \text{ then } u \text{ is } C_1.$$
$$\text{If } x_1 \text{ is } A_2 \text{ and } x_2 \text{ is } B_2, \text{ then } u \text{ is } C_2.$$

where A_i, B_i, C_i are linguistic labels (small, medium, large, etc.) and $(i = 1, 2)$. In this example, the fuzzification and rule reasoning networks are constructed using neural networks with two hidden layers and n is chosen as 6.

It is obvious that the computation process of the integrated neural network resembles the original fuzzy control system. The knowledge of the fuzzy rules has been represented by the weights in the individual neural networks in the structure. This integrated neural network can also be viewed as a parallel, neural-network implementation of the fuzzy control system. Its performance should be similar to the performance of the original fuzzy control system. It should be noted that the integrated neural network is not fully connected.

4 Training of the Network

In this section, a learning algorithm to train the three sub-networks will be described. The objective is to train the neural network so that it will refine the membership functions of the fuzzy systems. A set of learning rules similar to the backpropagation learning algorithm can be derived from the gradient descent method.

The back-propagation learning algorithm developed for standard multilayer feedforward neural networks can be extended for the training of the neuro-fuzzy network. Here, we assume that a set of desired or optimal input/output pairs is available. Let u^d be the desired control value. The error function is defined as

$$E = \frac{1}{2}(u^d - u)^2 \tag{6}$$

During the training process, we would like to maintain the defuzzification algorithm and hence the β are calculated from the equation using μ_i. The rule for updating w_{ij} in layer 3 can be shown to be :

$$w_{ij}(t+1) = w_{ij}(t) + \epsilon\alpha_j(u^d - u)(\beta_i - \sum_{l=1}^{n} u_l\mu_l/\gamma^2) \tag{7}$$

Note that $w_{ij}(t+1)$ only gets update if μ_i is within the range [0,1]. Also, ϵ is a learning rate between 0 and 1. The error term e_j that is back-propagated to the output of the rule reasoning network is :

$$e_j = (u^d - u)\sum_{i=1}^{n} w_{ij}(\beta_i - \sum_{l=1}^{n} u_l\mu_l/\gamma^2) \tag{8}$$

This error term is used to adjust the weights of the fuzzification and rule reasoning networks using the back-propagation algorithm.

When the training process has been completed, the new membership functions and conjunction operators can be recovered from individual sub-network. From the fuzzification network, we can obtain the modified membership functions for each linguistic label of input x_i. From the rule reasoning network, we obtain the modified conjunction operators of the rules. Different fuzzy control rules may now have different conjunction operators. To obtain the modified membership functions for the output linguistic labels, we have to set γ_i to 1 (where $i = 1, ..., r$), all others to zero, and obtain μ_i (where $i = 1, ..., n$).

5 Control Design Example

The control of an inverted pendulum is described in this section. In the classical inverted pendulum problem, the aim is to control the motor so as to balance an inverted pendulum in two dimensions (see Figure 8). The system's state variables are defined by θ and $\dot{\theta}$. u denotes the control variable or the value of the motor torque. A well-defined mathematical model can be obtained for the nonlinear system and well established nonlinear control system design technique can be used to obtain an excellent controller. Hence, the simulation results using the neuro-fuzzy approach can be verified and compared with the conventional design method.

The mathematical equation of a single link manipulator in a vertical plane can be given as :

$$MgLsin(\theta) - u - k\theta - \mu\dot{\theta} = I\ddot{\theta} \tag{9}$$

where θ is link rotation, I is link inertia, M is the link mass, L is half link length, g is gravity constant, μ is velocity constant, k is friction constant and u is the input torque. If all the parameters are known, excellent control result can be obtained using the computed-torque method :

$$u = K_v\dot{e} + K_pe + MgLsin(\theta_d) - k\theta_d - \mu\dot{\theta}_d - I\ddot{\theta}_d \tag{10}$$

where θ_d is the desired trajectory, $e = \theta_d - \theta$ is the tracking error, and K_v, K_p are the velocity and position gain constants. For $K_v = 8$ and $K_p = 28$, the simulation result is given in Figure 9, which shows the zero steady state position error and short settling time of the controlled system. The inverted pendulum is released from an angle of 30 degrees from the vertical.

For fuzzy control, let the term set for each fuzzy variable θ, $\dot{\theta}$ or u be $\{NM, NS, ZE, PS, PM\}$ where NM denotes Negative Medium, NS denotes Negative Small, ZE denotes Zero, PS denotes Positive Small and PM denotes Positive Medium.
The rules are defined as follows :

If θ is ZE and $\dot{\theta}$ is ZE, then u is ZE
If θ is PS and $\dot{\theta}$ is ZE, then u is NS
If θ is PM and $\dot{\theta}$ is ZE, then u is NM
If θ is NS and $\dot{\theta}$ is ZE, then u is PS
If θ is NM and $\dot{\theta}$ is ZE, then u is PM
If θ is ZE and $\dot{\theta}$ is NS, then u is PS
If θ is ZE and $\dot{\theta}$ is NM, then u is PM
If θ is ZE and $\dot{\theta}$ is PS, then u is NS
If θ is ZE and $\dot{\theta}$ is PM, then u is NM
If θ is PS and $\dot{\theta}$ is NS, then u is ZE
If θ is NS and $\dot{\theta}$ is PS, then u is ZE

All variables are normalized into the interval [-1,1]. The algebraic product is used as the conjunction operator. Figure 10 shows the result of fuzzy control of the inverted pendulum from the set of fuzzy rules defined above. Next, the fuzzy neural network is trained using the data from the computed torque method. The result is shown in Figure 11 and the modified membership functions are shown in Fig. 12.

6 Conclusions

This paper has discussed a method of implementing fuzzy rule-based control systems using artificial neural networks. The idea is based on constructing a neural network which is functionally equivalent to the original fuzzy system. Back-propagation-like algorithm is then used to train the neural network with the objective of improving the performance of the fuzzy system. During the training process, the weights of the neural network are changed and the modified membership functions of the fuzzy control system can be retrieved. Therefore, the method combines fuzzy knowledge representation with the association and learning functions of neural networks.

References

[1] J.S.R. Jang, "Self-learning fuzzy controllers based on temporal back propagation", IEEE Trans. on Neural Networks, Vol. 3, No. 5, pp. 714-723, 1992.

[2] H.R. Berenji, "Learning and tuning fuzzy logic controllers through reinforcements", IEEE Trans. on Neural Networks, Vol. 3, No. 5, pp. 724-740, 1992.

[3] S. Horikawa, T. Furuhashi and Y. Uchikawa, "On Fuzzy Modeling Using Fuzzy Neural Networks with the Back-Propagation Algorithm", IEEE Trans. on Neural Networks, Vol. 3, No. 5, pp. 801-806, 1992.

[4] H. Ishibuchi, R. Fujioka and H. Tanaka, "Neural networks that learn from fuzzy if-then rules", IEEE Trans. on Fuzzy Systems, Vol. 1. No. 2, pp. 85-97, May 1993.

[5] T. Yamakawa, "A neo fuzzy neuron and its applications to system identification and prediction of chaotic behaviour",in *Computational Intelligence : Imitating Life*, J.M. Zurada, R.J. Marks II, C.J. Robinson (eds.), pp. 383-395, IEEE Press, 1994.

[6] C.T. Lin and C.S. G. Lee, "Neural-network fuzzy logic control and decision system", IEEE Trans. on Computers, Vol. 40, No. 12, pp. 1320-1336, Dec 1991.

[7] C.T. Lin and C.S. G. Lee, "Reinforcement structure/parameter learning for neural-network-based fuzzy logic control systems", IEEE Trans. on Fuzzy Systems, Vol. 2, No. 1, pp. 46-63, Feb 1994.

[8] F.Y. Wang and H.M. Kim, "Implementing adaptive fuzzy logic controllers with neural networks : a design paradigm", to appear in Journal of Intelligent and Fuzzy Systems.

[9] J. Nie and D.A. Linkens, "Neural network based approximate reasoning : principles and implementation", in *Advances in Intelligent Control*, C.J. Harris (eds.), pp. 321-336, Taylor & Francis, 1994.

[10] T. Yamaguchi et al., "Self organizing control using fuzzy neural networks", in *Advances in Intelligent Control*, C.J. Harris (eds.), pp. 337-361, Taylor & Francis, 1994.

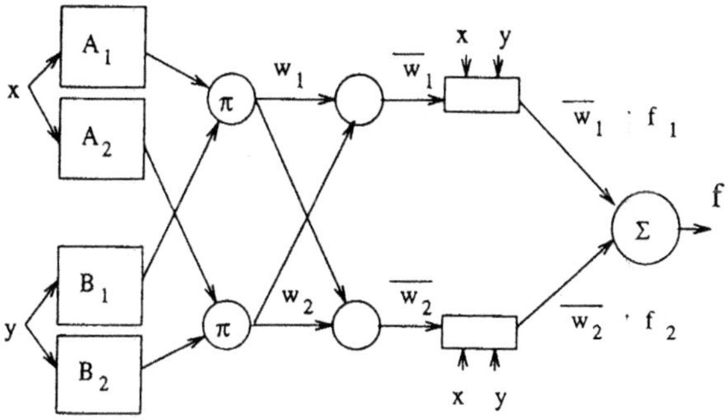

Fig. 1 : ANFIS

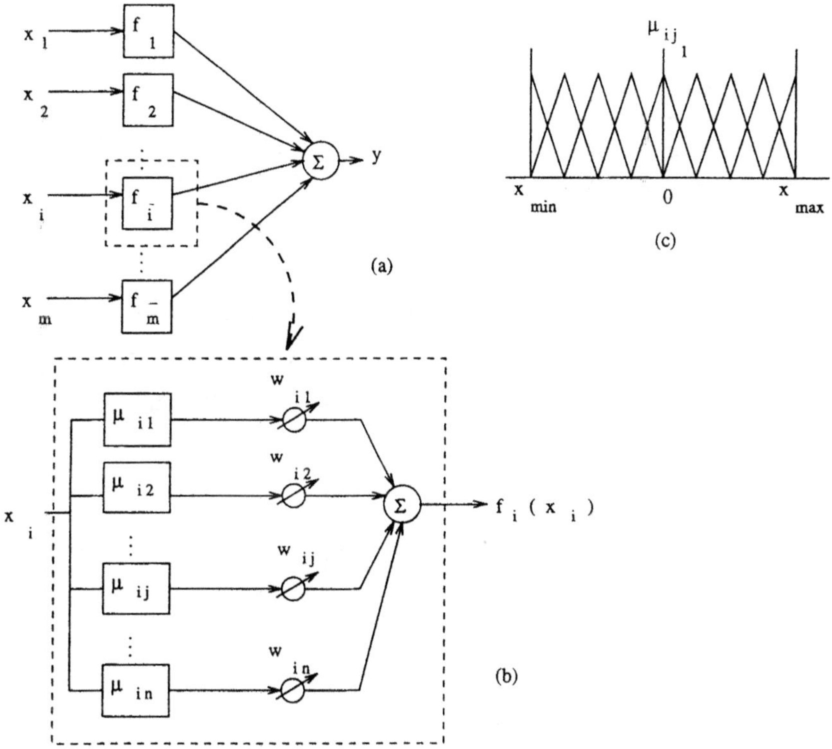

Fig. 2 : (a) Structure of the neo-fuzzy neuron, each synaptic characteristics of which is represented by a nonlinear function f_i (b) Structure of the nonlinear synapse which is described with a set of IF-THEN rules including singletons in consequents.
(c) Triangular and complementary membership functions assigned for the fuzzy segments

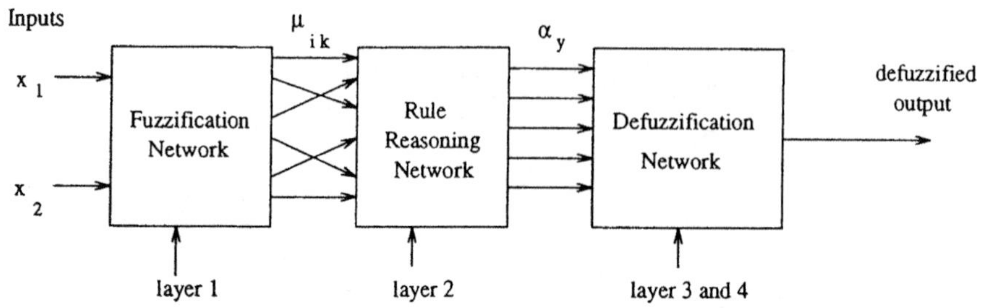

Fig. 3 : Structure of a neuro-fuzzy system

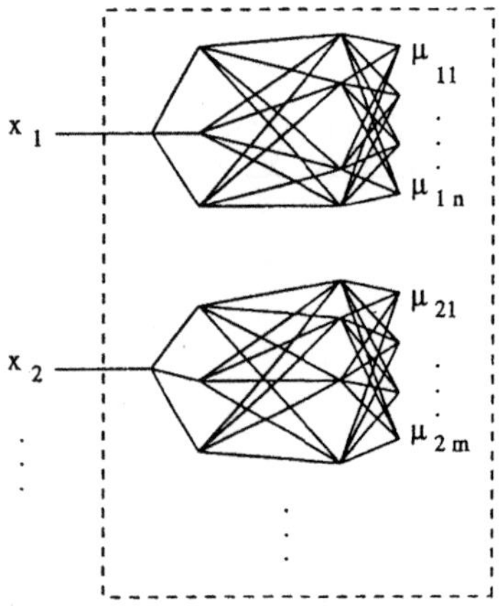

Fig. 4 : Fuzzification Network

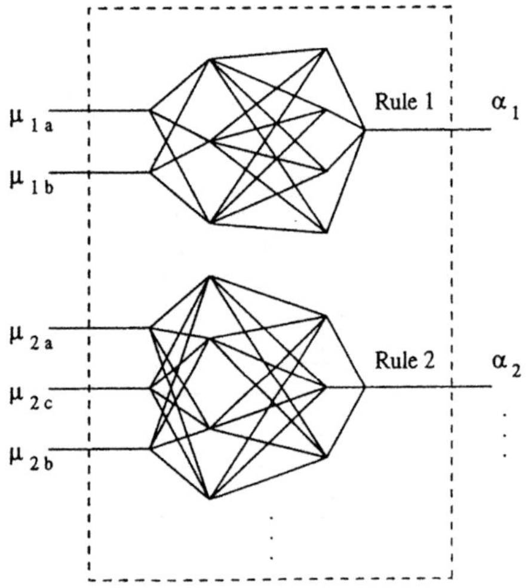

Fig. 5 : Rule Reasoning Network

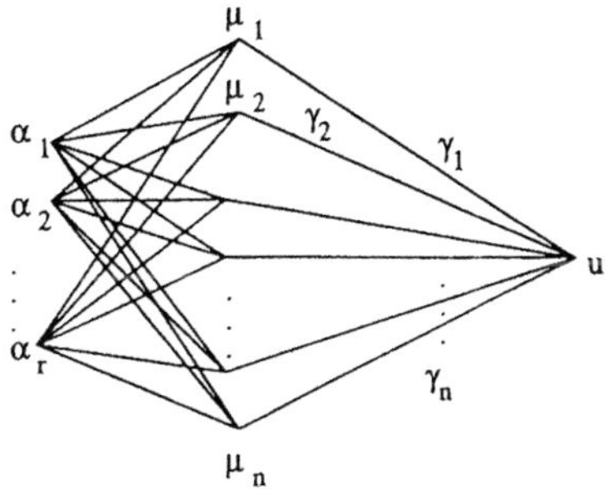

Fig. 6 : Defuzzification Network

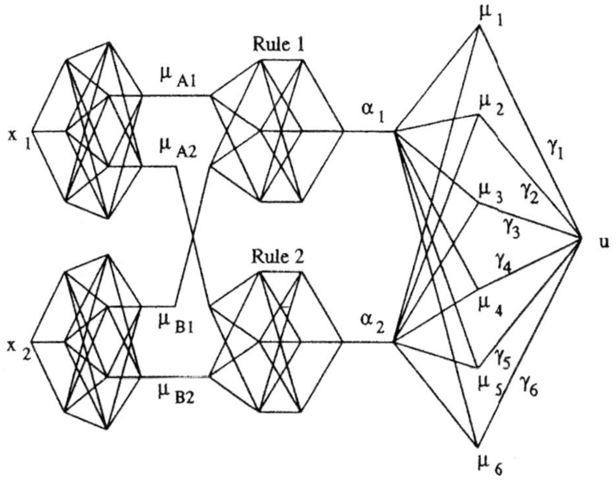

Fig. 7 : An example of a neural fuzzy network

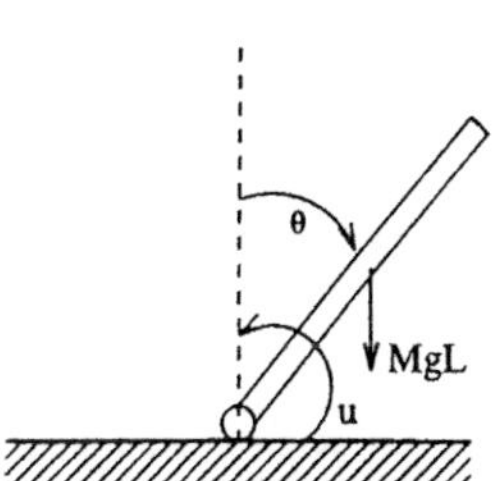

Fig. 8 : An inverted Pendulum

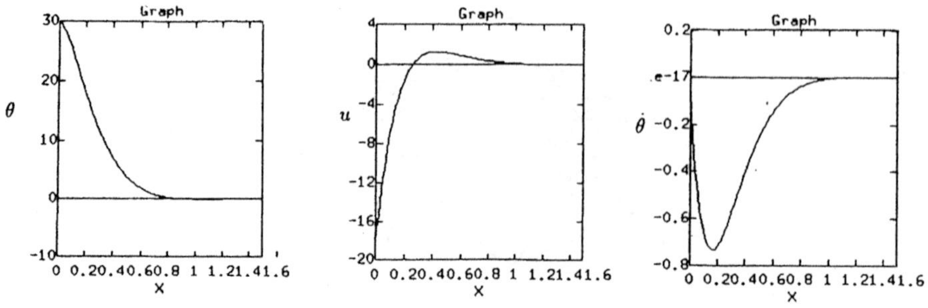

Fig. 9 Performance of the system using the computed torque method

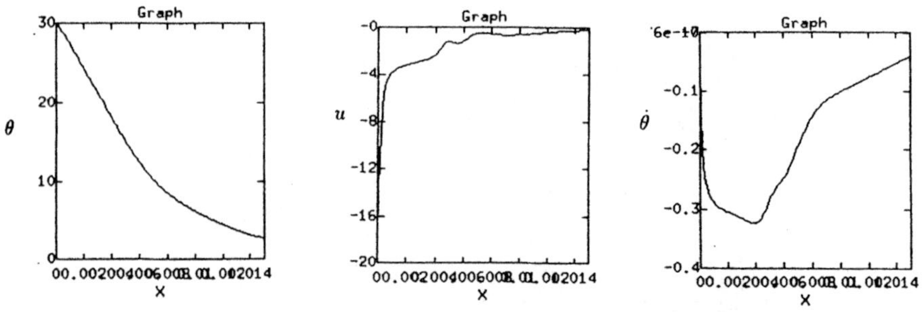

Fig. 10 Performance of the system using fuzzy control (before training)

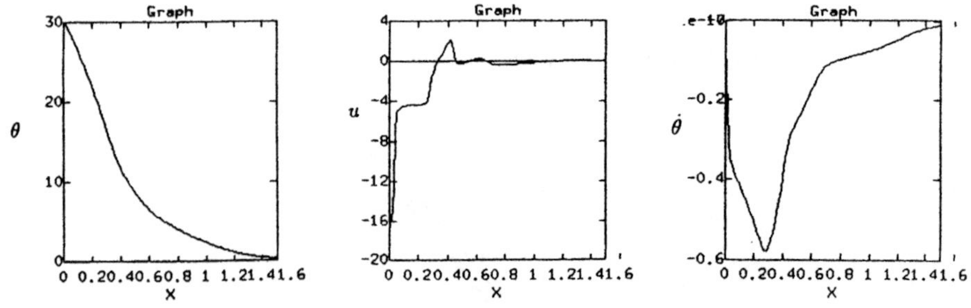

Fig. 11 Performance of the system using fuzzy control (after training)

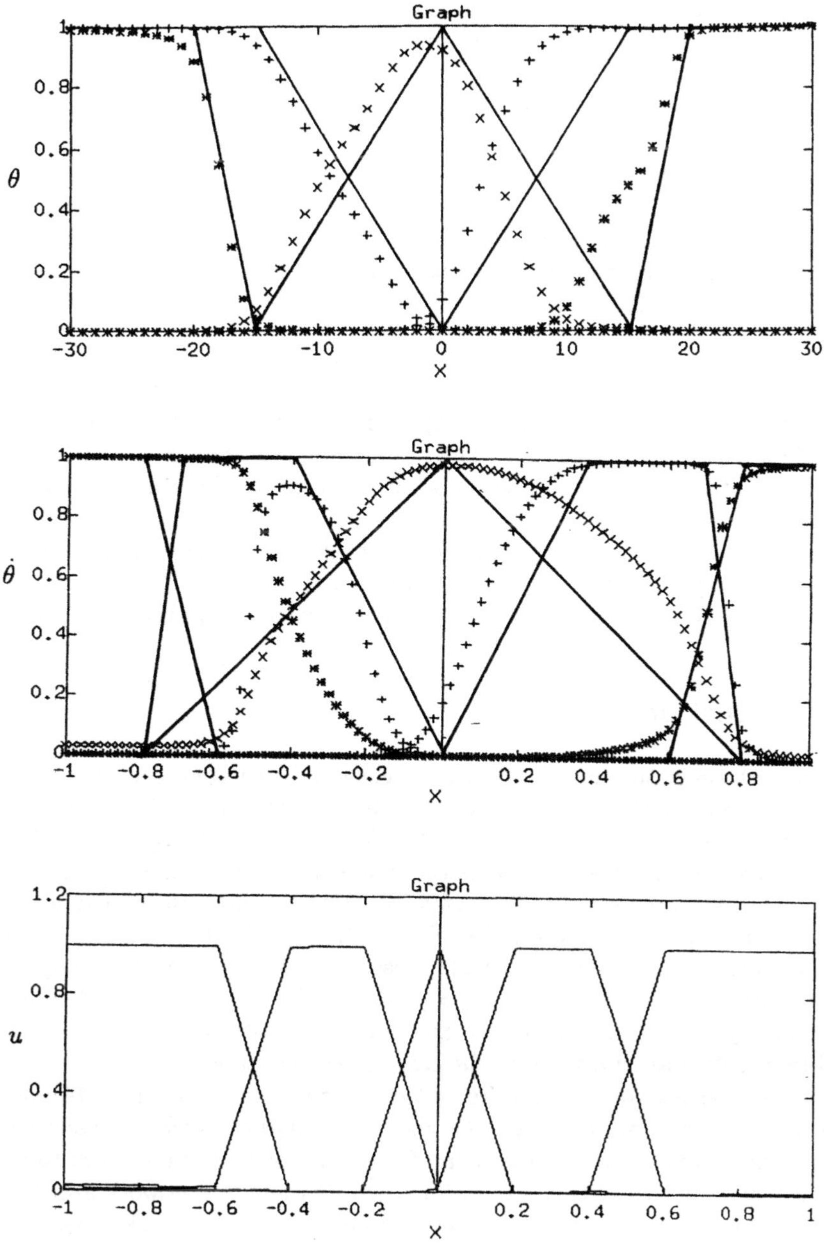

Fig 12 Modified membership functions of θ, $\dot{\theta}$ and u

(The solid line is the original definition of the membership function)

Efficient Neural Network Techniques for Discrete Optimization Problems

Andreas N. Stafylopatis

National Technical University of Athens
Department of Electrical and Computer Engineering
157 73 Zographou, Athens, Greece

Abstract. Some efficient neural network techniques are presented for the solution of several classes of discrete optimization problems including binary constraint satisfaction problems. The proposed approaches are based on the development of computational schemes, such as group updates and multiscaling, that improve the performance of discrete Hopfield–type networks. The appropriateness of the above techniques is established by means of theoretical results and experimental performance evaluation.
Keywords. Hopfield networks, discrete optimization, constraint satisfaction

1 Introduction

Discrete Optimization Problems in their general formulation can be defined in terms of a tuple (S, S', f_c), where S denotes the state space of the problem, $S' \subseteq S$ denotes the set of *feasible* states, i.e., those satisfying the problem constraints, and $f_c : S \rightarrow \Re$ denotes the function that determines the cost of each state. The problem is to find a feasible state for which the cost function is optimal, i.e., to find a state $s' \in S'$ such that $f_c(s')$ is optimal in S' [1]. The conventional approach to tackle these problems is to regard the given optimization problem as a tuple (S, S, f') where the function f' has the form $f' = f_p + f_c$, with the function f_p encoding the constraints of the problem and taking its optimum value in the case of feasible states.

Most neural network techniques applied to the solution of discrete optimization problems are based on the above formulation which expresses the constraints through the penalty term f_p. An alternative approach is to directly encode the set of constraints (or part of it) into the network architecture so that constraint satisfaction is guaranteed by the network operation. The most widely studied neural network models used in solving optimization problems are based on the Boltzmann Machine and the analog Hopfield network [4, 16]. As far as the efficiency of the above approaches is concerned, computation times can be very long as the problem size increases, whereas it is not easy to avoid getting stuck in local minima during the searching process.

This presentation concerns some neural network techniques for the solution of discrete optimization problems, that exhibit very good behaviour with respect to the above mentioned difficulties. The general framework relates to the notion of *group update* in binary Hopfield-type networks (Boltzmann machines): instead of selecting a single unit to update at each time step, we consider a group of units which are selected and updated simultaneously. A first approach concerns the application of *multiscaling* to binary networks [10, 11]. According to the multiscale principle, a smaller network is constructed by grouping the units of the original network and iterations are performed either in the original fine-scale network or in the smaller coarse-scale network. The operation of multiscaling in binary networks is closely related to the notion of group updates. We consider the solution of two optimization problems, for which an efficient grouping of units can be devised based on an original mapping of the problems to a network architecture. As a second application of group updates, an efficient technique is developed for the solution of binary constraint satisfaction problems. An operation scheme based on double updates leads to a faster and more effective search of the problem state space and significantly improves performance, as indicated by experimental results concerning several instances of typical constraint satisfaction problems. In the following, proofs are omitted for the sake of conciseness, thus emphasizing the application aspect of the work.

2 Group Updates

Consider a binary Hopfield-type neural network with n units. Let w_{ij} denote the connection weights of the network (with $w_{ii} = 0$) and θ_i the threshold values of the units. The energy corresponding to state $\vec{y} = (y_1, \ldots, y_n)$ ($y_i \in \{0, 1\}$ for every i) is the following [5, 6]:

$$E(\vec{y}) = -\frac{1}{2}\sum_{i=1}^{n}\sum_{j=1}^{n} y_i y_j w_{ij} - \sum_{i=1}^{n} \theta_i y_i \tag{1}$$

The operation of the network is *asynchronous*. At each time instant one unit is selected randomly and the change in the network's energy that will result if the selected unit i changes state is computed:

$$\delta E_i(\vec{y}) = (2y_i - 1)(\sum_{j=1}^{n} w_{ji} y_j + \theta_i) \tag{2}$$

If $\delta E_i(\vec{y}) < 0$ then the change is accepted, otherwise it is rejected. We shall use the term *trial* to denote the above operation. Changing the state of a unit will be referred to as *update*. Hence, a trial is eventually followed by an update. The notion of *group update* constitutes a generalization of the notion of single update. Instead of selecting a single unit to consider for update, we select a group containing a number of units and calculate the difference in the energy that would result if the state of *all* units in the group were altered. Then a decision must be made as to whether the states of *all* these units must change or not. Clearly, group operation differs from *synchronous* operation [1], which consists of simultaneously performing individual trials and updates and also from *block-sequential* operation [7]; according to the latter operation mode, the state vector is divided into subvectors which are sequentially updated, where within each subvector individual updates are performed simultaneously.

Let $\vec{y} = (y_1, \ldots, y_n)$ be the state of the network at some time instant during operation; we consider a group G consisting of m units and perform a trial on this group. If the trial is successful, a group update takes place and the new state $\vec{z} = (z_1, \ldots, z_n)$ of the network will be such that

$$z_i = \begin{cases} y_i + u_i & \text{if } i \in G \\ y_i & \text{if } i \notin G \end{cases} \tag{3}$$

where $u_i = 1 - 2y_i$, $i \in G$. Equation (3) simply expresses the fact that units in G change state.

It can be shown that the change in the energy of the network caused by a group update of group G from state $\vec{y}$ is given by

$$\Delta E_G(\vec{y}) = \sum_{i \in G} \delta E_i(\vec{y}) - \frac{1}{2} \sum_{i \in G} \sum_{j \in G} u_i u_j w_{ij} \tag{4}$$

where $\delta E_i(\vec{y})$ is the change in the network's energy in case unit i is selected for a *single* update from state $\vec{y}$.

The change in the network's energy due to a group update from state $\vec{y}$ as given by (4) is equal to the sum of the individual energy changes that would result if the units of the group sequentially changed state starting from $\vec{y}$ and following an arbitrary order. This provides a more intuitive interpretation of the above result and can be seen as a consequence of energy conservation.

In order for an operation scheme based on group updates to be successful, appropriate groups of units must be identified, which represents a highly problem dependent task. Grouping schemes can be distinguished into *state dependent* and *state independent*. In the case of state independent grouping it is possible to reduce the state space through multiscaling and construct a coarse scale network having the property that single updates in that network simulate group updates in the original fine scale network. On the other hand, application of the group update technique in cases of state dependent grouping schemes can significantly accelerate the solution of constraint satisfaction problems.

3 Multiscaling in Binary Networks

The idea of *multiscaling* originates from the methods of numerical analysis and particularly the multigrid methods. The above idea is used in [12] to optimize a general Hopfield objective function of *continuous* neural variables. A quite general methodology is suggested, that cannot be applied in the case where the fine and coarse scale vectors have binary elements. The multiscale method discussed here uses a transformation that depends on the current state of the fine-scale network and is closely related to group updates. We first proceed to the description of the method and then establish its relationship with group updates.

Consider again a Hopfield-type neural network with n binary units (n-net), symmetric connection strengths w_{ij} (with $w_{ii} = 0$) and threshold values θ_i. Suppose now that we wish to switch to a smaller network with p units (p-net) that is also assumed to be a Hopfield-type neural network with binary units. The mapping from the n-net to the p-net is based on partitioning the units of the n-net into p disjoint groups, such that each group G_k, $k = 1, \ldots, p$, corresponds to a unit k of the p-net. We shall use the notation $\mathcal{P}$ to represent the partition. If the state of the n-net is $\vec{y} = (y_1, \ldots, y_n)$ the corresponding energy function $E(\vec{y})$ can be written:

$$\begin{aligned} E(\vec{y}) &= -\frac{1}{2}\sum_{i=1}^{n}\sum_{j=1}^{n} y_i y_j w_{ij} - \sum_{i=1}^{n} \theta_i y_i \\ &= -\frac{1}{2}\sum_{G_k \in \mathcal{P}}\sum_{G_l \in \mathcal{P}}\sum_{i \in G_k}\sum_{j \in G_l} y_i y_j w_{ij} - \sum_{G_k \in \mathcal{P}}\sum_{i \in G_k} \theta_i y_i \qquad (5) \end{aligned}$$

Suppose that while being in state $\vec{y}$ we decide to switch to the small network. Suppose also that we have a way of appropriately constructing the p-net. At some later time, after operation of the small network, we decide to switch back to the n-net. Then, if the final state of the p-net is $\vec{\xi} = (\xi_1, \ldots, \xi_p)$, the following transformation is used for obtaining the new state $\vec{z} = (z_1, \ldots, z_n)$ of the original network:

$$z_i = y_i + u_i \xi_k, \quad i \in G_k \qquad (6)$$

for all $G_k \in \mathcal{P}$, where the quantities u_i are defined as $u_i = 1 - 2y_i$ for $i = 1, \ldots, n$. An interpretation of the above transformation is that, for each group $G_k \in \mathcal{P}$, if the final state of the corresponding unit k of the p-net is 1 all units belonging to group G_k change their state with respect to the state vector $\vec{y}$, whereas if the resulting value ξ_k is 0, the units belonging to group G_k retain the state they had before switching to the p-net.

The connection weights x_{kl} of the p-net can be computed as follows. Using the transformation equation (6) the energy of the new state $\vec{z}$ of the n-net is:

$$E(\vec{z}) = -\frac{1}{2}\sum_{G_k\in\mathcal{P}}\sum_{G_l\in\mathcal{P}}\sum_{i\in G_k}\sum_{j\in G_l}(y_i+u_i\xi_k)(y_j+u_j\xi_l)w_{ij} - \sum_{G_k\in\mathcal{P}}\sum_{i\in G_k}\theta_i(y_i+u_i\xi_k) \tag{7}$$

Taking into account the energy definition, after some algebra equation (7) yields $E(\vec{z}) = E(\vec{y}) + \hat{E}(\vec{\xi})$, where the quantity $\hat{E}(\vec{\xi})$ is given from the following:

$$\begin{aligned}\hat{E}(\vec{\xi}) &= -\frac{1}{2}\sum_{G_k\in\mathcal{P}}\sum_{G_l\in\mathcal{P}}\xi_k\xi_l\Big(\sum_{i\in G_k}\sum_{j\in G_l}u_iu_jw_{ij}\Big)\\ &\quad -\sum_{G_k\in\mathcal{P}}\xi_k\Big(\sum_{i\in G_k}\theta_iu_i + \sum_{G_l\in\mathcal{P}}\sum_{i\in G_k}\sum_{j\in G_l}u_iy_jw_{ij}\Big)\end{aligned} \tag{8}$$

We can easily observe that the quantity $\hat{E}(\vec{\xi})$ represents the energy of a network of size p whose state is given by the vector $\vec{\xi} = (\xi_1, \ldots, \xi_p)$ and whose connection weights x_{kl}, $k = 1, \ldots, p$, $l = 1, \ldots, p$ take the form:

$$x_{kl} = \sum_{i\in G_k}\sum_{j\in G_l}u_iu_jw_{ij} \tag{9}$$

To be consistent with the energy definition (1) the quantities x_{kk} should be equal to zero. We can achieve this by making a slight change to equation (9). We add the terms $x_{kk} = \sum_{i\in G_k}\sum_{i'\in G_k}u_iu_{i'}w_{ii'}$ to the threshold values ω_k of the p-net and set $x_{kk} = 0$, $k = 1, \ldots, p$. Since the ξ_k values are binary and hence $\xi_k^2 = \xi_k$, we find:

$$\hat{E}(\vec{\xi}) = -\frac{1}{2}\sum_{k=1}^{p}\sum_{l=1}^{p}x_{kl}\xi_k\xi_l - \sum_{k=1}^{p}\omega_k\xi_k \tag{10}$$

where

$$x_{kl} = \begin{cases}\sum_{i\in G_k}\sum_{j\in G_l}u_iu_jw_{ij} & \text{if } k\neq l\\ 0 & \text{if } k = l\end{cases} \tag{11}$$

and the threshold ω_k of each unit k of the small network is:

$$\omega_k = \sum_{i\in G_k}\theta_iu_i + \sum_{G_l\in\mathcal{P}}\sum_{i\in G_k}\sum_{j\in G_l}u_iy_jw_{ij} + \frac{1}{2}\sum_{i\in G_k}\sum_{i'\in G_k}u_iu_{i'}w_{ii'} \tag{12}$$

It is important to note that the p-net has also symmetrical weights (i.e., $x_{kl} = x_{lk}$ for all $k = 1, \ldots, p$, $l = 1, \ldots, p$), as we can easily observe from equation (11). This fact ensures convergence of the p-net when used in Hopfield-type operations. Another observation is that in order to compute the weight x_{kl} only the weights w_{ij} connecting the units belonging to groups k and l are involved.

The operation of the coarse-scale network emulates the interaction among the groups of units of the original network. According to the result of this interaction all units participating in a group either change or retain their state. As a matter of fact, the multiscale method is closely related to the notion of group update not only in an intuitive, but also in a formal way. Suppose that while being in state $\vec{y}$ of the n-net, a p-net is created. Then, it can be shown that the change in the n-net's energy in state $\vec{z}$ caused by a group update concerning the units of the k-th group, is equal to the change in the p-net's energy caused by the single update of unit k in state $\vec{\xi}$, provided that the two networks are in consistent states, i.e., $z_i = y_i + u_i\xi_k$ if $i \in G_k$.

In the next section we consider optimization problems for which the employment of multiscaling is very beneficial as they can be characterized by a naturally arising fixed (state independent) grouping scheme.

4 Applications of Multiscaling

We shall now examine the appropriateness of the multiscale method in solving optimization problems by considering the *set partitioning* and *set covering* problems. For these two problems we have developed original mappings to the corresponding fine-scale Hopfield network (Boltzmann machine), which are of interest by themselves and allow an effective exploitation of the advantages offered by multiscaling.

4.1 Set Partitioning Problem

The formulation of the Set Partitioning Problem (SPP) has as follows [3]: Given a finite set S containing M elements and a collection C of subsets S_i $(i = 1, \ldots, N)$ of S, find the minimum subset F of C such that the subsets belonging to F constitute a partition of S, i.e., they are disjoint and their union is equal to S.

In order to deal with this problem we construct the following neural network architecture. Let k_i $(i = 1, \ldots, M)$ denote the number of occurences of the i-th element of S in the subsets belonging to C. The units of the neural network are arranged in M rows, such that row i corresponds to the i-th element of S and contains k_i units, each corresponding to an occurence of that element in the subsets belonging to C. Thus, the unit (i, j) corresponds to the j-th occurence of the i-th element of S in the subsets belonging to C. If in the final state of the network a unit (i, j) is in the 'on' state, the subset S_l corresponding to the j-th occurence of the i-th element of S must be incorporated in the solution set.

The problem constraints require that every element of S should be covered and the subsets belonging to the solution set should be disjoint. This implies that in the final equilibrium state there must be exactly one 'on' unit in each row. We shall construct the network so that every equilibrium state fulfils the above requirement. It is obvious, however, that this property of equilibrium states solves the problem of disjointness only partially, as it is still possible to have 'on' units in different rows corresponding to non-disjoint subsets. This problem can be solved if we enforce the network to reach equilibrium points where either all units corresponding to the same subset are in the 'on' state or all are in the 'off' state. We shall call these equilibrium points *feasible* since they correspond to feasible (valid) solutions of the original problem. It must be noted that the problem of finding even one valid solution (not nessecarily the optimal) of the SPP is NP-hard.

Let us consider connections and threshold values. We assign the positive value $\delta_1/|S_r|$ to the strength of the connection imposed between any pair of units (i,j) and (k,l) (with $i \neq k$) such that the j-th occurence of the i-th element of S and the l-th occurence of the k-th element of S correspond to the same subset S_r in C. The threshold value of unit (i,j) is determined as follows:

$$\theta_{ij} = \alpha - \frac{\delta_1(|S_r| - 1 + \delta_2)}{2|S_r|} \tag{13}$$

where S_r is the member of C corresponding to unit (i,j) and $|S_r|$ denotes the cardinality of the set S_r. The value of the parameter $\alpha > 0$ can be arbitrarily specified, while the values of δ_1 and δ_2 will be determined from the requirements of the problem as will be described next.

Since in an equilibrium state each row contains exactly one 'on' unit, the network energy has the form:

$$E = -M\alpha - \sum_{r=1}^{N} \delta_1 \frac{v_r(v_r - |S_r| - \delta_2)}{2|S_r|} \tag{14}$$

where v_r $(r = 1, \ldots, N)$ denotes the number of 'on' units that correspond to subset S_r. The value of δ_2 can be determined from the requirement that every feasible equilibrium state should correspond to lower energy than any non-feasible equilibrium state. Let us define the function $f_s(v) = v(s - v + \delta_2)/s$, with $f_s(0) = 0$ and $f_s(s) = \delta_2$, where s is a positive integer and v is an integer such that $0 \le v \le s$. Using the above function, equation (14) takes the form:

$$E = -M\alpha + \frac{\delta_1}{2} \sum_{r=1}^{N} f_{|S_r|}(v_r) \tag{15}$$

The energy corresponding to a feasible equilibrium state (i.e., either $v_r = |S_r|$ or $v_r = 0$ for all r) is given by $E = -M\alpha + |F|\delta_1\delta_2/2$, where $|F|$ is the size of the solution set F. The maximum value that can be assigned to $|F|$ is N, which means that the maximum energy of a feasible equilibrium state is $-M\alpha + N\delta_1\delta_2/2$. On the other hand, it can be readily verified that, if $0 < \delta_2 < 1$, all integer values of v in the interval $(0, s)$ (with $s > 1$) yield values of f_s greater than δ_2. Also the minimum of $f_s(v)$ in $0 < v < s$

$(s > 1)$ occurs for $v = 1$ and has the value $1 - (1 - \delta_2)/s$. Hence, a lower bound of $\sum_{r=1}^{N} f_{|S_r|}(v_r)$ for $0 < v_r < |S_r|$ is the quantity $1 - (1 - \delta_2)/s_{\min}$, where $s_{\min}$ denotes the minimum value of $|S_r|$ $(r = 1, \ldots, N)$. Therefore, the minimum energy of a non-feasible equilibrium state cannot be less than $-M\alpha + [1 - (1 - \delta_2)/s_{\min}]\delta_1/2$. Consequently, the requirement that every feasible equilibrium state should correspond to lower energy than any non-feasible equilibrium state implies that $1 - (1 - \delta_2)/s_{\min} > N\delta_2$, which yields

$$\delta_2 < \frac{s_{\min} - 1}{N s_{\min} - 1} \tag{16}$$

The value of the parameter δ_1 can be determined from the requirement that the threshold value θ_{ij} should be greater than zero for all units (i, j). The following inequality is readily obtained:

$$\delta_1 < 2\alpha \frac{s_{\max}}{s_{\max} - 1 + \delta_2} \tag{17}$$

where $s_{\max}$ denotes the maximum value of $|S_r|$ $(r = 1, \ldots, N)$.

After having determined positive connection weights and threshold values, it remains to specify the negative weight between each pair of units belonging to the same row, which must ensure that it is not possible to have more than one 'on' unit in each row in an equilibrium state. We define the *positive potential* of a unit in a given state as the sum of the threshold of that unit (which is assumed positive) and the positive excitation it accepts from other units. If unit (i, j) corresponds to subset S_r, its maximum positive potential over all states is $b_{ij} = \theta_{ij} + \delta_1(|S_r| - 1)/|S_r|$. We shall consider a common value b for the strength of all negative connections of the network, which is taken greater than the maximum b_{ij} over all units (i, j).

It is apparent that the employment of multiscaling is very attractive for this problem, since there is a state independent grouping of the units of the network. According to this grouping, all units corresponding to the same subset in C should participate in the same group and one unit of the coarse-scale network is allocated to each of these groups. If creation of the small network takes place at a state of the original network, where all units belonging to the same group are in the same state, then the created network has the additional important feature that the feasible equilibrium points of the fine-scale network correspond to equilibrium points of the coarse-scale network.

The equilibrium points of the coarse-scale network that correspond to feasible equilibrium points in the fine-scale network will be referred to as *feasible coarse equilibrium points*. All other equilibrium points of the coarse-scale network will be referred to as non-feasible coarse equilibrium points. It is possible to extend the above result by showing that feasible coarse equilibrium points correspond to lower energy than any non-feasible coarse equilibrium point.

It follows that the feasible solutions of a given instance of the Set Partitioning Problem correspond to the low energy states of the coarse-scale network.

Consequently, the construction of the p-net allows us to approach (or even attain) the optimal solution despite the fact that the operation takes place at a coarse level.

4.2 Set Covering Problem

The Set Covering Problem (SCP) has the following definition [3]: Given a finite set S containing M elements and a collection C of subsets S_i ($i = 1,\ldots,N$) of S, find the minimum subset F of C such that the subsets belonging to F cover the set S, i.e., their union is equal to S.

The neural network architecture constructed to solve the problem features analogy to the one described previously. We denote by k_i ($i = 1,\ldots,M$) the number of occurences of the i-th element of S in the subsets belonging to C and consider a network with its units arranged in M rows. Each row $i = 1,\ldots,M$ is composed of two collections of units. The first collection contains k_i units, with unit (i,j) corresponding to the j-th occurence of the i-th element of S in the subsets belonging to C. If in the final state of the network a unit (i,j) of the above type is in the 'on' state, the subset S_l corresponding to the j-th occurence of the i-th element of S must be incorporated in the solution set F. As every element of S must be covered, there must be *at least one* unit in the 'on' state in each row in the final state of the network, i.e., there must be at least one subset covering each element. Satisfaction of this requirement can be achieved by letting each row i ($i = 1,\ldots,M$) contain a collection of $k_i - 1$ additional units and by imposing the constraint that in every equilibrium state there must be exactly k_i 'on' units in row i. A simple way to do this is to set a negative connection of strength α between each pair of units belonging to the same row and specify the threshold values and the strength of the positive connections of the network so that the maximum positive potential of each unit (i,j) will be less than αk_i and the minimum positive potential will be greater than $\alpha(k_i - 1)$. The additional units shall be called *slack units* [15] and will be distinguished from the originally considered first collection of units, which correspond to subsets and shall be called *actual units*. The above construction scheme ensures that all equilibrium states of the network are feasible since they satisfy the covering constraint.

A positive connection of strength $\delta_1/|S_r|$ is imposed between any pair of actual units (i,j) and (k,l) (with $i \neq k$), such that the j-th occurence of the i-th element of S and the l-th occurence of the k-th element of S correspond to the same subset S_r in C. Equilibrium points having the property that either all units corresponding to the same subset in the 'on' state or all of them in the 'off' state shall be called *complete*. The weights and thresholds of the network are specified so that complete equilibrium states correspond to lower energy than any other equilibrium state.

The threshold value of unit (i,j) is specified as follows. If unit (i,j) is an actual unit corresponding to subset S_r then

$$\theta_{ij} = \alpha(k_i - \delta_3) - \frac{\delta_1(|S_r| - 1 + \delta_2)}{2|S_r|} \tag{18}$$

where $\delta_1 > 0$ and $0 < \delta_2, \delta_3 < 1$. If unit (i,j) is a slack one, then $\theta_{ij} = \alpha(k_i - \delta_3)$. The specification of the parameters δ_1 and δ_2 follows a procedure analogous to the case of set partitioning. The desired network characteristics are obtained if the following inequalities are satisfied:

$$\delta_2 < \frac{s_{\min} - 1}{N s_{\min} - 1} \tag{19}$$

$$\delta_1 < 2\alpha\delta_3 \frac{s_{\max}}{s_{\max} - 1 - \delta_2} \tag{20}$$

$$\delta_1 < 2\alpha(1 - \delta_3) \frac{s_{\max}}{s_{\max} - 1 + \delta_2} \tag{21}$$

Therefore, first the values of $\alpha > 0$ and $0 < \delta_3 < 1$ must be arbitrarily specified, then the value of δ_2 must be determined so as to satisfy (19) and, finally, the value of δ_1 must be determined according to the inequalities (20) and (21).

In analogy with the SPP, it can be proved that the states of the resulting coarse-scale network that correspond to complete equilibrium points of the fine-scale network are also equilibrium points. In addition, it can be shown that the complete coarse equilibrium points are of lower energy than any non-complete coarse equilibrium point. As a result, it is possible to find complete equilibrium states of the original network (and therefore near-optimal problem solutions) just by operating at the coarse-scale level.

4.3 Experiments

In order to evaluate the efficiency of our approach a series of experiments has been conducted regarding both the Set Partitioning and the Set Covering problems.

To generate the instances of the SPP first the number M of elements of the set S and the number N of subsets in the collection set C were specified. Then, in order to ensure the existence of at least one valid solution, we first constructed a number of subsets of S by allocating each of the elements of S to one and only one of these subsets so as to create a disjoint solution. The remaining subsets were randomly constructed by deciding with probability q whether an element of S would belong to a particular subset. By varying the value of q the density of the resulting problem instance could be adjusted. The value of q was chosen in the range from 0.05 (sparse instances) to 0.4 (dense instances). For the generated problem instances, first the optimal solution was found using exhaustive search. Then, the fine-scale architecture was used to solve the problem and finally the multiscale approach was applied.

As far as the fine-scale architecture is concerned, we tested its effectiveness on small and medium (N smaller than 100 and M smaller than 50) problem instances since for large problem instances the size of the resulting neural network was very high (over 1000 units). Both sparse and dense instances were considered. In what concerns the annealing schedule, we have adopted the logarithmic scheme $T_n = T_{n-1}/(1+\log f(n))$, where $f(n) = f(n-1)(1+r)$ (with $f(0) = 1$) and r is the parameter that adjusts the speed of the schedule [16]. All experiments with the fine-scale architecture were conducted starting from temperature $T_0 = 10.0$ and using a temperature reduction rate $r = 10^{-7}$. Also the value of the parameter α was fixed to 10.0 for all experiments. At each temperature, a number of trials equal to $2n$ (where n is the number of network units) was performed and the annealing stopped if for 10 consecutive temperature values none of the units changed its state. In all tested instances, first the fine-scale network was constructed and then the creation of the p-net took place at the zero state of the n-net, i.e., $y_i = 0$ for all i. The operation of the p-net started from temperature $T_0 = 20.0$ and followed a temperature reduction rate $r = 10^{-7}$. After the operation of the p-net was terminated, the corresponding state of the n-net was found and the fine-scale network operated as a pure Hopfield network (with zero temperature) to provide the final solution to the problem.

As a metric of the quality of a feasible solution obtained for a given instance I of the problem we use the *approximation ratio* which is defined as $h(I)/opt(I)$, where $h(I)$ denotes the cardinality of the obtained feasible solution and $opt(I)$ denotes the cardinality of the optimal solution [13]. For small and medium size instances the percentage of feasible solutions obtained using the coarse scale network was 100% in almost all cases and over 75% even in cases where the fine scale network practically failed. For big size instances (N from 200 to 500 and M greater than 50) the percentage of feasible solutions obtained using the coarse scale network was over 80% in most cases. The average approximation ratio was between 1.1 and 1.4 in all cases.

Similar experiments were performed for the Set Covering problem. Due to the existence of slack neurons, the size of the resulting networks was bigger in this case, thus the experiments were performed for problem instances with N between 30 and 100 and M between 10 and 50. Moreover, both sparse and dense problem instances were examined. The value of α was fixed to 10 and the value of δ_3 was fixed to 0.1 for all the experiments. The construction of a problem instance was similar to the SPP case, except that we were not concerned to create a disjoint solution. Care was taken so that every element participated in at least one subset. Experiments with the fine-scale network have shown that the quality of the obtained solutions was poor, even for small problem instances. On the contrary, the multiscale method provided results of good quality. The multiscale scheme used in the experiments had the same characteristics as in the SPP case, except for the fact that the initial temperature of the p-net was set equal to 0.5. The annealing rate of the p-net was $r = 10^{-7}$. The obtained results are quite satisfactory, since the average case approximation ratio for instances of various sizes was less than 1.3.

5 Solving Constraint Satisfaction Problems

A constraint satisfaction problem (CSP) consists of finding an assignment to a collection of variables that satisfies a set of constraints among them [9]. In the following we shall assume that constraints are binary, i.e., each constraint specifies the values that cannot be taken on simultaneously by a given pair of variables. We consider a general binary CSP involving a set of N variables $X_1, \ldots, X_N$ and a set of binary constraints C. Each variable X_k ($k = 1, \ldots, N$) assumes values from a finite domain D_k. Let d_{kl} denote the l-th value in the domain D_k. The set of binary constraints C contains pairs of d_{kl} values that are incompatible with each other.

The representation of a CSP in terms of a binary Hopfield network can be done very easily [2]. The network units are arranged in a matrix structure with a matrix row corresponding to each of the variables. The number of units in each row k is equal to the number of values that can be assigned to variable X_k, i.e., it is equal to the size $|D_k|$ of the domain set D_k. Under this representation, a unit (k, l) corresponds to the value d_{kl} and the assignment of d_{kl} to X_k is represented by $y_{kl} = 1$, where y_{kl} is the state of unit (k, l). The connection weights $w_{kl,mn}$ and the threshold values θ_{kl} are chosen so that the network states corresponding to solutions of a given CSP are equilibrium states of minimum energy (global equilibria):

$$\theta_{kl} = \beta \tag{22}$$

$$w_{kl,mn} = \begin{cases} -\alpha & \text{if } (d_{kl}, d_{mn}) \in C \\ -\eta & \text{if } k = m \text{ and } l \neq n \\ 0 & \text{otherwise} \end{cases} \tag{23}$$

where β, α and η are positive constants and $\alpha, \eta > \beta$. The above specification of weight and threshold values ensures that a state corresponding to a problem solution is an equilibium point of minimum energy $-N\beta$. If the pure binary Hopfield network is used to solve this problem, it will be found that, in most of the runs, the network settles into equilibrium states corresponding to local minima of the energy function. Such local minimum states are not useful since there exist rows with no unit in the 'on' state.

5.1 Double Updates

We next describe an operation mode of the Hopfield network based on the group update technique, which leads to an effective search of the problem state space. We start from an initial network state having the property that there is exactly one unit in the 'on' state in each row. We shall refer to states having the above property as *consistent* states. In order to perform state transitions between consistent states only, we consider that at every time instant a *double update* takes place, i.e., a group update concerning a group of two units.

Consider that the network is in a consistent state $\vec{y}$ and let $(i, i^\star)$ denote the index of the unit that is in the 'on' state in each row i $(i = 1, \ldots, N)$. We randomly select a row k and a unit (k, l) in that row that is different from $(k, k^\star)$, i.e., $y_{kl} = 0$. The selected unit along with unit $(k, k^\star)$ constitute the group that will be tested for update. If this trial is successful, the network will move to a consistent state and we will set $k^\star = l$. The energy difference caused by the group update can be derived by direct application of equation (4):

$$\Delta E_G(\vec{y}) = \delta E_{kk^\star}(\vec{y}) + \delta E_{kl}(\vec{y}) + w_{kk^\star,kl}\ , \qquad G = \{(k, k^\star), (k, l)\} \qquad (24)$$

We have $\delta E_{kk^\star}(\vec{y}) = \beta + \sum_{m \neq k} w_{mm^\star,kk^\star}$ and $\delta E_{kl}(\vec{y}) = -\sum_{m \neq k} w_{mm^\star,kl} + \eta - \beta$. Since $w_{kk^\star,kl} = -\eta$, the energy difference takes the form:

$$\Delta E_G(\vec{y}) = \sum_{m \neq k} w_{mm^\star,kk^\star} - \sum_{m \neq k} w_{mm^\star,kl}\ , \qquad G = \{(k, k^\star), (k, l)\} \qquad (25)$$

We define an *equilibrium point under the double update scheme* as a network state such that for each unit (k, l) with $l \neq k^\star$ it is valid that $\Delta E_G(\vec{y}) \geq 0$, where $G = \{(k, k^\star), (k, l)\}$.

It can be shown that the states of the network that correspond to valid problem solutions are global minimum equilibrium points under the double update scheme.

The advantage of always moving between consistent states, in addition to favouring an efficient search, has an immediate impact on the computational complexity of the approach. In order to perform a single update in the single update scheme a number of multiplications must take place which is equal to the number of units in the network (in our case equal to $\sum_{k=1}^{N} |D_k|$). On the other hand, the double update scheme ensures that there is exactly one 'on' unit in each row. Therefore, the knowledge of the $k^\star$ index for each row k, reduces the number of computations required to perform a double update to $2N - 1$.

5.2 Experimental Results

To examine the effectiveness of the proposed double update technique in solving large scale problems, we have conducted experiments concerning the two most popular CSPs, namely the N-Queens problem and the 3-Graph Coloring problem.

The N-queens problem [14] constitutes a well-known benchmark problem that has been used for examining the efficiency of various constraint satisfaction methods. It deals with the placement of N queens on a $N \times N$

chessboard so that no queen threatens another. We can represent this problem as a binary CSP with N variables representing the chessboard rows and each variable assuming one of N values representing the position in the row in which the corresponding queen may be placed. A series of experiments have been conducted for several values of N, starting with small problem instances ($N = 8$) and progressively increasing the dimension of the state space up to the value $N = 512$ which corresponds to a very large problem instance. For each value of N with $N \leq 128$, we performed a series of 50 runs using different seed values and initial network states. For larger problem instances 20 runs were performed for each value of N. In what concerns the characteristics of the Boltzmann machine, the logarithmic annealing schedule described previously has been adopted with initial temperature $T_0 = 0.1$ and using a rate $r = 10^{-4}$. Moreover, the value of the parameter α was fixed to 2.0 for all the experiments. (The values of β and η need not be specified in our apprach.) At each temperature, a number of trials equal to $2n$ (where n is the number of network units) was performed and the annealing stopped when a state with energy $-N\beta$ was found. All experiments were performed on a SPARCstation ELC with 12Mb of main memory. The obtained results concerning the time to find an acceptable solution, which ranged from a few seconds for medium size problems (N up to 128) to some hundreds of seconds for large size problems (N greater than 512), have shown that the proposed method is considerably fast.

The 3-graph coloring problem constitutes a difficult constraint satisfaction problem that belongs to the class of NP-complete problems [3]. Given a graph $G = (V, E)$ the problem is to assign one of three colors to each graph node, so that there are no neighbouring nodes having the same color. Each graph node k can be considered as a variable assuming one of three values. Thus, a network of $3N$ units is required where N is the size of the graph. Each unit (k, l) ($k = 1, \dots, N$, $l = 1, \dots, 3$) is connected through an inhibitory connection with units (m, n) where either $k = m$ and $l \neq n$ or $(k, m) \in E$ and $l = n$. In the experiments, random instances were generated so as to ensure the existence of at least one valid solution. To achieve this, we have used the graph construction technique presented in [2]: First we partition the graph nodes into three groups each containing $N/3$ nodes (approximately) and we color the nodes of each group with the same color. Second, we add K edges between nodes of different colors so that there is at least one edge incident to each node. Two values of K were considered: $K = N(N-1)/4$ (easy instances) and $K = 2N$ (hard instances). A series of experiments have been conducted regarding instances of several sizes in the range between $N = 30$ and $N = 300$, for both choices of K. For easy instances solutions have been obtained very fast, so results are presented only for the case $K = 2N$. For each value of N, 50 experiments were performed with each experiment corresponding to a different problem instance. The annealing shedule and the parameter values were the same as in the N-Queens approach, except for the initial temperature which was set equal to $T_0 = 0.3$. The average time required to reach an acceptable solution was less than 1 minute for graphs of size up to 300 nodes. Thus, the double update approach has proved very effective and can provide solutions to large instances in reasonable time.

6 Conclusions

We have presented the notion of group update in the context of binary Hopfield-type neural networks, and we have shown that the multiscale approach is an easy and fast way for performing group updates. A single update in the coarse-scale network is equivalent (in terms of the resulting energy difference) to the corresponding group update in the original fine-scale network. This fact makes the multiscale approach very attractive, especially in problems where state independent grouping schemes can be devised. We have exploited the idea of multiscaling in solving optimization problems, such as the Set Partitioning and Set Covering problems. An original mapping of these problems to a binary Hopfield neural network architecture has been constructed, which is characterized by an order preserving energy function. Then, by appropriately grouping the units of this network, a coarse-scale network was created having the property that the low energy equilibrium points correspond to feasible equilibrium points of the original network. Thus, feasible solutions can be obtained by operating exclusively at the coarse level, allowing for both faster and more effective search of the problem state space. In such a case, multiscaling can be considered as a method for developing a coarse representation of the problem without losing any significant information.

We have then presented a very effective method for solving general binary constraint satisfaction problems that is based on the discrete Hopfield neural network operating under the double update mode. The generality of the method stems from the fact that it assumes no knowledge about the problem domain other than value and value-pair inhibitions. Therefore, it can be easily employed for any problem that assumes finite domains and binary constraints between pairs of domain values. The double update approach can also be employed in the case of more general optimization problems that involve, in addition to constraint satisfaction, optimization of a cost function.

References

1. Aarts, E., Korst, J.: Simulated Annealing and Boltzmann Machines, A Stochastic Approach to Combinatorial Optimization and Neural Computing. John Wiley & Sons Ltd. 1989
2. Adorf, H., Johnston, M.: A Discrete Stochastic Neural Network Algorithm for Constraint Satisfaction Problems. Proceedings Int. Joint Conf. on Neural Networks, San Diego 1990. 3, 917–924. 1990

3. Garey, M., Johnson, D.: Computers and Intractability: a Guide to the Theory of NP-completeness. San Francisco: W.H. Freeman and Company 1979
4. Herault, L., Niez, J.: Neural Networks and Combinatorial Optimization: a Study of NP-complete Graph Problems. In: E. Gelenbe (ed.) Neural Networks: Advances and Applications. North-Holland 1991
5. Hopfield, J.: Neural Networks and Physical Systems with Emergent Collective Computational Abilities. Proceedings of the Nat. Academy of Sciences USA. 79, 2554–2558 (1982)
6. Hopfield, J., Tank, D.W.: Neural Computation of Decisions in Optimization Problems. Biological Cybernetics 52, 141–152 (1985)
7. Kamp, Y., Hasler, M.: Recursive Neural Networks for Associative Memory. John Wiley & Sons 1990
8. Kirkpatrick, S., Gellat Jr., C.D., Vecchi, M.P.: Optimization by Simulated Annealing. Science 220, 671–689 (1983)
9. Kumar, V.: Algorithms for Constraint Satisfaction Problems: A Survey. The AI Magazine 13, 32–44 (1992)
10. Likas, A., Stafylopatis, A.: Multiscaling in binary Hopfield-type neural networks: application to the Set Partitioning Problem. Int. Journal of Neural Networks 3, (1992)
11. Likas, A., Stafylopatis, A.: Group Updates and Multiscaling: An Efficient Neural Network Approach to Combinatorial Optimization. IEEE Trans. on Systems, Man, and Cybernetics. To appear
12. Mjolsness, E., Garret, C., Miranker, W.L.: Multiscale Optimization in Neural Nets. IEEE Trans. on Neural Networks 2, (1991)
13. Papadimitriou, C.H., Steiglitz, K.: Combinatorial Optimization: Algorithms and Complexity. Prentice-Hall 1981
14. Stone, H.A., Stone, J.M.: Efficient Search Techniques - An Empirical Study of the N-Queens Problem. IBM Journal on Research and Development 31, 464–474 (1987)
15. Tagliarini, G.A., Page, E.W.: Learning in Systematically Designed Networks. Proceedings IJCNN, Washington D.C. 1989. 1, 497–502. 1989
16. Zissimopoulos, V., Paschos, V., Pekergin, F.: On the Approximation of NP-complete Problems by Using Boltzmann Machine Method: The Cases of Some Covering and Packing Problems. IEEE Trans. on Computers. 40, 1413–1419 (1991)

Neural Network Based Extrapolation Strategies in Structural Analysis and Design

Z. Peter Szewczyk

Research Scientist, The Center for Computational Structures Technology, NASA LaRC, Hampton, VA 23681

Abstract. This paper presents three hybrid strategies for enhancing prediction capabilities of feedforward neural networks and two numerical studies employing these strategies for predicting partial information about nonlinear systems behavior. The strategies are: (1) limited data excess strategy in which scarce additional training data are added to the existing training set, (2) weights regression strategy in which network parameters are modified based on linear regression, and (3) data expansion strategy in which additional training data are generated based on Taylor series expansion with predicted coefficients. In the first numerical study predictions of the maximum deflection and the strain energy of a composite panel under uniform pressure are considered. The problem of collecting training data for analysis of such nonlinear systems via a perturbation technique is also addressed in this study. The second study treats the design space of an energy absorbing composite structure. In addition to expanding the knowledge about the effect of geometric variables on energy absorbing capacity, predictions of the value of a Tsai-Wu number are also attempted.

Keywords. Function extrapolation, BPN network, extrapolation strategies, structural analysis, design, Padé approximants, energy absorbing structures, Tsai-Wu number, curved composite panel.

1 Introduction

Function extrapolation arises naturally in many areas of scientific and engineering applications, and examples can be found in forecasting, commodity markets, sociological analyses, safety and reliability, design, and other disciplines. In extrapolation one attempts to predict values of a given relation outside of a region for which knowledge of its behavior exists. Such a modeling problem cannot be solved in general, and any formulation of an extrapolation problem to be of a practical value requires some qualifiers. Despite inherent limitations and the necessity of tailoring prediction techniques to a given class of problems, the advantages of extrapolation encourage research in this field.

Important applications for extrapolation exist in analysis of response of nonlinear structural systems and their design spaces since gathering information about behavior of such systems, via numerical simulations with analytical models or experimenting on empirical models, tends to be very expensive. The use of a reliable extrapolation tool may alleviate problems associated with collecting information, and consequently, save computational resources and reduce the time required to complete a design process.

Nonparametric estimation is a model-free approach to data representation in which no assumptions are made about the data structure. Allowing a model and parameters to be determined alleviates a "bias-variance" dilemma [1] and results in achieving the best performance for any nontrivial estimation problem provided appropriate limits. A growing interest in the approach in recent years has spawned a number of methods such as Parzen windows, recursive partitioning, multivariate adaptive regression splines, and also feedforward neural networks, commonly known as backpropagation neural (BPN) networks. The latter method has drawn much attention for its straightforward implementation and its suitability for parallel and analog computational methods.

There are several neural networks that are suitable for data fitting. A list includes feedforward neural networks, radial basis function networks, vector quantization networks and some other hybrid architectures. All exhibit very good interpolation properties, and several researchers (for example [2]) have shown that the interpolation capabilities of BPN networks exceed that of regression techniques, and therefore BPN networks have been used extensively on function interpolation problems in structural analysis (for example [3]). Only feedforward networks, however, appear to possess extrapolation capabilities that result from the product of nonlinearities (for instance, sigmoidal functions) imbedded in the network. Despite a critical paper [4] showing limitations of the multilayer feedforward network in extrapolating arbitrary functions, this particular architecture is still the most promising. Reference [5] reports predictions of flight data in the range of 90-95% accuracy obtained using such networks. Nonlinear behavior of composites can be predicted satisfactorily in a short range of variations of a load parameter as shown in Reference [6]. The most recent studies ([7] and [8]), however, indicate a need for hybrid methods of extrapolation to extend the range and accuracy of predictions. In such methods a neural network is only a component of a computational scheme and is complemented by regression, series expansions, analytical and heuristic modeling, and other techniques.

In this work three hybrid strategies for enhancing the extrapolation capabilities of a BPN network are presented, namely (1) limited data excess, (2) weights regression, and (3) data expansion strategies. All these strategies exploit the premise that additional training data improve network predictions and provide a correction to a weight matrix to be used for extrapolation. Adding a correction to the original weight matrix preserves generalization capabilities of the network in the interpolation region. This approach, however, requires associating the correction with the spatial orientation of the extrapolation region, and such a mapping is obtained herein via a feature-sensitive neural network [9].

The limited data access strategy assumes some access to data from outside the interpolation region and adds a correction to a weight matrix to improve predictions for a prescribed design variable. This strategy uses a weights analysis [10] to estimate extrapolation error for an arbitrary variable. The underlying assumption of some access to data directly addresses a class of problems for which collecting additional training data for a given variable may be computationally prohibitive. In structural design, a typical example of such a problem can be found in the shape sizing of a structural component. Any design variable which requires a new finite element model would be more expensive to explore than a variable related to member cross-sectional dimensions only. In the weights regression strategy additional information is extracted from existing data to generate training samples outside an interpolation region. The strategy utilizes linear regression to modify network parameters, interconnection weights and biases, within a fixed network topology. Modifications are determined by applying regression to a sequence of weight matrices produced by training performed on subsets of training data. In the data expansion strategy, a given function is expanded into a Taylor series with derivatives predicted by a neural network. New training samples are added to the existing training set and used to retrain a network. A detailed description of these strategies follows in the next section.

2 Hybrid Strategies for Extrapolation

2.1 Objective of Extrapolation Strategies

To state the objective of extrapolation strategies, assume the existence of a neural network model $\mathbf{W}$ of a given function, $f : \mathbf{x} \in \mathbf{R}^n \mapsto \mathbf{y} \in \mathbf{R}^m$, for the function current domain, or the interpolation region, $\mathbf{D} \subset \mathbf{R}^n$. The weights matrix $\mathbf{W}$ containing interconnection weights and biases is obtained, for instance, by a backpropagation training algorithm (for alternatives see Reference [11]) operating on a discrete training set $C \subset \mathbf{D}$. The size of the interpolation region $\mathbf{D}$ can be defined as $\delta = \max\|\mathbf{x}\|$. Define also an extrapolation region $\hat{\mathbf{D}}$ of the size $\hat{\delta}$ as adjacent to $\mathbf{D}$ for which approximated values of the function are to be determined. In the extrapolation region error of predictions, e_e, may be allowed to grow, but its upper bound, B, should be a function of the interpolation error e_i.

The objective of the extrapolation problem is to find a necessary correction to the weights matrix, $\Delta\mathbf{W}$, so that for a given bound on error the size of the extrapolation region increases. Thus,

$$\text{For } e_e \leq B \text{ determine } \Delta\mathbf{W} \text{ such that } \hat{\delta}(\mathbf{W}) < \hat{\delta}(\mathbf{W} + \Delta\mathbf{W})$$

The problem is solvable provided that access to the training data from $\hat{\mathbf{D}}$ is unrestricted. This conclusion is based on the representation theorem for a network with two hidden layers, or even with one under mild conditions [12]. Effective solution of the problem requires network retraining on additional training data drawn from $\hat{\mathbf{D}}$ which is in essence an interpolation problem for an extended function domain. However, any restrictions on the access to data from $\hat{\mathbf{D}}$ neccessitates limits on its size to control e_e. Since the accuracy of estimation is sensitive to the weight matrix, the problem formulation in which corrections to the existing weight matrix are sought avoids degradation of approximations in the interpolation region.

Defining interpolation on a hypercube or hypersphere, depending upon a selected norm, implies that a training set contains sufficient amount of information about the function behavior. For multi-dimensional mappings gathering that much information may not be feasible resulting in undersampling certain regions. It is therefore more practical to define an interpolation problem on a cluster of data and extrapolation on its complement. In numerical studies presented in this paper the problem of collecting a sufficient amount of data is addressed to avoid the existence of undersampled regions.

2.2 Strategy with Limited Access Data

This section contains a strategy which improves the quality of extrapolation by modifying parameters of a feedforward neural network within a fixed network topology. Weights matrix corrections are obtained via additional training of the network performed on limited data available outside the current function domain. Corrections are associated with a spatial orientation of samples, and the feature-sensitive network is used to assign an appropriate correction. In addition, a weights analysis is used in this strategy to impose limits on the size of extrapolation regions in directions for which performing numerical simulations or experiments would be impractical.

The assumption of limited access to the training data from $\hat{\mathbf{D}}$ puts obvious restrictions on an extrapolation range, but the size $\hat{\delta}$ depends on the actual effect an independent variable has on an output quantity. To determine directions that put stricter limits on the size of the extrapolation region or require additional training data to bound the error, an analysis of the interconnection weights can be performed. A transition matrix $\mathbf{T}_{m \times n} = [t_{li}]$

for a network with one hidden layer is defined as follows [10]:

$$\tilde{t}_{li} = \sum_{j=1}^{h_1} \frac{| w_{ji}^{(h)} |}{\sum_{r=1}^{n} | w_{jr}^{(h)} |} \times | w_{lj}^{(o)} |, l = 1, \ldots, m, i = 1, \ldots, n \tag{1}$$

and after normalization

$$\mathbf{T}_{n \times m} = [t_{li}] = \tilde{t}_{li} / \sum_{s=1}^{m} \tilde{t}_{ls} \tag{2}$$

Superscripts are introduced to distinguish between the weights of hidden and output layers, and it is assumed that there are h_1 neurons in the hidden layer. The ratio which appears in Eq.1 expresses the contribution of the ith input to a total input directed to the jth neuron in the hidden layer. After computing the transition matrix $\mathbf{T}$, inspection of rows allows identification of dominant components of the input vector $\mathbf{x}$. There are two cases to consider.

In the first case, the component x_s of an input vector exists which does not affect an output quantity y_j. Thus, the correction $\Delta\mathbf{W} = \mathbf{0}$, and the model $\mathbf{W}$ is sufficient for extrapolation of y_j. In other words, the error remains bounded regardless of the extent of extrapolation in the s direction. In the second case, extrapolation is required in the direction d that affects an output quantity y_j, and the correction matrix $\Delta\mathbf{W}$ is to be found. It is assumed that additional training data $\mathbf{p}_n \in \hat{\mathbf{D}}$, $n = 1, \ldots, N$, have been collected in the d direction and that $\mathrm{Card}(C) \gg N$. Performing additional training on $\mathbf{p}_n$ yields a matrix correction $\Delta\mathbf{W}$. To estimate the size of $\hat{\mathbf{D}}$, it is assumed that the size in one of the d directions can be determined through simulation. An initial claim is made that the bound in all other directions can be computed as proportional to corresponding elements of the transition matrix. Thus for all j

$$h_i = t_{jd}/t_{ji} \times h_d, i = 1, \ldots, n, i \neq d \tag{3}$$

It is expected, based on extensive experience, that the CPU time required for additional training will be insignificant compared to the full training. The orientation of data drawn from $\hat{\mathbf{D}}$ can be determined using a clustering routine to assure that additional training data and testing data coincide in the space. By introducing appropriate corrections to the weight matrix, the original matrix can still be used for interpolation.

2.3 Weights Regression Strategy

In the weights regression strategy an attempt is made to extract additional information from existing training data by using linear regression to process changes in a weights matrix caused by introducing new training data from the domain.

Assume that the training set C contains data representative of the function and that improved predictions are required for some input quantity x_d. Training data from C can be ordered in a succession such that $C_1^{(d)} \subset C_2^{(d)} \subset \ldots \subset C_\nu^{(d)} \subseteq C$. Each subset $C_\nu^{(d)}$ gives rise to a weight matrix after completion of training to some prescribed training error τ_t. Applying linear regression to these matrices

$$\mathcal{L}(\mathbf{W}_1^{(d)}, \ldots, \mathbf{W}_\nu^{(d)}) \Rightarrow \mathbf{W}_{\nu+1}^{(d)} \tag{4}$$

produces a new matrix which then can be used to generate data in the adjacent region $\hat{\mathbf{D}}$. Obtaining a correction $\Delta\mathbf{W}^{(d)}$ requires additional training on these data and subtracting the original weight matrix from it. The correction is only valid in the direction d, and this procedure must be repeated for all input quantities of interest.

It is intuitively understood that this strategy yields the best results when, during each successive training, key patterns are added to the current training subset. This results in meaningful changes to a weight matrix, and regression should produce a modified matrix with better predicting capabilities. Also, network parameters can be initialized only at the beginning of the training process to assure continuity in weight matrices, and training should terminate when training error τ_t has been satisfied.

2.4 Data Expansion Strategy

In the data expansion strategy additional training data in $\hat{\mathbf{D}}$ are generated based on a Taylor series expansion of a given function. Since the accuracy of approximations yielded by the expansion depends on the number of available function derivatives and the distance from the current point, the range of validity of such approximations can be extended by utilizing predicted values of coefficients in the series. Coefficients (function derivatives) are predicted by neurocomputing models of a given function assuming an unrestricted access to data from the function current domain.

The approach begins with obtaining a neural network model $\mathbf{W}$ of a given data for any $\mathbf{x} \in \mathbf{D}$

$$\mathbf{y} \in \mathbf{R}^m \approx \mathbf{W}(\mathbf{x}) \tag{5}$$

Similarly, neurocomputing models of function derivatives can also be obtained provided the unrestricted access to data in $\mathbf{D}$, and it is clear that such models simply transform discrete sets of training data into continuous functions. Assuming that for a small stepsize into an extrapolation region network predictions of the function and its derivatives are of the comparable accuracy, it can be shown that the formula

$$\mathbf{y} = \mathbf{W}(\mathbf{x}) + \mathbf{W}'(\hat{\mathbf{x}})h + \frac{1}{2!}\mathbf{W}''(\hat{\mathbf{x}})h^2 + \ldots + \frac{1}{n!}\mathbf{W}^n(\hat{\mathbf{x}})h^n \tag{6}$$

where $\hat{\mathbf{x}} \in \hat{\mathbf{D}}$, produces better results than those obtained by using the neurocomputing model $\mathbf{W}$ directly for predicting function values. By adding data generated via Eq.6 to the existing training set, a network can be retrained to determine the necessary weights matrix correction.

The effectiveness of the strategy depends upon the accuracy of function derivatives, and better predictions can be obtained with just first few derivatives. Despite the fact that neurocomputing models are very efficient at function evaluations, they lack the necessary accuracy to produce improved results via Eq.6. Higher order finite difference formulas and network differentiation can only produce the first derivative with the satisfactory accuracy, and therefore in one of the numerical studies presented below a perturbation technique is used to approximate a function and its derivatives.

2.5 Illustrative Example

To demonstrate the effectiveness of the extrapolation stategies, an explicit vector function $\mathbf{y} = \mathbf{y}(\mathbf{x})$ defined by:

$$\begin{aligned} y_1 &= x_1^2 - \exp x_2 + \sin x_3 \\ y_2 &= \sin x_2 + \ln x_3 \end{aligned}$$

is considered. For this problem $x_i \in [0.3, 1.7]$, and, hence $y_1 \in [-4.64, 2.38]$, and $y_2 \in [-0.84, 1.53]$. Sample data of this vector function are randomly generated and normalized so that $x_i, y_j \mapsto [0.1, 0.9]$.

In the first simulation, the limited data access strategy is tested. For interpolation, samples from the interval $[0.2, 0.8]$ are used, and the extrapolation problem is defined on the complement of the normalized domain. To select the most suitable BPN network architecture, the transition matrix is computed, and it is observed that for the 3-4-2 architecture trained on 60 samples, elements of the matrix

$$\mathbf{T} = \begin{pmatrix} 0.2682 & 0.5920 & 0.1398 \\ 0.0628 & 0.5787 & 0.3585 \end{pmatrix}$$

reflected causal relations between $\mathbf{x}$ and $\mathbf{y}$. The network yields estimates with the mean relative error of 0.93% on 65 new testing data from $\mathbf{D}$.

The use of BPN networks for extrapolation requires a correction to the weight matrix. Such a correction can be determined assuming some access to the adjacent region $\hat{\mathbf{D}}$, and it is conjuctured that adding just one training pattern improves network extrapolations. To support this claim, simulations are performed in which testing patterns are drawn from $\hat{\mathbf{D}}$. All patterns are clustered to determine their spatial orientation, and the cluster examplar is used to further train the network. In all cases training converges rapidly within a few iterations. The results, the mean value of relative error expressed in %, are summarized in Table 1 for five distinct clusters of size n.

Table 1: The mean relative error (%) of predictions for clustered testing data.

cluster	1	2	3	4	5
n	25	12	15	6	15
e_W	4.22	4.49	2.28	2.42	1.31
$e_{\Delta W}$	3.45	1.65	1.91	1.74	1.11

Table 2: The mean relative error (%) of predictions for different relations

relation	$f_1(x_1)$	$f_1(x_2)$	$f_1(x_3)$	$f_2(x_2)$	$f_2(x_3)$
n	129	102	130	102	130
e_W	10.50	25.53	4.1	13.44	2.64
$e_{\Delta W}$	4.20	9.54	1.73	6.35	1.57

The results obtained in this simulation show that a very small correction obtained by further training on just one additional pattern can significantly improve extrapolation of a BPN network (note very good predictions of the network for the stepsize up to 0.175 of unscaled values). However, network performance is highly sensitive to any disturbance introduced to the weight matrix. This may result in very poor predictions if no clustering is performed to assure that the correction is computed for a pattern that is close in the space to the testing data (this is essentially, an additional mapping which relates a correction matrix to data for extrapolation).

In the second simulation, the data expansion strategy is used to improve predictions of a network. The interpolation range is redefined as [0.1, 0.5], and the extrapolation range is a complement of this normalized interval. Two networks of 3-3-1 architecture are selected to model $y_1 = y_1(\mathbf{x})$, and $y_2 = y_2(\mathbf{x})$. Both networks are trained independantly on 145 training samples so that the training error (MSR) is reduced to 1%. The mean relative error of interpolation on 180 testing data is 2.2% and 1.8% for y_1 and y_2, respectively. Obtaining models of the first and second partial derivatives using FD formulas is also attempted, and it is observed that a 1.9% error for the interpolation range has grown to 7.2% and 15.1% for the first and second derivatives in the [0.5, 0.7] interval, respectively. Therefore, exact values of derivatives are extented by neural networks into the extrapolation range. Using a Taylor series expansion with predicted values of coefficients, nine additional training patterns are computed assuming that only one independent variable lies in $\hat{\mathbf{D}}$. The networks are further trained and used to predict function values in the extrapolation range. Table 2 summarizes results of this simulation; n denotes the number of testing data for each relation.

Results of this simulation illustrate that the data expansion strategy improves network predictions, and in this particular simulation the error is reduced by more than 50%. Its utility, however, depends largely on obtaining accurate models of the function derivatives, and in this example exact values are used in network training. It is noted that the derivatives are only extended to the middle of the extrapolation range, and the whole extrapolation range is covered only in predicting the function itself. This way the continuity of a function is utilized better. The weights analysis

$$\mathbf{T} = \begin{pmatrix} 0.3217 & 0.4712 & 0.2071 \\ 0.0202 & 0.6871 & 0.2427 \end{pmatrix}$$

indicates that in the function domain the dependencies change as the interpolation interval shifts and that the dependencies appear to be reflected in the growth of the error. The importance of spatial orientation has been demonstrated again by using a modified matrix for interpolation in the case $y_1(x_3)$; the error has grown to 3.3%.

3 Numerical Studies

3.1 Nonlinear Analysis of a Panel with a Cut-out

The availability of very powerful modeling techniques leads to complex analyses requiring vast computational resources. Despite rapid development of computational techniques and platforms (for example [13]), the resources and time required for such analyses may not always be available. In addition, the amount of information resulting from these analyses may be excessive and difficult to manage. In many practical situations only partial information about system behavior may suffice to make decisions on project development. Consequently, there is an ever-growing need for efficient approximation techniques for performing partial analyses [14].

3.1.1 Padé approximants

Power series representation of a function is an approximation technique to provide insights into a mathematical aspect of physical problems, and, for example, the Taylor series has been extensively used for approximations. The accuracy of such polynomial approximations depends upon the number of available derivatives, and in practice the validity of approximations, or radius of convergence, is limited. It is proved by Padé (Reference [15]) that polynomial approximations are asymptotically equivalent to rational fractions and shown, for example in Reference [16], that a significant increase in the radius of convergence may be achieved by a simple transformation. However, neither of these approximation techniques allow partial problem analysis. In this numerical study a backpropagation neural network is used to extend solutions to a nonlinear system obtained via Padé approximations.

Consider a Taylor series expansion of the function

$$f(x) = c_0 + c_1 x + c_2 x^2 + c_3 x^3 + \dots \tag{7}$$

with various order derivatives as the coefficients $c_0, c_1, \cdots$. Padé approximation to this function is simply the ratio of two polynomials

$$[L/M] \doteq \frac{a_0 + a_1 x + \dots + a_L x^L}{b_0 + b_1 x + \dots + b_M x^M} \tag{8}$$

whose coefficients are determined by the coefficients of the Taylor series. It is usually assumed that $b_0 = 1$, and therefore the ratio in (8), contains $L + M + 1$ unknowns to be determined from $c_0, \cdots, c_{L+M}$. The Taylor series has to be known to the nth term, where $n = L + M$. A detailed algorithm to compute Padé approximants can be found in Reference [15]. The radius of convergence may be determined by selecting a control parameter which limits the spread of values of diverging approximants. By denoting $P_{[L/M]}$ a Padé approximation for a given parameter and assuming that the allowed tolerance is τ_P, q^{max} satisfies the condition

$$\tau_P = \frac{P^{max}_{[L/M]}(q) - P^{min}_{[L/M]}(q)}{0.5(P^{max}_{[L/M]}(q) + P^{min}_{[L/M]}(q))} \tag{9}$$

Approximations for any $q \leq q^{max}$ can be computed as the simple average, and an arbitrary number of these estimates can be generated as training samples for a neural network.

3.1.2 Numerical simulations

In this study a response of a doubly-curved composite panel with a cut-out under a uniform pressure is analyzed. The panel is a shallow square, $0.254m \times 0.254m$, with a circular hole of $0.0381m$ radius and consists of 16 plies of $[[45, -45, 0, 90]_2]_s$ orientation; thickness of each ply is $0.000132m$. Material properties are as follows: $E_L = 0.15168 \times 10^{12} Pa, E_T = 0.965 \times 10^{10} Pa, \nu_{LT} = \nu_{TT} = 0.32, G_{12} = G_{13} = 0.593 \times 10^{10}$, and $G_{23} = 0.365 \times 10^{10} Pa$. The panel is modeled with 320 9-node shell elements for a total of 6860 degrees-of-freedom; nodal displacements include three translations u, v, w, and two rotations θ_1, θ_2. The panel is clamped along one side and simply supported on the opposite side.

In the first simulation, predictions of displacements of an element with the maximum deflection located on the edge of the cut-out are obtained. The displacements of a corner node and the center node are selected as representative for the element. A full system analysis is first performed at the load factor $q = 0$ to generate twenty basis vectors representing derivatives of a displacement vector with respect to load. To develop Padé approximants, coefficients in the Taylor series are computed, and approximate solutions are obtained for several different combinations of L and M. By setting the value of τ_P to 0.01, q^{max} is found to be in the range from 10,000 to 15,400 depending on a specific DOF. Training sets are next generated for $q \in [0, q^{\max}]$, and data are normalized so that each vector component is in the interval $[0.1, 0.3]$ or $[0.9, 0.7]$ to allow fourfold increase or decrease in a function value before a neural network saturation effect would start distorting results.

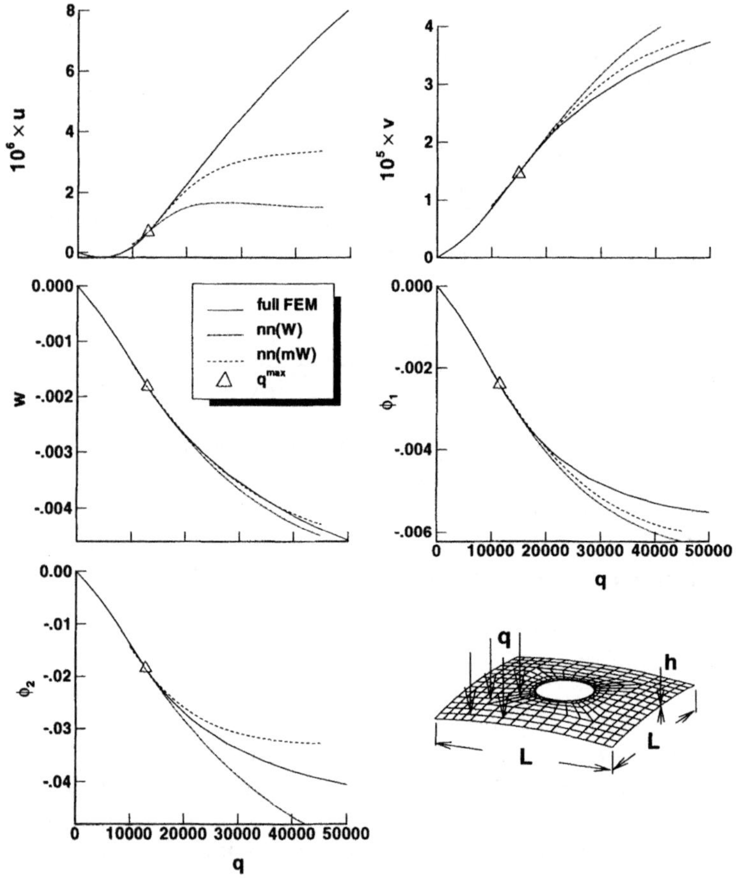

Figure 1: Comparison of predictions of the center node displacements

A BPN network architecture with eight neurons in the hidden layer is selected for extrapolation. The networks, one for each DOF, are trained so that the training error τ_t is reduced to 0.001. Predictions of these networks, labeled "nn(W)", for the center node are shown in Figure 1, and displacement predictions for a corner node are shown in Figure 2. For comparison, the full system solution (a solid line) is also shown. A hollow triangle on each plot shows the interpolation range, $q^{\max}$, for each DOF.

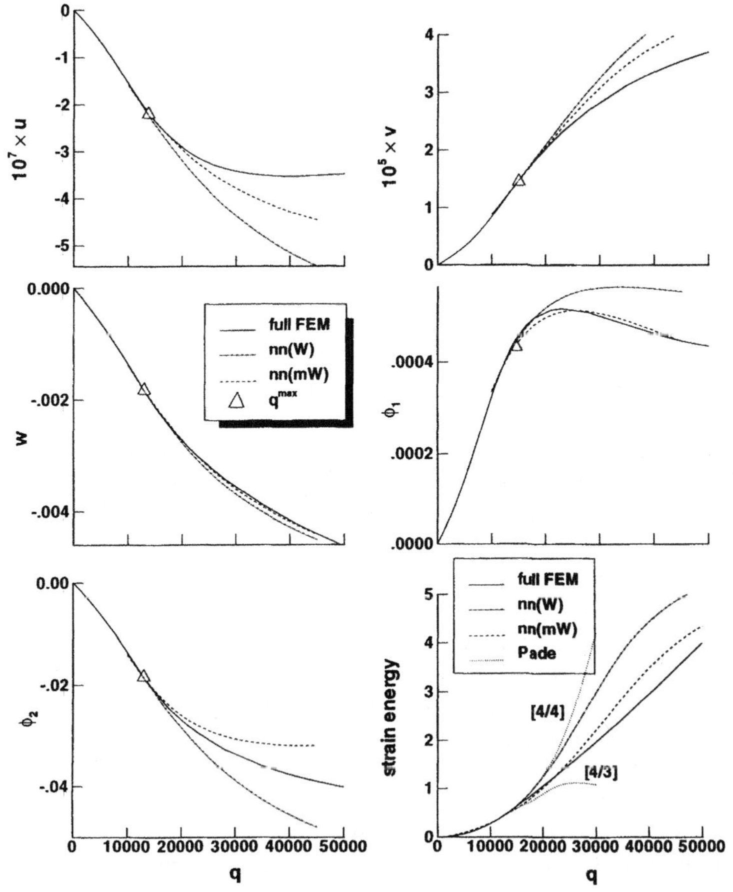

Figure 2: Predictions of the corner node displacements and the strain energy

To improve the accuracy and range of predictions, the data expansion strategy is used to generate additional training data. The first and the second derivatives of a displacement with respect to the load are computed and extended to the load factor of 20,000 by trained neural networks. Using Eq.6, additional training data are computed, and original networks are retrained to the same training error. Predictions of these networks, labeled "nn(mW)" are also shown in Figures 1 and 2.

In the second simulation the strain energy of a deforming panel is predicted. In this case nine basis vectors (the strain energy derivatives with respect to the load) are computed and used to obtain several Padé approximants. For a given τ_P, valid samples are found for load factors up to 13,700. A network with four neurons in the hidden layer is trained, and its predictions are shown in Figure 2 labeled "nn(W)". On the same plot, shapes of two approximants, [4/3] and [4/4], are also shown.

Additional training samples are generated using the weights regression strategy. The interpolation range is divided in three subsets, and a network is trained in succession to produce three weight matrices. After linear regression (a quadratic fit) is applied to elements of these matrices, the resulting model is used to generate additional training data. The original network is retrained on this extended training set, and its predictions, labeled "nn(mW)", are also shown on the strain energy plot of Figure 2.

Results of this numerical study indicate that by the use of various approximation techniques long-range predictions of behavior of a composite panel are feasible. The costly full system analysis is performed only once at zero load, and polynomial approximations and neural network extensions yield valid solutions up to a load of 50,000 in the case of a vertical displacement, the largest output quantity. The hybrid strategies tested in this study improve network predictions. The weights regression strategy is useful because there is almost no additional computational cost of applying this strategy. The network training is performed on subsets of training data with rapid convergence at each stage. It is, however, imperative to add key patterns at each stage, otherwise no meaningful change in network parameters is obtained. The case of the strain energy is amenable to this approach as available data scattered over the interpolation range define training subsets in a natural way. Displacement predictions are treated with the data expansion strategy because the large number of basis vectors allows evaluation of derivatives by Padé approximants with shifted basis vectors. This strategy, although more general than weights regression, requires additional computational effort because neurocomputing models of derivatives must be developed through training of networks.

3.2 Design Space of an Energy Absorbing Cylinder

In the preliminary design of energy absorbing composite structures, it is necessary to determine the most promising lines of investigation based on very limited data that relates structure behavior to static and dynamic loads. Such data are expensive to collect since either numerous experiments with changing design variables and load conditions or numerical simulations on complex finite element models have to be performed. Despite a steadily growing body of knowledge on the subject of design with safety as the dominant objective (References [17] and [18]), designers still rely heavily on experience and intuition when designing energy absorbing structures. A description of the design space is important, so that such decisions can be made in a rational and possibly automated fashion.

In this study the data for training neural networks are generated via an axisymmetric finite element model developed for structures with an arbitrary number of homogeneous, hyperelastic and anisotropic layers undergoing finite deformations subject to quasi-static loads ([19]). Motion of the body is restricted by a rigid obstacle, and a frictionless contact model is developed to enforce the contact constraints. Data are collected by solving a sequence of linear equations obtained by simplifying nonlinear equations by means of an incremental/iterative solution technique. The iterative procedure is terminated once the residual force is less than some given tolerance, and failure is assumed when the Tsai-Wu (TW) number reaches one for the first ply in the laminate. It is noted that this solution technique is computationally very intensive since it requires a fine mesh to correctly model contact, and very small load increments are necessary for the algorithm to converge and produce meaningful results.

3.2.1 Numerical simulations

In simulations performed for this study the design space description for the cylinder shown below is considered. The cylinder top surface is fixed, and the bottom surface is free and on contact with a rigid plate. An initially stress-free cylinder is compressed by moving the bottom plate in the axial direction, and for a certain displacement failure occurs. In general, failure of a cylinder depends on material properties, ply orientation, and its geometry. Here, it is assumed that the material properties remain unchanged, and there is only one ply orientation $[(0/0/0)_n]_s$. The design variables are only r (inner radius), H (wall thickness), and L (length), and any changes in their values require remeshing. Range of variations of the design variables is also shown below.

Material properties:

$E_L = 0.181 \times 10^{12}$Pa
$E_T = 0.103 \times 10^{10}$Pa
$\nu_{LT} = \nu_{TT} = 0.32$
$G_{12} = G_{13} = G_{23} = 0.717 \times 10^{10} Pa$

Design variable ranges:

$0.05\text{m} \leq \text{r} \leq 0.1\text{m}$
$0.005\text{m} \leq H \leq 0.01\text{m}$
$0.1\text{m} \leq L \leq 0.17\text{m}$

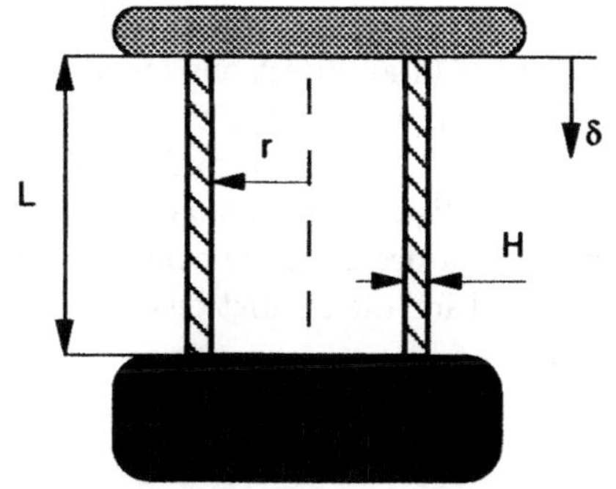

The first simulation shows that the incremental/iterative procedure can be more efficient if extrapolation strategies are used to predict the displacement for which the TW number reaches one. The numerical algorithm is executed for several increments of δ and after convergence the mapping $\delta \mapsto TW$ is created. Then network training progresses in succession on additional samples of the relation yielded by the algorithm once they become available.

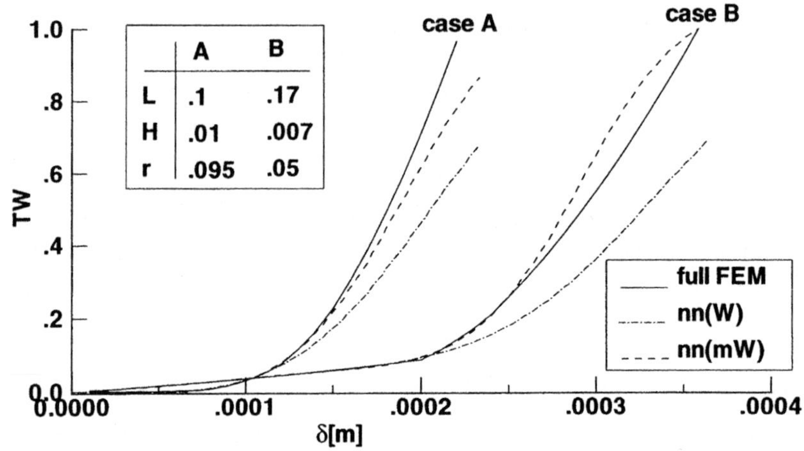

Figure 3: Predictions of the TW number for two cases

Linear regression is applied to a sequence of weight matrices, and predictions are made as to when $TW = 1$. Figure 3 shows predictions for two different sets of the design variables. In the second simulation a mapping L, r, $H \mapsto TW$ is sought. A network of 3-4-1 architecture is trained on 31 samples of this relation generated via numerical simulations. The weight analysis $\mathbf{T} = (\, 0.4437 \quad 0.1477 \quad 0.4087\,)$ indicates that of the three design variables the inner radius affects TW the least. Therefore, in Figure 4 a three-dimensional plot of the design space is shown for a given radius of 0.075m; plots for different values of r are very similar. The description is extended in all directions for at least 50% of the interpolation range. A numerical simulation indicates that the maximum error is about 3% (predicted value is $4.48 \times 10^{-4}m$ and the exact is $4.6 \times 10^{-4}m$). Results of this simulation indicate that a small number of samples may suffice to satisfactorily describe a design space containing mild nonlinearities when neural networks are used for the predictions.

4 Closing Remarks

Previous studies have shown that without additional modifications, backpropagation neural networks produce poor generalization results when presented with an input vector that falls out of the range of training data; therefore, the present work describes hybrid strategies to enhance the extrapolation capabilities of such networks. The strategies explore function continuity and the premise that network predictions improve when additional data from outside the interpolation range are used to train a network. They differ, however, in the way such data are generated. The data expansion strategy is shown to work well on number of testing cases, but it requires an additional computational effort to produce models of funtion derivatives. The weights regression strategy, although very efficient, may not improve predictions unless key patterns are added at each stage of training. The limited data access strategy coupled with the weights analysis is shown to improve the accuracy of predictions as expected, but its forte is shown to be the ability to estimate error for computationally intensive cases.

In problems whose solution requires an iterative/incremental procedure, networks can be used to predict a convergence point thereby increasing the efficiency of numerical algorithms. Nonlinear structures can also be analyzed more efficiently when neural networks are employed to further extend solutions yielded by other approximation techniques. It is also shown that fewer design points may suffice to describe a space if extrapolation strategies are coupled with numerical techniques. For the considered class of problems using BPN network predictions in lieu of the exact analysis has merit, and by careful selection of hybrid strategies significant improvements in the range and accuracy of predictions can be obtained. However, the need for hybrid strategies indicates that a BPN network functions as a data processing technique rather than the operator of a given relation. Further development of neural network based strategies for extrapolation may require incorporating in the modeling process other sources of information such as, for example, qualitative reasoning and heuristics. Both studies show that a model-free approach to data proccessing based on neural networks provides a promising tool for predicting nonlinear system response.

References

[1] S. Geman, E. Bienenstock, and R. Doursat. Neural networks and bias-variance dilemma. *Neural Computation*, 4(1):1–58, 1992.

[2] W.C. Carpenter and J.F.M. Barthelemy. A comparison of polynomial approximations and artificial neural nets as response surfaces. *Structural Optimization*, 5:1–15, 1993.

[3] P. Hajela and L. Berke. Neural networks in analysis and design; an overview. *Computing Systems in Engineering*, 3(1-4):525–538, 1992.

[4] P. J. Haley and D. Soloway. Extrapolation limitations of multilayer feedforward neural networks. In *Proceedings of IJCNN*, volume 4, pages IV–25 to IV–30, Baltimore, MD, 1992.

[5] A. B. Cook and et. al. The prediction of nonlinear dynamic loads on helicopters flight variables using artificial neural networks. In *Proceedings of DGLR/AIAA Aeroacoustics Conference*, pages 971–979, 1992.

[6] J. Ghaboussi, J. H. Garrett, and X. Wu. Knowledge-based modeling of material behavior with neural networks. *Journal of Engineering Mechanics*, 117(1):132–153, 1991.

[7] M. L. Thompson and M. A. Kramer. A hybrid modeling methodology to combine prior knowledge and neural networks. In *Proceedings of IJCNN*, volume 4, Nagoya, Japan, 1993.

[8] Z. P. Szewczyk and P. Hajela. Function extrapolation in structural reanalysis using neural networks. In *Proceedings of the 35th SDM Conference*, number AIAA - 94 -1599, pages 2247–2257, Hilton Head, SC, 1994.

[9] Z. P. Szewczyk and P. Hajela. Feature-sensitive neural networks in structural response estimation. In Dagli and Berke, editors, *Intelligent Engineering Systems Through Neural Networks*, pages 159–164. ASME Press, 1992.

[10] P. Hajela and Z. P. Szewczyk. Neurocomputing strategies in structural design - on analysing weights of feedforward neural networks. *Structural Optimization*, 8(4):236–242, 1994.

[11] W.C. Carpenter. Newnet user's guide. University of South Florida, Tampa.

[12] K. Funahashi. On the approximate realization of continuous mappings using neural networks. *Neural Networks*, 2:183–192, 1989.

[13] B. C. Watson and A. K. Noor. Nonlinear structural analysis on distributed-memory computers. *Computers & Structures*, 1995. to be published.

[14] Z. P. Szewczyk and A. K. Noor. A hybrid neurocomputing/numerical strategy for nonlinear structural analysis. *Computers & Structures*, 1995. to be published.

[15] G. A. Baker and P. Graves-Morris. *Padé Approximants*. Addison-Wesley Publishing Company, 1981.

[16] B. Cochelin, N. Damil, and M. Potier-Ferry. Asymptotic-numerical methods and padé approximants for nonlinear elastic structures. *International Journal of Numerical Methods in Engineering*, 37:1187–1213, 1994.

[17] G. L. Farley and R. M. Jones. Prediction of the energy-absorption capability of composite tubes. *Jounal of Composite Materials*, 26:388–403, 1992.

[18] J. C. Simo and T. A. Laursen. An augmented lagrangian treatment of contact problems involving friction. *Computers & Structures*, 42(1):97–116, 1992.

[19] L. Karaoğlan and G. S. Springer. Axial impact of composites. *Computers & Structures*, 49(5):807–823, 1993.

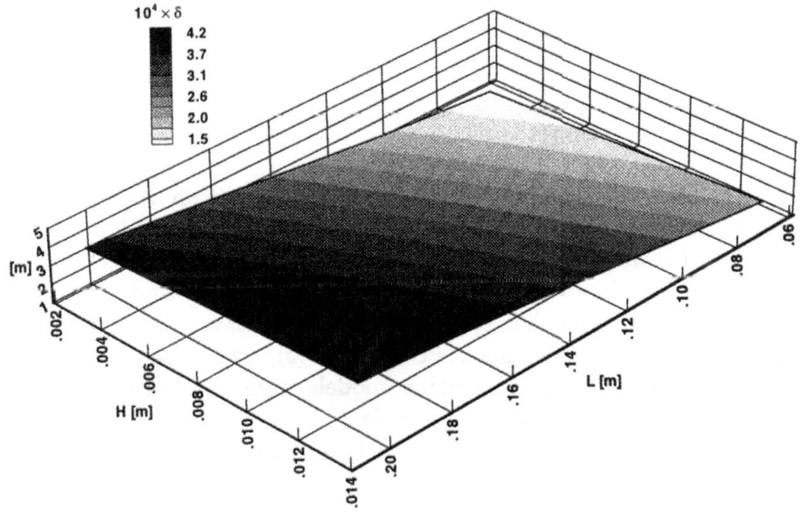

Figure 4: A design space for the compressed cylinder (r=0.075m)

Neural Networks for Learning and Optimization

B.H.V. Topping† and A. Bahreininejad‡
† Professor of Computational Mechnanics
‡ Postgraduate Student
Department of Mechanical and Chemical Engineering
Heriot-Watt University, Riccarton,
Edinburgh, EH14 4AS, United Kingdom

Abstract This paper describes a parallel processing implementation for backpropagation neural networks and its application to finite element mesh design. The software used was NETS 2.01 [1] which is based on the backpropagation algorithm. The paper also describes the application of unsupervised recurrent Mean Field Annealing Networks for solving optimization problems, in particular, the domain decomposition of adaptive unstructured finite element meshes.

1 Introduction

There has been an upsurge in the interest in Artificial Neural Networks (ANNs) in recent years. The first wave of interest emerged after the introduction of simplified neurons by McCulloch and Pitts in 1943 [2]. These neurons were presented as models of biological neurons and as conceptual components for circuits that could perform computational tasks. After the publication in 1969 of the book *Perceptrons* [3], by Minsky and Papert in which the defficiencies of perceptron models were described, very few researchers continued their work in this field. However, after the discovery of backpropagation networks and the development of new hardware, interest re-emerged again.

ANNs can be characterized as computational models with particular properties such as the ability to learn, generalize, cluster or organize data for which the operation is based on parallel distributed processing (PDP) [4]. This paper attempts to demonstrate the appropriate application of ANNs in structural computations. It begins with the efficient implementation of the backpropagation network on a network of transputers. The application of the backpropagation network to finite element mesh generation is described. The paper continues to describe the unsupervised Mean Field Annealing network

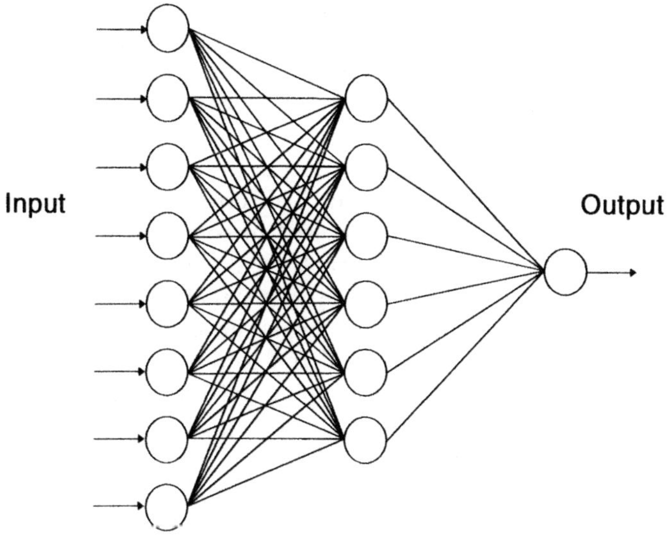

Figure 1: A three-layered fully connected network.

and its application to subdomain generation of adaptive unstructured finite element meshes.

2 Backpropagation Neural Networks

A typical multi-layered network shown in Figure 1 consists of a set of neurons or units organized into a sequence of layers with full or patterned connecting weights between the layers. Biases may also be employed which act as imaginary weights from a neurons which have an output of one at all times. Typically only two layers terminate a network: the input layer for which data is presented to the network, and the output layer where the response of the network to a given input is received. Layers other than these two are called hidden layers.

Neural networks recognise patterns after training by the presentation of a number of input-output (i-o) patterns. During the training the weights are adjusted using a training algorithm. Such networks learn the mapping task by example and hence no predefined knowledge of the problem is required.

The backpropagation (BP) algorithm is based on the *Generalized Delta Rule* [4, 5] which may be used to carry out association between input and output patterns by means of adjusting the weights of the connections between the processing units of the corresponding layers. This is done when a set of inputs is presented to the input layer units. The inputs are propagated forward through to the other layers where they are multiplied by the associated weights. The sum of these products within each unit is subjected to a transfer function. The result is an output which becomes the input to the next layer.

At the output layer, the computed output is compared with the desired output and an error value is computed. This error value together with the derivative of the transfer function are used to compute an error signal for the output layer units. The error signals from the output layer units are propagated back in order to calculate the error signals for the previous hidden layer(s). Once all the appropriate error signals have been computed they are used to calculate the delta-weights and the corresponding weight changes. The process is repeated for all the training patterns until a desired Root Mean Square (RMS) error level over all training patterns is achieved [6, 7]. The RMS value of the training is given by:

$$RMS_{err} = \sqrt{\frac{1}{PN_{output}} \sum_{p}^{P} \sum_{output}^{N_{output}} (t_p - o_p)^2} \tag{1}$$

where P is the total number of training patterns, p is a single training pattern, N_{output} is the number of units in the output layer, t_p is the desired output in the training data and o_p is the output of the network.

3 A Parallel Implementaion of the BP Algorithm

Backpropagation is a time-consuming algorithm. This is particularly true in the cases where either the size of the net is large and/or the number of the training patterns is large. The training process may take many iterations in order that the net reaches a desired accuracy. There have been many analytical approaches [8, 9, 10] to speed up the algorithm but the greatest scope for improving the computational efficiency lies with the parallel nature of the algorithm. This nature enables the algorithm to be implemented on parallel architectures. There are a number of parallel processing architectures including variations of:

- **SIMD** single instruction multiple data.
- **MIMD** multiple instruction multiple data.

The work was carried out using a transputer TMB 08 motherboard mounted in a PC and a transputer TMB 12 motherboard mounted in a Transrack [11]. Up to fourteen transputers connected in a pipeline were used. The 3L Parallel C compiler [12] was used for these studies. This architecture represents a MIMD -multiple processor distributed memory system. Parallel implementation of multi-layered neural networks on an array of transputers may be carried out in a number of forms including:

- Distribution of the BP algorithm among the processors;
- Distribution of the network by dividing the units and or layers amongst the processors;
- Distribution of units by representing each unit with a single processor; and
- Distribution of i-o data patterns among the processors

The first method implies that the entire BP learning algorithm is to be distributed over the processors. This means, for example, that the forward propagation would be carried out on one processor and the results transferred to the next processor in the pipeline to calculate errors and deltas and so forth.This form of paralllelisation is sometimes referred to as algorithmic parallelisation. The second method suggests that layers and/or units are distributed amongst the processors. The first two methods are particularly efficient if the constructed network consists of many units and layers but if during the learning a large quantity of data has to travel through the pipeline then the method will be inefficient.

The third paralellisation method is to represent each unit with an individual transputer. Except for small network structures an efficient implementation of this method may not be possible. This is due to the limited hardware links available in a transputer. Any point to point routing implementation will not contribute to the efficiency because the communication overhead would significantly increase.

If the network structure is not large but a large number of training patterns is to be used the third method is likely to be more efficient. The implementation is essentially a task farming algorithm that requires no or minimal inter processor communication which is undertaken by distributing the data amongst the Worker processors that usually only communicate with the ROOT processor. As a result of the large amount of data involved in structural problems, it was decided to adopt the data distribution method in the parallelisation process, as illustrated in Figures 2 and 3.

A limited amount of inter processor communication was carried to pass the delta values up the pipeline to the root by progressively adding the delta values to those of the previous processor in the pipeline.

The i-o data was distributed by the Master task on the ROOT processor amongst the Worker processes on the Worker and Master processors. These data patterns were then stored in the memory of the receiving Worker and used for subsequent computations.

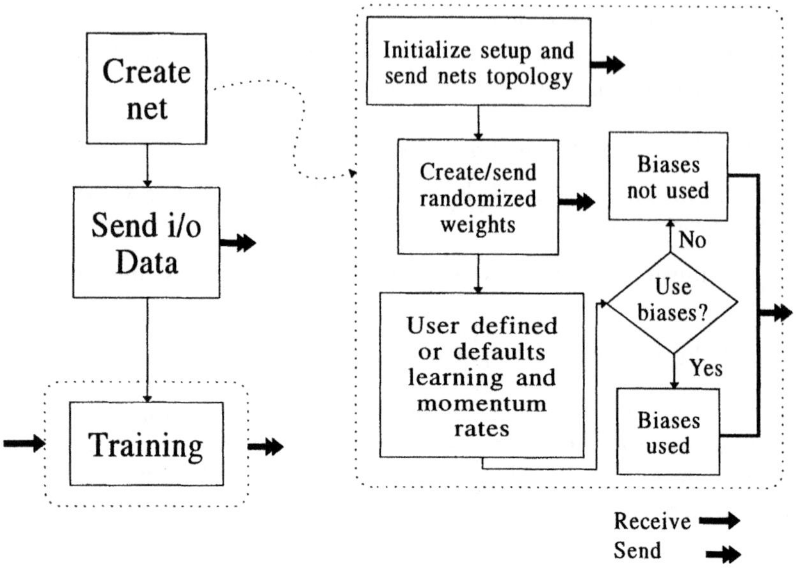

Figure 2: The Master task operation on the ROOT processor

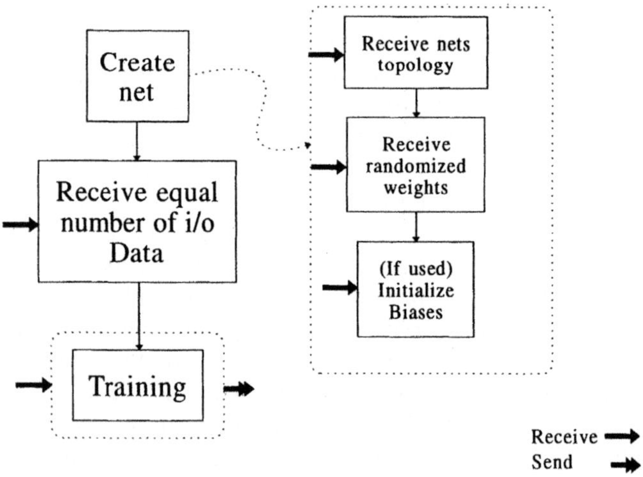

Figure 3: The Worker process task operation on the ROOT and Worker processors

3.1 Transputer Implementation of the BP Algorithm

The Master task creates a neural network model after receiving information concerning the network topology from a user defined file from the host computer. The Master then sends this information to the Worker processors for which they create their own local neural networks. Also at this stage the ROOT processor receives information from the host concerning the maximum and minimum limits for which a random weight values may be generated. Throughout the training process the ROOT keeps a copy of the current weights. At initialisation the ROOT sends a copy of the initial weights to each of the Workers.

Each processor initializes values such as delta-weights and delta-biases to zero. As the process of network creation continues, the learning and momentum rates are received by the ROOT to be applied during the training. These values may be user defined or defaults which are generated by the program. If the user wishes to apply or not apply biases, the ROOT informs the Workers of this decision.

Hence the ROOT and all the Worker processors share the same network configuration. Once the complete network topology is set-up on all processors, i-o data are then distributed equally among the Workers. No part of the i-o data need to be saved on the ROOT processor. The equal distribution of data between the Workers processes ensures load balancing in the network of processors. This load balancing must be exact to ensure that the communication scheme for passing delta values to the ROOT works. Although this places a constraint on the number of training patterns which must be an exact multiple of the number of Workers it ensures that interprocessor communication concerning the delta-weights is kept to a minimum. Further the additional task of adding the delta-weights is not undertaken on the ROOT processor.

The next stage is the main part of the neural computations and that is the learning or the training process. The ROOT receives the *Constraint Error* variable from the host computer which has already been specified by the user. This variable is used to stop the learning operation. NETS 2.01 uses the constraint error as a terminating limit of the maximum difference between the desired error and the observed error $(t-o)_{max}$, however the RMS error was employed in this research to halt the training. The RMS value for the parallel training is given by:

$$RMS = \sqrt{\frac{1}{PN_{output}} \sum_{proc}^{N_{proc}} \sum_{p}^{P_{proc}} \sum_{output}^{N_{output}} (t_p - o_p)^2} \tag{2}$$

where N_{proc} is the number of Worker processors and P_{proc} is the number of patterns per Worker processor.

The NETS 2.01 software operates on the basis of *Single Pattern* or on-line training where the weights are changed after each i-o pattern has been subjected to the net. *Batch Pattern* or epoch training is another popular method in which the weight changes occur after the presentation of a batch or all i-o patterns [6, 13]. In general, different results are obtained depending which method is used, but provided the *learning rate* is sufficiently small, single pattern training is shown to be a very close approximation to gradient descent in terms of the sum-squared error [4, 5].

3.1.1 The Parallel Single Pattern Training

The learning process which is shown in Figures 4 and 5 is divided between the ROOT and the Worker processors in the following manner:

1. The ROOT instructs the Workers to start the learning process. To ensure efficiency a Worker task was also kept on the ROOT processor.
2. All the Workers propagate their input data through the neural network to compute the observed output(s) in the output layer.
3. The $(t-o)$ error, $(t-o)^2$ and delta-signals are computed and a back-propagation is made to calculate the delta-signals in the hidden layer(s).
4. Workers then calculate the delta-weights for a single pattern and after summation to the delta-weights from processors lower in the pipeline the partially summed delta-weights are passed to the next processor. At the ROOT the total weight change is given by:

$$\Delta w_{ROOT} = \sum_k \delta_k o_k \tag{3}$$

where k denotes a Worker processor on which the pattern is processed, δ_k is the delta-signal, o_k is the output and Δw_{ROOT} is the total computed weight change for all the Worker processors.

The total weight change, which is not determined until all the delta values have been progressively passed up to the ROOT processor, is calculated using the appropriate learning and momentum rates and is given by:

$$\Delta w_{total} = \eta \Delta w_{ROOT} + \alpha \Delta w_{old} \tag{4}$$

where η is the learning rate, α is the momentum and Δw_{old} is the old delta-weight.

This value is used to calculate the new weights on the ROOT processor and is also sent to all Workers so that weight modifications take place on all processors. The next learning sequence continues unless a satisfactory RMS condition has been reached. The process of changing the bias values (if they are used) is carried out in exactly the same manner as changing the weights.

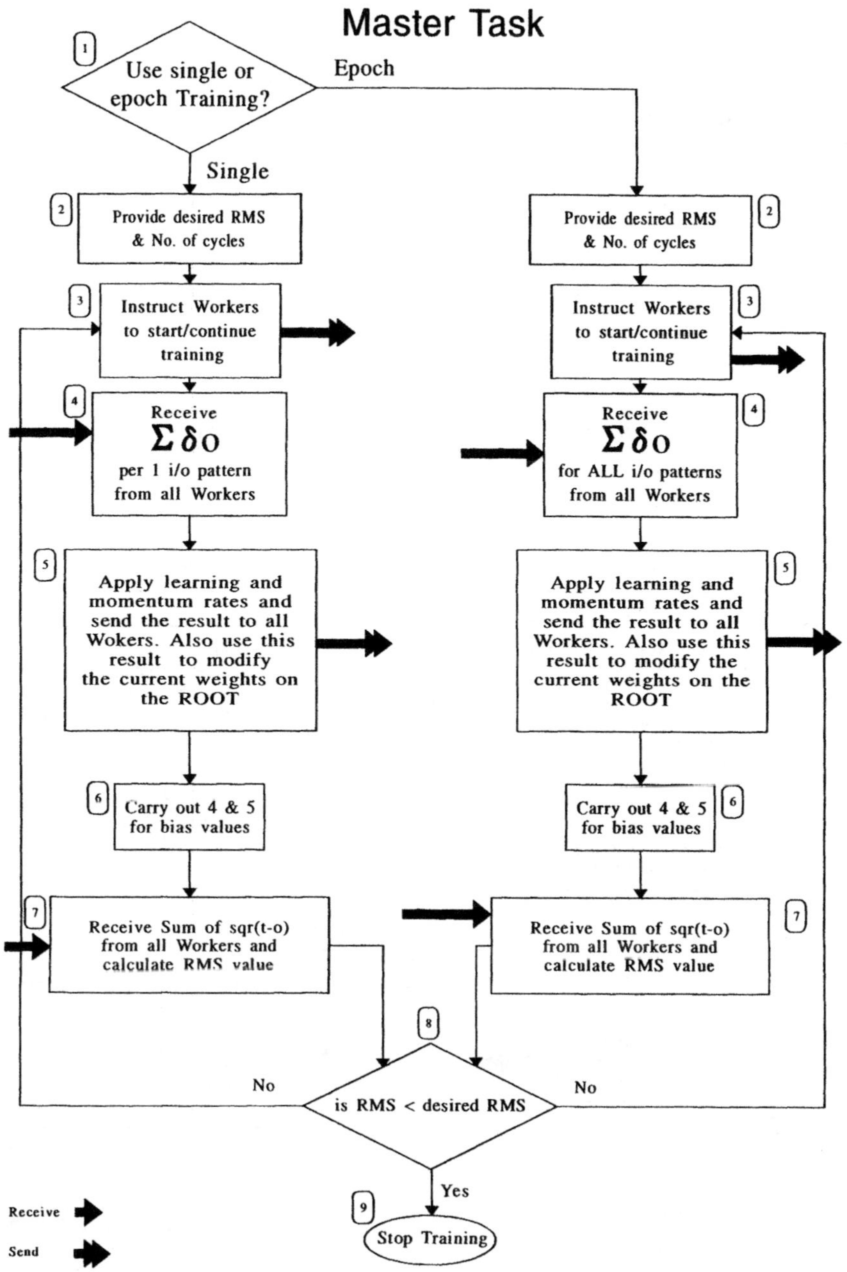

Figure 4: The ROOT's Master task operation during the training

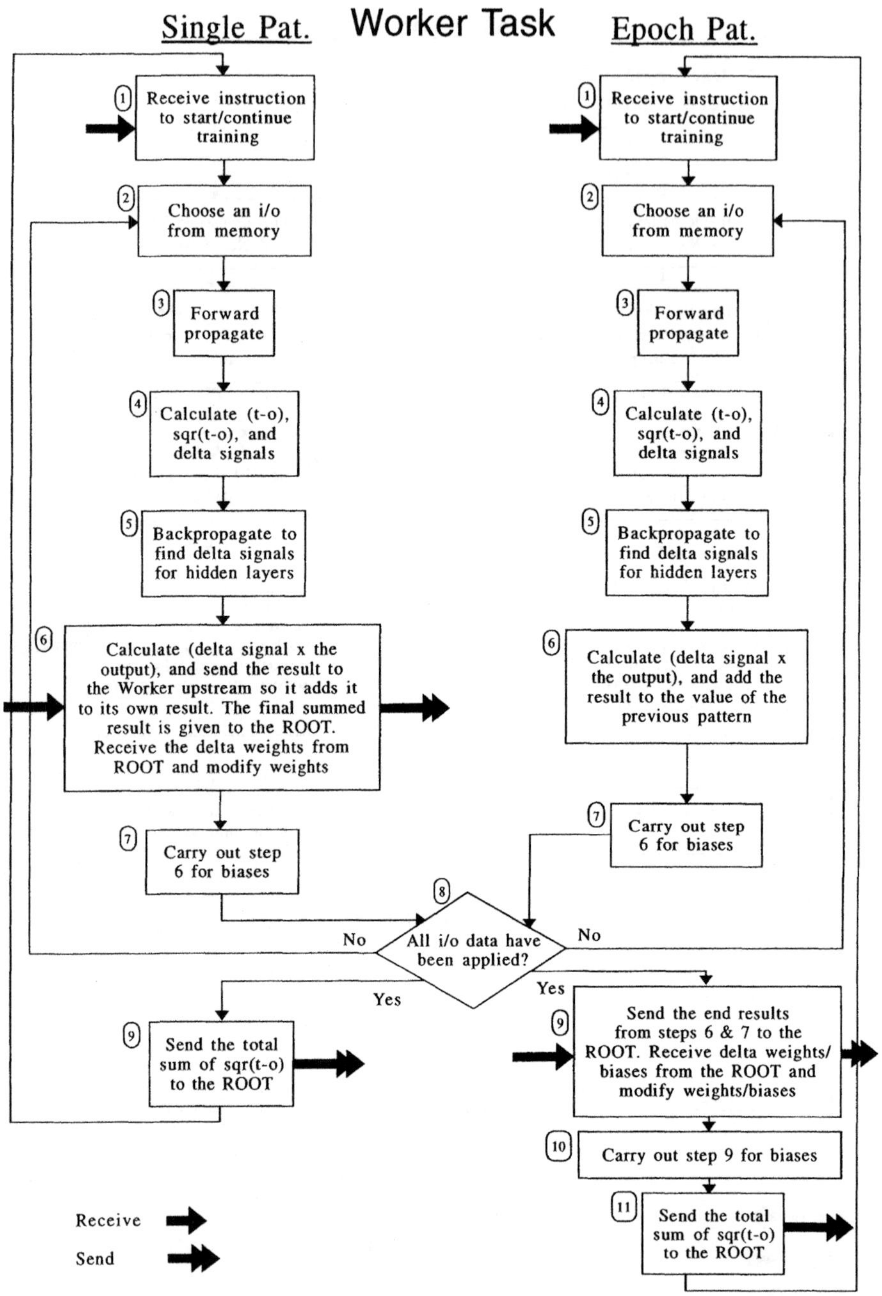

Figure 5: The Worker process task operation during the training

All the above operations are repeated for all the i-o data and the sum-squared error and the maximum error difference, $(t - o)_{max}$, are sent back to the ROOT. Using equation 2, a root mean square operation on the sum-squared errors over all the i-o patterns is carried out on the ROOT processor. The ROOT compares this RMS with the user defined permitted constraint error and if RMS is less than the permitted constraint error, the learning process is stopped, otherwise the ROOT instructs the Workers to resume training using the same i-o data.

3.1.2 The Parallel Epoch Training

When several training patterns are presented to the neural network before changes to the weights are made, this process is called epoch or batch pattern training [14]. In sequential training of the BP method, an epoch consists of a portion or all the i-o patterns. However in this paper an epoch consists of all the i-o patterns allocated to a single Worker processor.

In the parallel single pattern training method described in the last section a form of epoch training existed in that an epoch consisted of a single pattern from each of the processors. The sequential version of NETS 2.01 did not support epoch training. It was therefore decided to implement a complete epoch pattern training algorithm by modifying the original source codes and comparing it with single pattern training. The overall method is quite similar to the single pattern training with the following exceptions:

- the accumulated sum of delta-weights is communicated after the complete i-o patterns have been trained once on all the Workers; and
- the same operation is carried out for the bias values.

The performance of the parallelised NETS 2.01 is demonstrated for single pattern training in Figure 6 and for epoch training in Figure 7. Here an epoch is given by the number of trining sets divided by the number of processors. The high performance of epoch training is due to the low number of communications between the ROOT and Worker processors. Epoch training however requires low values of learning and momentum rates therefore a much greater number of iterations for the training of a given neural network to a given error state may have to be carried out. The network's *computing-speed* ψ may be computed by:

$$\psi = \frac{NPC}{T} \tag{5}$$

where T is the time for training P number of i-o patterns and C the number of neural network connections to carry out N number of iterations where C is defined for a fully connected network as:

$$C = N_i N_h + N_h N_o \tag{6}$$

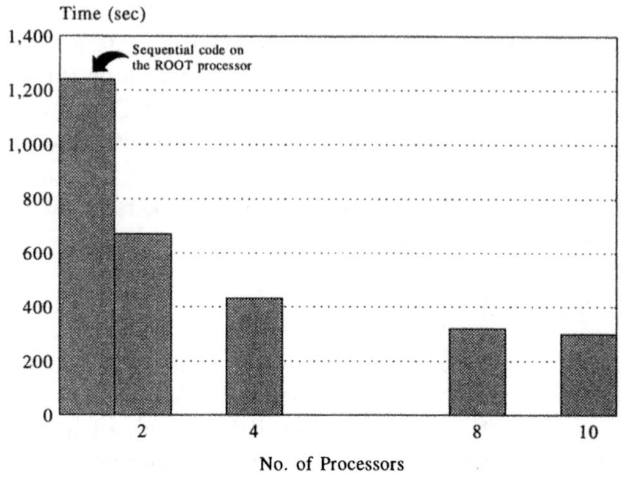

Figure 6: Parallel-Sequential run-time performance for single pattern training for 1000 iterations of 400 i-o patterns

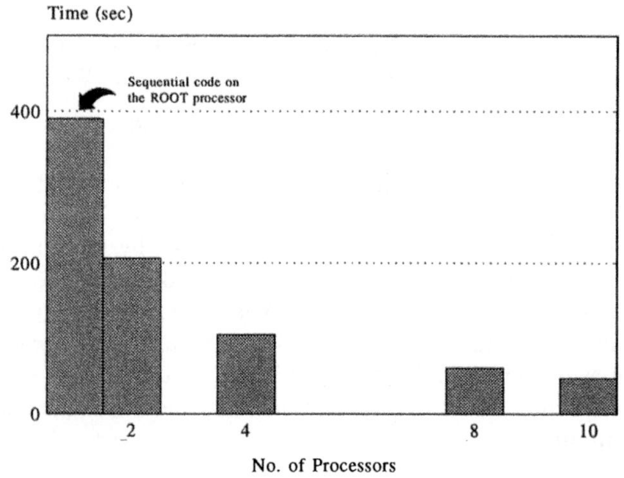

Figure 7: Parallel-Sequential run-time performance of epoch training for 1000 iterations of 400 i-o patterns

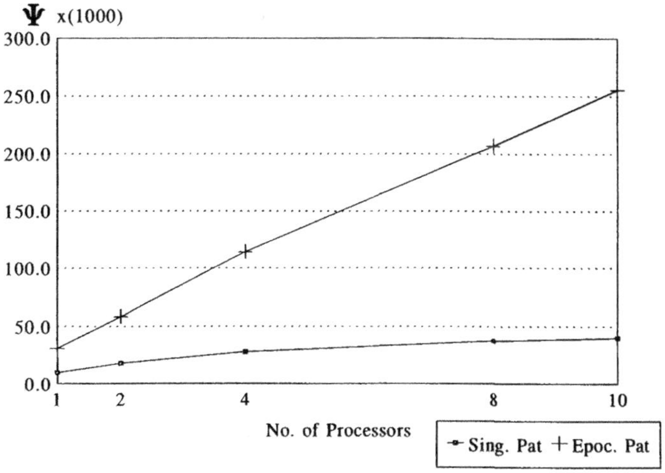

Figure 8: Computing-speed of the neural network using single and epoch training for 1000 iterations of 400 i-o patterns

N_i is the number of input units, N_h is the number of hidden units and N_o is the number of output units. The computing-speed represents the speed of training per second. In terms of the transputer implementation, the ***speed-up*** and *efficiency* are calculated using:

$$S = \frac{T_{sequential}}{T_{parallel}} \tag{7}$$

$$E = \frac{S}{N_{proc}} \tag{8}$$

Figure 8 and 9 illustrates the computing-speed and efficiencies of the neural network training using various numbers of Worker processors.

4 BP Training for Adaptive Finite Element Mesh Generation

In the Subdomain Generation Method [15, 16] it is important to know the number of elements that will be generated within a single element of a coarse mesh that is to be remeshed when using an unstructured adaptive parallel mesh generator. This permits load balancing of the subdomains in the refined mesh for parallel finite element analysis when using explicit time stepping dynamic analysis procedures.

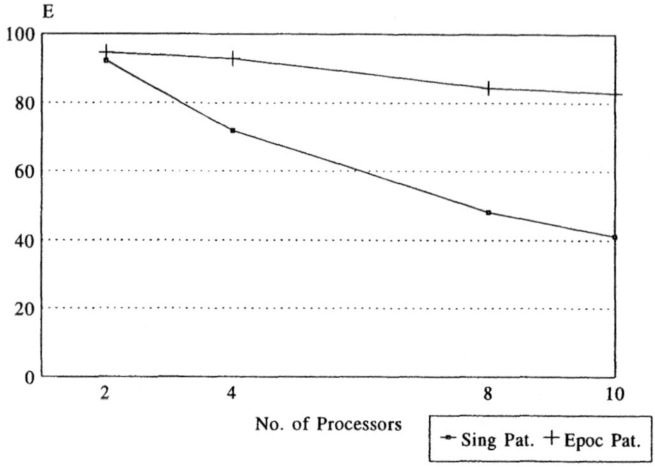

Figure 9: Efficiency of the neural net in different training conditions for 1000 iterations of 400 i-o patterns

BP neural nets have adaptive pattern recognition capability which may be used to approximately predict the number of elements that will be generated within each element of the initial background mesh after remeshing.

In order to carry out the training, several background coarse meshes were chosen and these meshes were analyzed using different point loads [6, 7, 15]. An input-output training data was created and applied in the training procedure which consisted of:

- the data regarding the geometry of the individual elements;
- the nodal mesh parameters; and
- the number of elements generated in the refined adaptive mesh

The geometry of a triangular element was represented by the side lengths and the three internal angles. It was noted that the geometry of a triangle may be defined by the length of its sides, thus the three internal angles of the triangular elements need not have been included in the training data-file [6, 7]. The three nodal mesh parameters actually represent the size of the triangle to be generated and they were scaled down to two. Therefore, the input data consisted of three side lengths, three internal angles and two scaled mesh parameters of each element. The output data consisted of the number of generated triangles in the refined mesh. A network was trained and once a desired trained network was achieved it was used to predict the

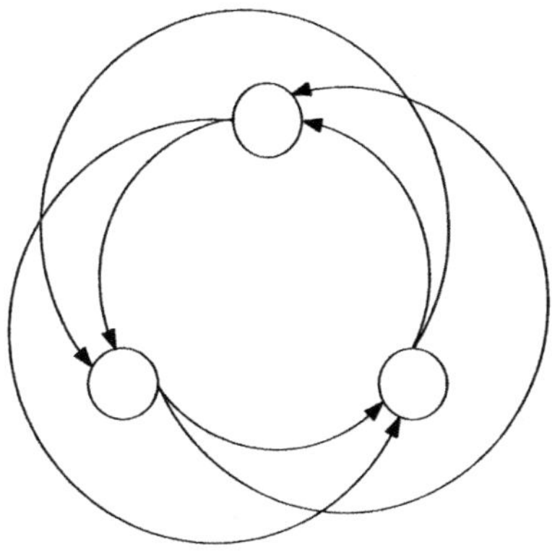

Figure 10: A simple representation of a feedback network.

number of elements which may be generated within an element of a coarse mesh.

5 Neural Networks for Optimization

Combinatorial optimization problems arise in many areas of science and engineering. Unfortunately, due to the NP (non-polynomial) nature of these problems, the computations increase with the size of the problem. Most computational methods that have so far been developed which generally yield good solutions to these problems rely on some form of heuristic. ANNs make use of highly interconnected networks of simple neurons or processing units which may be programmed to find approximate solutions to these problems [17, 18]. They are also highly parallel systems and have significant potential for parallel hardware implementation.

The origin of the optimization neural networks goes back to the work by Hopfield and Tank [19] which was a formulation of the Traveling Salesman Problem (TSP). The Hopfield network is a feedback type of neural network where the output(s) from a processing unit is (are) fed back as the input(s) of other units through their interconnections. This type of network structure is a nonlinear continuous dynamic system. Figure 10 illustrates a typical feedback neural network.

Following the poor performance of Hopfield networks in determining valid solutions to the TSP problem there followed a considerable research effort to improve the performance of the this type of network and to find ways of applying it to other optimization problems.

At about the same time of the emergence of Hopfield networks, a new optimization method called *Simulated Annealing* [20] was developed and researched. This technique provides a method for finding good solutions to most combinatorial optimization problems, however the algorithm takes a long time to converge. To overcome this problem, Mean Field Annealing (MFA) networks were proposed as an *approximation* to simulated annealing. MFA strikes a balance between computational time and quality of the resulting solution.

6 Mean Field Annealing

The roots of MFA are in the domain of statistical physics which combines the features of Hopfield neural network and simulated annealing. MFA is a deterministic approach which essentially replaces the discrete degrees of freedom in a simulated annealing problem with their average values as computed by the Mean Field Approximation.

6.1 Hopfield Network

NP problems may be mapped onto the Hopfield discrete state neural network using the energy function or Liapunov function (in statistical physics) or the Hamiltonian (in statistical mechanics) which is defined as:

$$E(\vec{S}) = -\frac{1}{2}\sum_{i=1}^{N}\sum_{j=1}^{N} t_{ij} s_i s_j + \sum_{i=1}^{N} I_i s_i \tag{9}$$

where: $\vec{S}$ represents the state of the network $\vec{S} = (s_1, s_2, ..., s_N)$; I is the bias; N is the number of processing units; and t_{ij} represents the strength of the synaptic connection between the units. It is assumed that the t_{ij} matrix is symmetric and has no self-interaction (i.e. $t_{ii} = 0$).

In order to move $E(\vec{S})$ downwards on the energy landscape, the network state is modified asynchronously from an initial state by updating each processing unit according to the updating rule:

$$s_i = sgn(\sum_{j=1}^{N} t_{ij} s_j - I_i) \tag{10}$$

where the output of ith unit is fed to the input of the jth unit by the connection t_{ij}. The symmetry of the matrix t_{ij} with zero diagonal elements enables $E(\vec{S})$ to decrease monotonically with the updating rule.

In optimization problems, the concept is to associate the Liapunov function given in equation 9 with the problem's objective function by setting the connection weights and input biases appropriately.

6.2 Simulated Annealing

Simulated annealing is a probabilistic hill-climbing algorithm which attempts to search for the global minimum of the energy function. It carries out uphill moves in order to increase the probability of producing solutions with lower energy. The method carries out random search in order to find new configurations using the Boltzmann distribution:

$$h(\vec{S}) = \frac{e^{-\frac{E(\vec{S})}{T}}}{Z} \tag{11}$$

where T is the temperature of the system and Z is the partition function of the form:

$$Z = \sum_{(\vec{S})} e^{-\frac{E(\vec{S})}{T}} \tag{12}$$

However simulated annealing involves stochastic search for generating new configurations. In order to reach good solutions, a large number of configurations may have to be searched which involves the slow lowering of the temperature and therefore is a very CPU time consuming process.

6.3 Mean Field Annealing Network

The purpose of the MFA approach is to approximate the stochastic simulated annealing by the average of the stochastic variables with a set of deterministic equations.

Peterson and Anderson in reference [21] showed that the discrete sum of the partition function 12 may be replaced by a series of multiple nested integrals over the continous variables u_i and v_i giving:

$$Z = C \prod_{j=1}^{N} \int_{-\infty}^{\infty} dv_j \int_{-i\infty}^{i\infty} du_j e^{-E'(\vec{V},\vec{U},T)} \tag{13}$$

where C is a constant and $\prod \int \int$ refers to multiple integrals. E' may be given in the form of:

$$E'(\vec{V},\vec{U},T) = E(\vec{V})/T + \sum_{i=1}^{N} u_i v_i - \log(\cosh u_i) \tag{14}$$

and the multiple integrals may be determined using a saddle point expansion of E' which involves the mean field approximation that is found in reference [22]. The saddle point positions are given by:

$$\frac{\partial E'}{\partial u_i} = 0 \tag{15}$$

$$\frac{\partial E'}{\partial v_i} = 0 \tag{16}$$

Hence using equations 14, 15 and 16, at the saddle points, gives:

$$v_i - \tanh u_i = 0 \tag{17}$$

and

$$\frac{\partial E(\vec{V})}{T\partial v_i} + u_i = 0 \tag{18}$$

thus combining equations 9, 17 and 18 and assuming $I_i = 0$, the MF equation is given by:

$$v_i = \tanh\left(-\sum_{j=1}^{N}\frac{t_{ij}v_j}{T}\right) \tag{19}$$

Furthermore, the continous variables, v_i are used as approximations to the discrete variables at a given temperature (i.e. $v_i \approx \langle s_i \rangle_T$), thus the final value of v_i approximates whether the value of s_i is +1 or -1.

7 Domain Decomposition Using MFA Network

In this paper, the partitioning of a mesh is based on the recursive bisection approach. The problem of bisecting meshes may be mapped onto a feedback type neural network and may be defined by an objective function in the form of the Hopfield energy function given in equation 9. This is carried out by assigning a binary unit of $s_i = +1$ or $s_i = -1$ to each element of the mesh defining which partition the element is to be assigned. The connectivity of the triangular elements is encoded in the t_{ij} matrix in the form of:

$$t_{ij} = \begin{cases} 1 & \text{if a pair of elements are connected by an edge} \\ 0 & \text{otherwise} \end{cases}$$

With this terminology the minimization of the first term of equation 9 will maximize the connectivity within a bisection while minimizing the connectivity between the bisections. This has the effect of maximizing the boundary. However using this term alone as the cost function forces all the elements to move into one bisection, thus a penalty term is applied which measures the violation of the equal sized bisection constraint [22]. This penalty term may be used to reduce Therefore the neural network Liapunov function for the mesh bisection is of the form of:

$$E(\vec{S}) = -\frac{1}{2}\sum_{i=1}^{N}\sum_{j=1}^{N} t_{ij}s_is_j + \frac{\alpha}{2}(\sum_{j}^{N} s_j)^2 \tag{20}$$

where α is the *imbalance* parameter which controls the bisection ratio and the cutsize.

The mean field equation for the mesh bisection is given by combining equations 17, 18 and 20 which gives:

$$v_i = \tanh\left(\sum_{j=1}^{N}(t_{ij} - \alpha)\frac{v_j}{T}\right) \tag{21}$$

Equation 21 is used to compute a new value of v_i asynchronously. Initial values for the vector $\vec{V}$ are assigned using small continous random values.

7.1 Selection of Parameters

A phase transitions occurs in materials as they cool down or heat up and they change from one state to another. Sharp changes in the properties of a substance are often visible as the material changes from one state to another. The transition from liquid to gas, solid to liquid or from paramagnet to ferromagnetic are common examples. There is a *critical temperature* such as the melting or boiling point where the state of a system suddenly changes from a high to a lower energy state.

Although there are some analytical methods for determining the critical temperature [21, 23, 24], it was decided, in this research, to determine an approximate value for the critical temperature by using the following procedure:

1. Determine an initial bisection using an initial temperature value (for example T $\geq$ 2).
2. During the bisection, the changes in temperature and the corresponding spins (v_i) values are recorded.
3. The iteration continues using a temperature reduction factor of 0.9. The programme is terminated either when the number of iterations exceeds 100 or when the system reaches a saturation state which is defined by the term:

$$\frac{v_i^2}{N} \leq 0.999 \tag{22}$$

   where N is the number of spins.
4. Once the program has terminated, the temperature where a sudden change to the spins has occurred is identified.

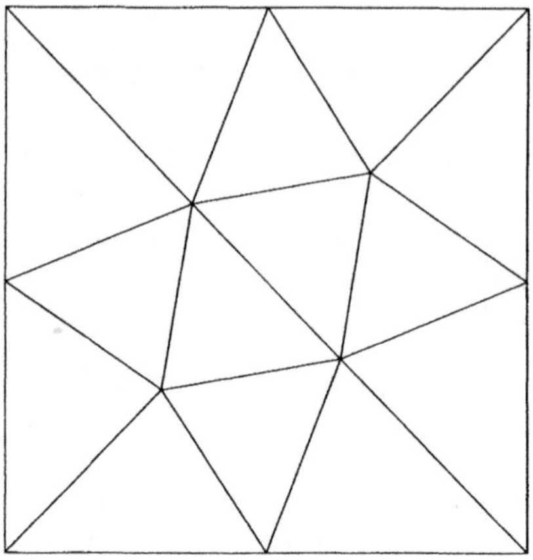

Figure 11: The background mesh before the bisection.

5. The MFA bisection is then repeated by initializing the temperature with a value slightly higher than the temperature identified.

The bisection of the mesh in Figure 11 is shown in Figure 12. The initial temperature was chosen as 3. The total number of iterations for the bisection was 25.

The effect of the decrease in temperature upon individual spins for this example is represented in Figure 13. This Figure shows that at high temperatures the spin average diverges to near zero for all the elements which indicates that the bisection is maximally disordered. As the temperature is decreased, the system reaches a critical temperature where each element starts to move significantly into one or another of two bisections. At sufficiently low temperatures the spins saturate at or near values of 1 or -1. Therefore, if a net is initialized with a temperature just above or equal to the critical temperature, the fastest convergence to a good global minimum should occur. Thus a bisection for the mesh shown in Figure 11 was carried out but this time with an initial temperature 0.8. The total number of iterations for this bisection was 11 which produced the same result shown in Figure 12. The temperature may be lowered using a cooling factor (0.9) after one complete iteration of equation 21.

The initial imbalance factor α is usually selected as 1.0 for a balanced bisection. This value in most cases ensures a balanced bisection but a minimum cutsize may not be produced. Therefore it was decided to carry out an

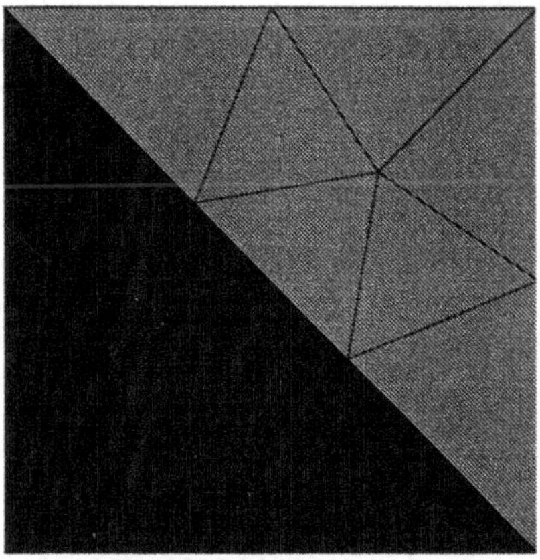

Figure 12: The mesh after the bisection.

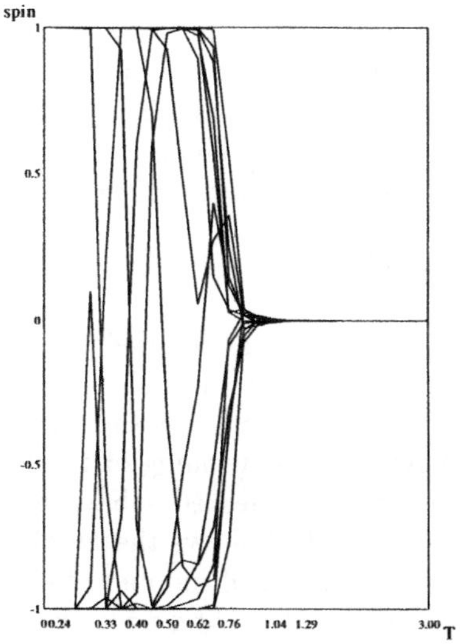

Figure 13: The variation of spin averages as the temperature is decreased.

inverse annealing or, in other words, incrementation of the initial value of α. This was carried out by selecting a small initial value for α (for example $\alpha \leq 0.1$) and once the neural network optimization has started, α is increased by a factor (for example 1.5) after one complete iteration of equation 21 until it reaches 1.0. The value of α is set at 1.0 for the rest of the optimization procedure. This dynamic implementation of α proved to be highly efficient in creating a good minimum bisection interfaces while producing two well balanced subdomains. This is due to the stronger emphasis which has been made upon the minimization of the cutsize at the early stages of the optimization.

7.2 The Predictive Mean Field Hamiltonian for Domain Decomposition

Equation 21 may be modified in order to accommodate the predicted number of elements which may be generated within an element of the coarse mesh. This equation was thus modified and the MFA bisection equation for using the predicted number of elements is given by:

$$v_i = \tanh\left(\sum_{j=1}^{N}(t_{ij} - (\alpha * N_{pred}))\frac{v_j}{T}\right) \tag{23}$$

where N_{pred} represents the predicted number of elements.

Equation 23 may be applied iteratively using the same procedure as the conventional asynchronous method described earlier by initializing v_j with small continous random values. The temperature may be lowered by a cooling factor (0.9) after each complete iteration of equation 23 .

7.3 Example Study

The domain shown in Figure 14 was selected for the example study and is uniformly meshed. This provided an initial mesh with 153 elements and is shown in Figure 15. This initial mesh was then decomposed using the MFA method. Figure 16 shows the partitioning of the initial mesh by the MFA method. Figure 17 shows the remeshed subdomains after the partitioning by the the MFA method. Table 1 shows the number of elements generated within each subdomain and the ideal number of elements required per subdomain. The total number of interfacing nodes between the partitions after the decomposition is 188. The total computation time for partitioning using a single transputer was 19.98 seconds using a single T800 transputer.

8 Concluding Remarks

The implementation of the BP algorithm on a transputer array computer with good performance has been demonstrated. The increase in training

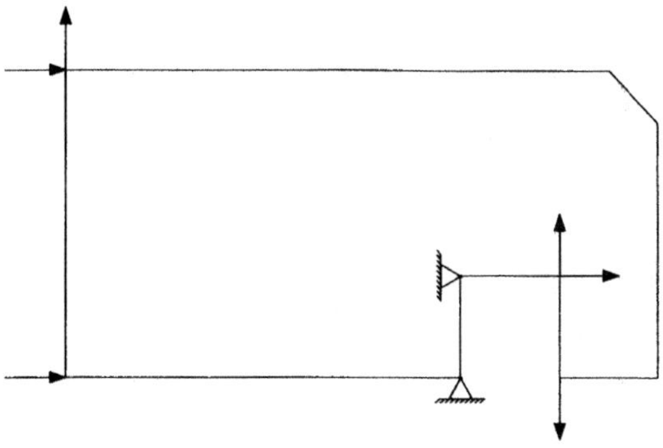

Figure 14: A domain with cut-out and chamfer.

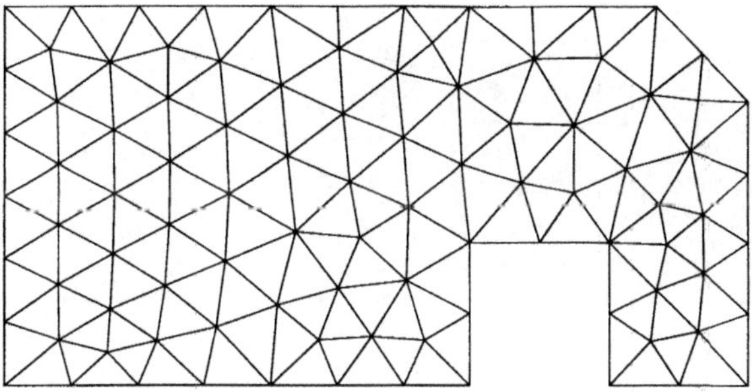

Figure 15: The initial mesh with 153 elements.

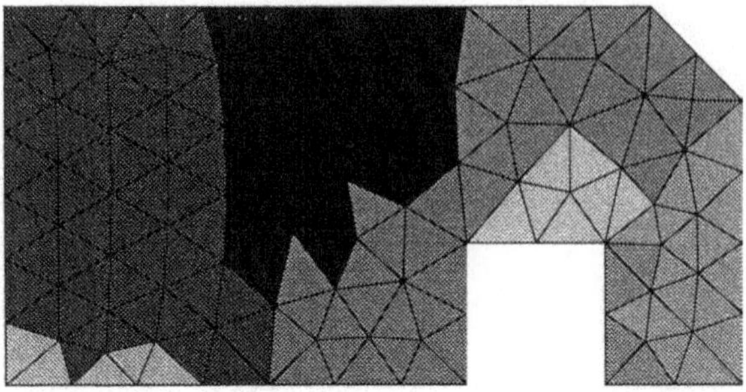

Figure 16: The initial mesh with 153 elements divided into eight partitions using the MFA method.

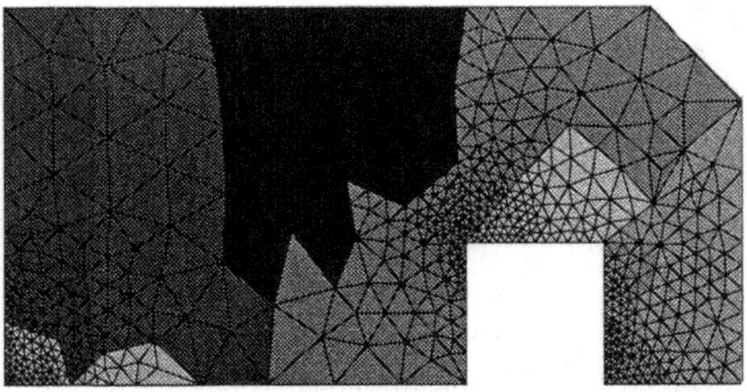

Figure 17: The remeshed subdomains with 1166 elements partitioned by the MFA method.

Subdomain No.	Generated Elements (Actual)	Required Elements	diff	%age diff
1	136	146.5	-10.5	-7.16
2	160	146.5	13.5	9.21
3	151	146.5	4.5	3.07
4	157	146.5	10.5	7.16
5	146	146.5	-0.5	-0.34
6	138	146.5	-8.5	-5.80
7	141	146.5	-5.5	-3.75
8	137	146.5	-9.5	-6.48

Table 1: Comparison of the actual number of generated elements per subdomain and the ideal number of elements per subdomain using the MFA method.

speed is significant but two factors should be noted:

1. The parallel single pattern training is relatively efficient because weights are changed after each pattern and faster convergence may be achieved using higher learning and momentum rates. The high rate of these values, however, does not guarantee a true gradient descent and therefore the net may not be able to learn efficiently. Continued training using low learning and momentum rates may assist in improving the quality of the gradient descent.

2. The parallel epoch pattern training is relatively fast in terms of the number of training cycles completed per second. This is because interprocessor communication is less of an overhead. With epoch pattern training the weight changes are modified after the propagation of all i-o patterns hence the learning and momentum rates may have to be much lower than those used for the single pattern training. Consequently convergence of a neural network to a given RMS value with epoch pattern training may be significantly slower than with single pattern training.

The effective and efficient use of BP neural networks for predicting the number of adaptively refined elements which are generated per coarse element was demonstrated. This information is of great value when subdomain generation and parallel finite element mesh generation is carried out [15, 16]. The efficient use of neural networks in the partitioning of finite element meshes has also been described. The partitioning using the predictive neural network model has a more complex energy criteria which may consist of many local minimas. Perhaps the only drawback of using the predictive MFA partitioning method is the high degree of parameter sensitivity (i.e. temperature and

α). In general, good partitions may be generated for both conventional and predictive MFA mesh partitioning with little computational expense. The method also has a high potential for parallel implementation which would increase the performance of the network in terms of convergence.

Acknowledgements

We wish to thank the Marine Technology Directorate Ltd for support of this work through the research contracts: "High Performance Computing for Marine Technology Research", (ref: GR/J22191); and "High Performance Adaptive Finite Element Computations for CAD of Offshore Structures using Parallel and Heterogeneous Systems", (ref: GR/J54017). The research described in this paper was also supported by the Systems Architecture Committee of the UK Engineering and Physical Sciences Research Council through the research contract: "Domain Decomposition Methods for Parallel Finite Element Analysis", (ref: GR/J51634). The support from British Gas Plc, Engineering Research Station, Killingworth, Newcastle upon Tyne is greatfully acknowledged.

The authors wish to thank other members of the Structural Engineering Computational Technology Research Group in the Department of Mechanical and Chemical Engineering, Heriot-Watt University for useful discussions.

References

[1] P. T. Baffes, *Nets User's Guide, version 2.01*, NASA, Lyndon B. Johnson Space Center, 1989

[2] W. S. McCulloch and W. Pitts, "A Logical Calculus of the Ideas Immanen in Nervous Activity", *Bulletin of Mathematical Biophysics* 5, 1943, 115-133.

[3] M. Minsky and S. Papert, *Perceptrons: An Introduction to Computational Geometry*, MIT Press, 1969.

[4] D. E. Rumelhart, G. E. Hinton, R. J. Williams, "Learning Internal Representation by Error Propagation", in D.E. Rumelhart and J. L. McClelland (Eds.) *Parallel Distributed Processing: Explorations in the Microstructure of Cognition, Vol. 1 : Foundations.*, MIT Press, 1986.

[5] Y. H. Pao, *Adaptive pattern recognition and neural networks*, Addison-Wesley Publishing Company, 1989

[6] A.I. Khan, B.H.V Topping, A. Bahreininejad, "Parallel Training of Neural Networks for Finite Element Mesh Generation" in B.H.V Topping and A.I. Khan (Eds.) *Neural Networks and Combinatorial Optimization in Civil and Structural Engineering*, 81-94, Civil-Comp Press, 1993.

[7] B.H.V. Topping, A. Bahreininejad and A.I. Khan, "Neural Networks for Adaptive Finite Element Mesh Generation", Submitted for publication in Engineering Applications of Artificial Intelligence, 1994.

[8] Y. Izui, A. Pentland, *Speeding Up Back Propagation* in M. Caudill, (Editor), *Theory Track Neural & Cognitive Sciences Track* of the Proceedings of the International Joint Conference on Neural Networks, IJCNN-90-WASH-DC, Lawrence Erlbaum Associates, Publishers, New Jersey, United States of America, vol.1, 639-642, 1990

[9] M. Hagiwara, *Accelerated Back Propagation using Unlearning based on Hebb Rule* in M. Caudill, (Editor), *Theory Track Neural & Cognitive Sciences Track* of the Proceedings of the International Joint Conference on Neural Networks, IJCNN-90-WASH-DC, Lawrence Erlbaum Associates, Publishers, New Jersey, United States of America, vol.1, 617-620, 1990

[10] S-B. Cho, J.H. Kim, *An Accelerated Learning Method with Backpropagation* in M. Caudill, (Editor), *Theory Track Neural & Cognitive Sciences Track* of the Proceedings of the International Joint Conference on Neural Networks, IJCNN-90-WASH-DC, Lawrence Erlbaum Associates, Publishers, New Jersey, United States of America, vol.1, 605-608, 1990

[11] A.I. Khan, B.H.V. Topping, *"Parallel adaptive mesh generation"*, Computing Systems in Engineering, vol. 2, no. 1, 75-102, 1991.

[12] 3L Ltd., *Parallel C User Guide*, 3L Ltd., Livingston, Scotland, 1988

[13] Russell C. Eberhart and Roy W. Dobbins, *Neural Network PC Tools*, Academic Press, Inc., 1990

[14] Ken L. Parker and Allison L. Thornbrugh, "Parallelized BP Training and Its Effectiveness", (Eds.) Maureen Caudill, *International Joint Conference on Neural Networks*, (Vol. 2), Lawrence Erlbaum Associates, Inc., 1990, Washington D.C.

[15] A.I. Khan, B.H.V. Topping, *"Subdomain Generation for Parallel Finite Element Analysis"*, Computing Systems in Engineering, Vol. 4, Nos 4-6, 473-488, 1993.

[16] B.H.V. Topping, A.I. Khan, *"Parallel Finite Element Computations"*, Saxe-Coburg Publications, Edinburgh, 1995.

[17] C. Peterson and T. Rognvaldsson, "An Introduction to Artificial Neural Networks", C. Verkerk (Ed.) *Proceedings of 1991 CERN School of Computing*, CERN Vol 92, No. 2, 113-170, May 1992.

[18] A. Bahreininejad, B.H.V. Topping and A.I. Khan, *"Subdomain Generation Using Multiple Neural Networks Models"*, in B.H.V. Topping, M. Papadrakakis, (Eds) "Parallel and Vector Processing for Structural Mechanics", Civil-Comp Press, Edinburgh, 1994.

[19] J. J. Hopfield and D. W. Tank, *"Neural Computation of Decisions in Optimization Problems"*, Biological Cybernetics, **52**, 141-152, 1985.

[20] S. Kirkpatrick, C. D. Gelatt Jr and M. P. Vecchi, *"Optimization by Simulated Annealing"*, Science, **220:4598**, 671-680, 1983.

[21] C. Peterson and J. R. Anderson, *"A Mean Field Learning Algorithm for Neural Networks"*, Complex Systems, **1**, 995-1019, 1987.

[22] C. Peterson and J. R. Anderson, *"Neural Networks and NP-complete Optimization Problems; A Performance Study on the Graph Bisection Problem"*, Complex Systems, **2:1**, 59-89, 1988.

[23] J. Hertz, A. Krogh and R. G. Palmer, *"Introduction to the Theory of Neural Computing"*, Addison-Wesley Publication Company, 1991.

[24] D. E. Van den Bout, T. K. Miller III, *"Graph Partitioning using Annealing Neural Networks"*, IEEE Transaction on Neural Networks, Vol. 1, No. 2, 192-203, June 1990.

Application of the Simulated Annealing Algorithm to Structural Design

Richard J. Balling
Department of Civil and Environmental Engineering
Brigham Young University
Provo, UT 84602

Abstract: The simulated annealing algorithm and its convergence properties are presented. Structural engineering applications are surveyed. Techniques for increasing the speed of the algorithm are examined including filtering and parallelization. The relationship of the simulated annealing algorithm to genetic algorithms and neural networks is investigated.

Keywords: structural optimization, simulated annealing, stochastic search

1 The Simulated Annealing Algorithm

The annealing algorithm originally described by [1] was first applied to discrete optimization problems by [2]. In that paper, good results were obtained for large NP-hard problems such as the Traveling Salesman Problem and an integrated circuit design problem. In the past decade, the use of simulated annealing strategies to solve optimization problems has mushroomed with applications in a wide range of disciplines.

Consider an optimization problem in which a scalar-valued objective function, f(x), is minimized where x is a set of discrete-valued design variables. Assume that the search space is finite, and let N be the total number of possible designs. Thus, the global optimum could be found by exhaustive search in N evaluations of the objective function f.

The simulated annealing algorithm has been shown, using Markov chains, to converge to the global optimum of the above problem with probability as close as desired [3-5]. The basic algorithm proceeds as follows:

1. Obtain a starting current design x_0, compute $f_0=f(x_0)$, specify a cooling schedule of M+1 temperatures $T_0, T_1, \ldots, T_M$, and set i=0.
2. Randomly generate a candidate design x_c which is a neighbor to the current design x_i, and compute $f_c=f(x_c)$.
3. Set x_{i+1} to x_c (and f_{i+1} to f_c) with probability

$$A=\min\{1, \exp((f_i-f_c)/T_i)\} \qquad (1)$$

and set x_{i+1} to x_i (and f_{i+1} to f_i) with the complementary probability.
4. Stop if i=M and return x_i as the optimal design, otherwise set i=i+1 and return to Step 2.

The key to the algorithm is the acceptance probability in Step 3. Note that if $f_c \leq f_i$, then $x_{i+1}=x_c$ with probability one. But if $f_c>f_i$, then $x_{i+1}=x_c$ with probability less than one. Thus, better candidates are always accepted, and worse candidates are sometimes accepted. The probability of accepting a worse design is greatest when f_c-f_i is small and/or T_i is high. This allows the simulated annealing algorithm to break out of local minima in the earlier iterations when the temperature is hot. Two parts of the algorithm require further elaboration: the random generation of candidates in Step 2, and the specification of a cooling schedule in Step 1. These tasks will now be discussed.

The random generation of candidate designs from current designs is application-dependent. However, the process must satisfy certain requirements. Let P_{ij} be the probability of generating candidate design x_j from current design x_i . One requirement on the generation of candidates is symmetry: $P_{ij} = P_{ji}$ or all possible designs x_i and x_j in the search space. Candidates must be generated within neighborhoods of the current designs. The neighborhood of current design x_i is the set of designs x_j for which $P_{ij} >0$, while $P_{ij}=0$ for all designs outside its neighborhood. Finally, the random generation of candidates must satisfy a connectedness requirement. This requirement states that for any two possible designs x_i and x_j in the search space, there exists a finite number of intermediate designs $x_1', x_2', x_3', \ldots, x_K'$ such that:

x_1' is in the neighborhood of x_i ;
x_2' is in the neighborhood of x_1';
x_3' is in the neighborhood of x_2';
...
x_K' is in the neighborhood of x_j .

The cooling schedule is a sequence of non-increasing temperatures, T_0, T_1, T_2,..., T_M, used at each iteration of the algorithm. Consider the case where all temperatures in the schedule are equal to the final temperature T_M. The convergence proofs show that one can get as close as desired to the global optimum if one sets T_M low enough and performs enough iterations to reach steady state. The number of required iterations for a given T_M is a function of the second largest eigenvalue of the transition matrix of the Markov chain which depends on the scheme used to generate random candidates. This number may be quite large, though finite, and is generally impossible to compute for large transition matrices of size NxN. In general, the number of required iterations is greater for smaller values of T_M. Best performance is achieved by starting with temperatures at high values and gradually decreasing to low values as the iterations proceed. The optimal cooling schedule is impossible to compute except in the most trivial of problems. Rarely in practice are enough iterations performed to reach steady state at end of the algorithm.

Many cooling schedules found in practice run at a constant temperature for a number of iterations and then drop the temperature by multiplying the current temperature by a specified constant which is less than one. A pragmatic cooling schedule is suggested by [5] which they claim has performed well on many simulations. Their schedule reduces the temperature at every iteration as follows:

$$T_0 >> U \qquad T_i = T_0/(i+1) \qquad M = (T_0/e)\ln(N) \qquad (2)$$

The parameter U is an estimate of the maximum f_j-f_0 where f_0 is the objective value of the starting current design, and f_j is the objective value of any other possible design in the search space. The parameter e is the acceptable error in the optimal objective value. Note that for large N, the number of iterations is clearly less than that required by exhaustive search where M=N.

The simulated annealing algorithm may be modified for constrained problems. This may be accomplished by multiplying the constraints by penalty parameters and adding the products to the objective function. Alternatively, the acceptance probability in Step 3 may be modified such that the probability of accepting infeasible candidates is zero. The latter approach requires a feasible starting design. The latter approach also requires that the connectedness requirement be modified such that any two feasible designs x_i and x_j are connected via neighborhoods by a finite number of feasible intermediate designs x_1', x_2', x_3',... x_K'.

The algorithm may be applied to problems involving continuous-valued design variables by discretizing continuous design space. Alternatively, the random generation of candidates may consist of the random perturbation of each design variable. The perturbation probability should be uniformly distributed ranging from minus to plus neighborhood size. Since N goes to infinity for continuous optimization problems, the number of iterations, M, is difficult to set. In general, gradient-based algorithms are much more efficient in solving continuous optimization problems unless the objective and constraint functions are fraught with multiple local minima, noise, or discontinuities.

1.1 Structural Engineering Applications

The simulated annealing algorithm has been applied to optimization problems involving the design of space trusses for astronomical installation. These trusses are comprised of many equal-length members, and they support telescopes and other scientific equipment. Typically, they must maintain a fixed geometric shape to micron-level accuracy when subject to a variety of static, dynamic, and thermal disturbances. The manufacturing process of the truss members may yield members whose lengths are in error. Nevertheless, the manufactured lengths can be accurately measured. The simulated annealing algorithm was used in [6] to determine the optimal locations of the measured truss bars in a given truss so as to minimize distortion and internal forces. Candidate designs were generated by randomly selecting two members in the current design and swapping. In another application [7], the simulated annealing algorithm was used to determine which members of a given truss should be replaced by a limited number of actively controlled truss members so as to maximize the cumulative energy dissipated over a finite time interval. Candidate designs were generated by randomly selecting an active member and an inactive member in the current design and swapping.

The author applied the simulated annealing algorithm to the discrete optimization of 3D steel framed buildings [8]. The members of such frames were grouped into beam and column groups, and the section for each group had to be selected from among standardized sections. The total frame weight was minimized subject to the satisfaction of various code-specified constraints on stress, buckling, and deflection under dead, live, and seismic load cases. A conservative starting design was obtained, and the acceptance criteria of the simulated annealing algorithm was modified so that infeasible designs were always rejected. Actual structural analysis of candidate designs was replaced by an approximate analysis about the starting design, and then used to explicitly evaluate stress and buckling formulas [9].

The standardized sections were ordered by increasing cross-sectional weight, and candidate designs were generated by randomly selecting a single member group in the current design and changing its section to either the next heavier or next lighter section. Note that this procedure did not require the construction of functional relationships between cross-sectional properties. In fact, although the section area was monotonically increasing in the ordering of the standardized shapes, other section properties such as strong and weak moments of inertia and section moduli were not. This led to the existence of "local minima" in the search strategy. Local minima in the search strategy also resulted from the fact that the strategy modified only one member group at a time, since occasionally the reduction of frame weight while maintaining feasibility requires the simultaneous increase and decrease in the weights of multiple member groups.

The simulated annealing algorithm produced results comparable to those achieved by enumerative techniques such as the linearized branch and bound technique [10]. One drawback of the algorithm is the excessive number of analyses (albeit approximate analyses) required. Techniques for speeding up the simulated annealing algorithm will be discussed in the next section.

The author and a graduate student have also recently experimented with applying the simulated annealing algorithm to the design of reinforced concrete frames (work not yet published). In this work the optimization problem was decomposed so that the gross section dimensions of the members were optimized at the system level while the reinforcement was optimized in each member individually. Such a decomposition is natural for reinforced concrete structures since the member weights and stiffness required by system-level frame analysis can be approximated directly from gross section dimensions without regard for reinforcement. However, the calculation of the objective and constraints at the system level requires the execution of frame analysis as well as the optimization of the reinforcement in each of the members.

The simulated annealing algorithm was applied at the system level to find optimal continuous values of the gross section dimensions. A gradient-based algorithm proved ineffective because the objective and constraint functions were woefully discontinuous. These discontinuities arose from the fact that evaluation of functions such as strength and cost required the discrete optimization of the reinforcement. Discrete reinforcement optimization involved determination of the number and sizes of bars, the bar layout, the cut-off and bend points, and the number stirrups and ties. Exhaustive search and heuristic search were used in this determination.

The simulated annealing algorithm made it possible to achieve good results. But again, the computational effort is excessive due to the large number of function evaluations required.

In summary, the simulated annealing algorithm has been used to solve discrete optimization problems and ill-conditioned in structural design. One might expect to see topological optimization applications in the future. The simulated annealing algorithm is suitable for topological optimization problems since it does not require gradients and it is well-adapted to discreteness. Researchers may find ways to use the algorithm to build up optimal truss and frame topologies in a general way. Creative ideas for how to randomly generate candidate neighboring topologies from the current topology in a fashion which is symmetric and connected are needed. Evaluation of candidate topologies may include the gradient-based optimization of member sizes and nodal coordinates. Creative ideas for streamlining topology evaluation are needed.

1.2 Speeding Up the Algorithm

In the case of structural design, computational effort is dominated by the structural analysis used to evaluate objectives and constraints, and the simulated annealing algorithm typically requires several such evaluations. How can the computational time be reduced? Two approaches will be examined in detail: filtering and parallelization.

A filter for the simulated annealing algorithm was proposed by [11]. The idea behind the filter was to use additional knowledge about the problem to filter out ridiculous candidate designs before they are evaluated. The additional knowledge came in the form of gradients evaluated at the starting design. Thus, the filter was applicable only to problems where gradients are available. This would rule out most topological optimization problems.

After a candidate was generated, a rapid approximation of the objective, f, and each constraint function, g, was evaluated. The approximate values of f and each g for a candidate design, f_c* and g_c*, were based on their values at the current design, f_i and g_i, and their gradients at the starting design, ∇f_0 and ∇g_0:

$$f_c^*=f_i+\nabla f_0 \bullet (x_c-x_i) \qquad g_c^*=g_i+,\nabla g_0 \bullet (x_c-x_i) \tag{3}$$

An estimate of the candidate "potential" was calculated as:

$$P=\max(0, f_c^*-f_i)+\max_{\text{over constraints}}[\max(0,g_c^*)] \tag{4}$$

The above formula assumes that the constraints are satisfied when less than zero, and that the objective and constraints are properly scaled to the same order of magnitude. The candidate was then passed on for exact function evaluation with probability S^P and rejected outright with the complementary probability. The parameter $0<S\leq 1$ is the filter sieve size. When S=1, the algorithm is basically unfiltered and all candidates are passed on for exact evaluation. As S is decreased, the number of candidates which are filtered out increases.

The filtered simulated annealing algorithm was applied to the discrete optimization of steel frames as described in the previous section. Recall that in that application, the "exact" evaluation of the objective and constraint functions was itself a special approximation. Nevertheless, this special approximation required significantly more computation time than the evaluation of the candidate potential described in this section. Figure 1 shows plots of the objective function for the current design vs. iteration. Note that as the sieve size is decrease, the jumps in the objective value for the current design are less erratic. Nevertheless, upward jumps in objective value are still possible even with a low sieve size. This is necessary to allow the algorithm to jump out of local minima.

The solution time of the algorithm was reduced over that required by the unfiltered algorithm by factors of 4, 8, and 13 for sieve sizes of 0.1, 0.01, and 0.001, respectively. This reduction was due to the fact that many candidate designs are filtered out before exact evaluation. For filter sizes of 0.1 and 0.01, the algorithm reached the same optimum design as was reached by the unfiltered algorithm in 3 out of 4 runs with different random number sequences. A design with a slightly higher objective value was reached for a filter size of 0.001.

At first glance, the simulated annealing algorithm may seem to be a sequential process which is not amenable to parallelization. However, an interesting parallel implementation based on the concept of speculative computation is given by [12].

Assume that the computational effort is dominated by the function evaluations of the objective and constraints. If one has three processors available, one can go ahead and speculate on the outcome of the accept/reject criteria and generate the candidates for both outcomes before any function evaluations take place. The function evaluations for the candidates corresponding to the three nodes of the balanced tree in Figure 2 can be performed in parallel. After the results are

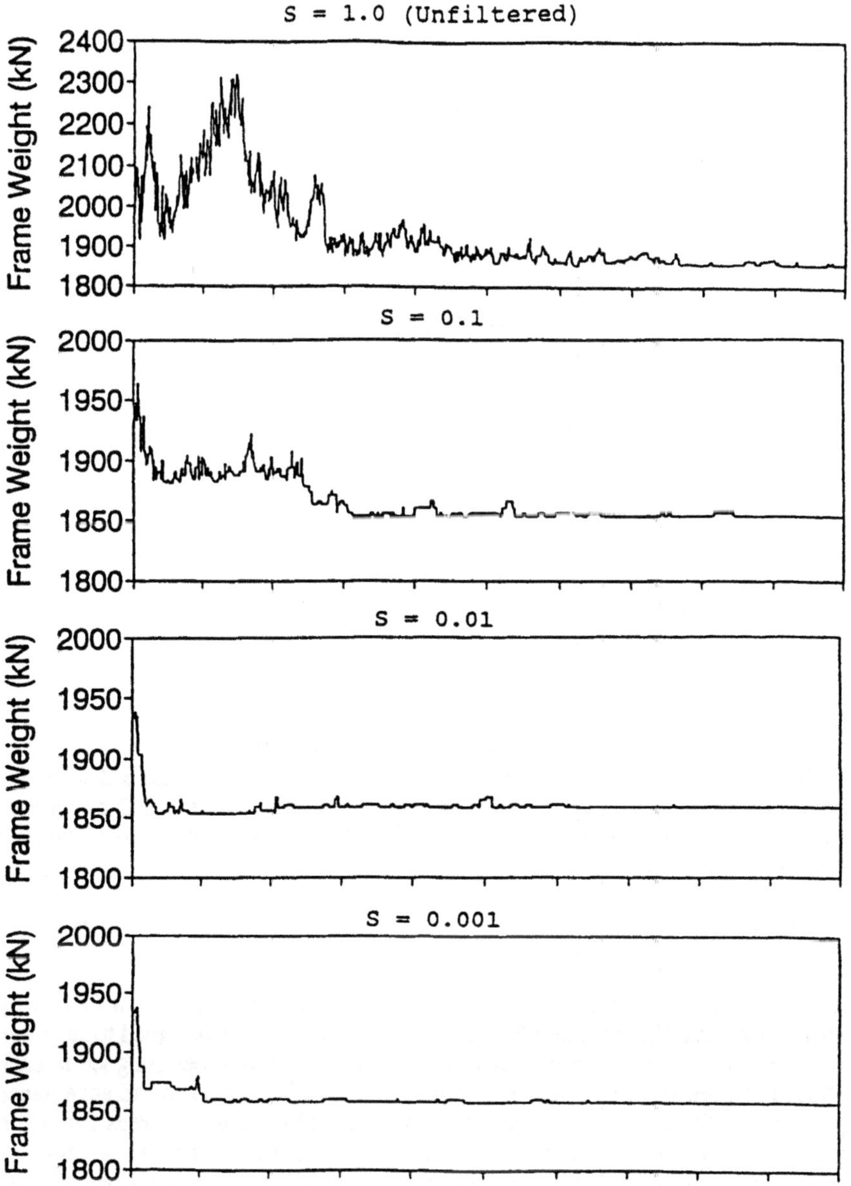

Figure 1: Objective vs. Iteration for Different Sieve Sizes

obtained, it will be recognized that one of the function evaluations was wasted effort. Nevertheless, two iterations of the algorithm will have been performed in the time it would have taken for one iteration of the serial implementation.

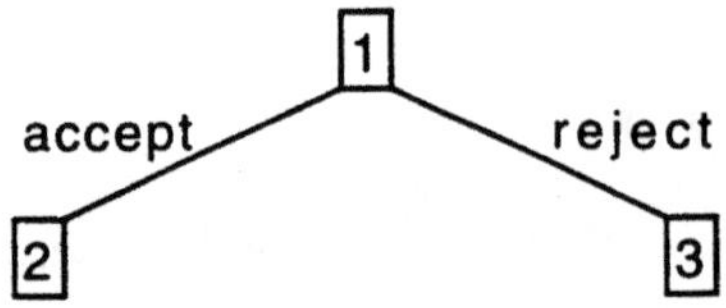

Figure 2: Three-Node Balanced Tree

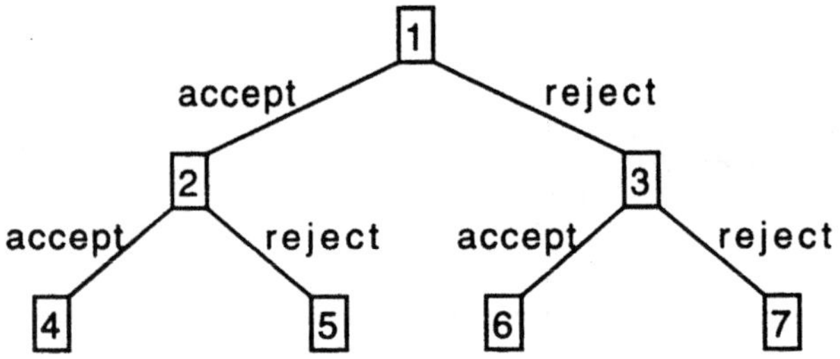

Figure 3: Seven-Node Balanced Tree

Using seven processors as shown in the balanced tree in Figure 3, one can perform three iterations of the parallel implementation in the time it takes for one iteration of the serial implementation. In general, if one has Q processors available, the speed up from the serial implementation to the parallel implementation is:

$$R=\log_2(Q+1) \tag{5}$$

The speed up of the parallel implementation can be increased even further by speculating that the same number of processors need not be assigned to rejection nodes as are assigned to acceptance nodes. In fact, the number assigned to each should change from iteration to iteration and should be tied to the acceptance probability for the current iteration. The optimal unbalanced trees for seven processors are shown in Figure 4 for different values of the acceptance probability, A. Of course, acceptance probability is a function of both temperature and the increase in objective value from current to candidate design (see Equation 1). Some average value of the latter would have to be used so that the tree could be specified before function evaluations are computed. The following formula for speed up due to parallelization is given by [12] assuming that the optimal unbalanced tree is continually updated as the simulated annealing algorithm proceeds:

$$R=(Q-\log_2 Q)/(\ln Q-\ln(\log_2 Q)) \tag{6}$$

As Q goes to infinity, R goes to Q/lnQ. Speed ups for a constant balanced tree and for optimally updated unbalanced trees are compared below:

Q	R, balanced tree	R, unbalanced trees
4	2.3	2.9
8	3.2	5.1
16	4.1	8.7
32	5.0	14.5
64	6.0	24.5

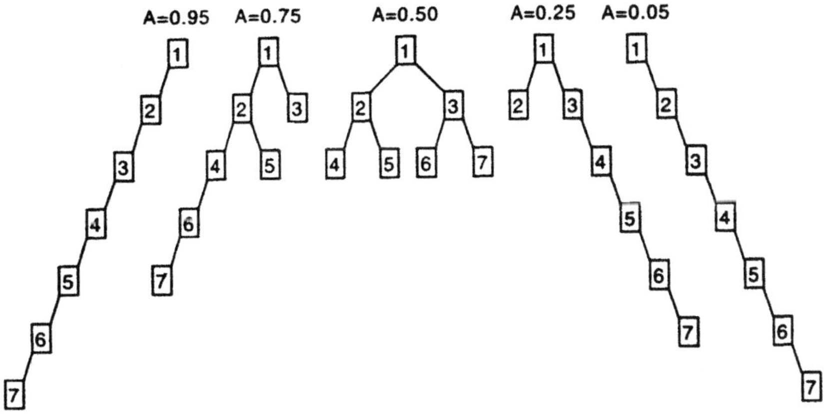

Figure 4: Seven-Node Unbalanced Trees

1.3 Relationship to Genetic Algorithms and Neural Networks

Genetic algorithms, like the simulated annealing algorithm, are a class of stochastic search strategies for optimization. They have many properties in common: they have the ability to converge to global optima; they are applicable to both continuous and discrete problems; they can treat nasty objective and constraint functions; they do not require function gradients. Genetic algorithms differ from the simulated annealing algorithm in that the current and candidate designs are replaced with entire populations of designs. In the simulated annealing algorithm, the process for randomly generating candidate designs is left open for the application. In genetic algorithms, the process for randomly generating the next population is application-independent and specifically involves encoding designs as chromosomes and the crossover and mutation operations. Another fundamental difference is the absence of acceptance/rejection criteria in genetic algorithms. Instead, genetic algorithms insure that progress towards the optimum is made in the reproduction operation by favoring as parents the most fit members of the current population.

A genetic annealing algorithm was proposed by [13]. In that algorithm, operations from genetic algorithms were used to generate a candidate population from the current population. A scalar objective for the population was evaluated as the sum of the objectives of members of the population. This scalar was used to evaluate an acceptance probability for the entire candidate population. The goal of the genetic annealing algorithm was to use genetic algorithms to improve to quality of candidate designs.

The simulated annealing algorithm has been used in conjunction with neural networks. In particular, it has been used to converge fully-connected Hopfield nets to global minimum states when starting from some specified input pattern. Such a network is referred to as a Boltzmann machine [14]. The Boltzmann machine can be used to solve combinatorial optimization problems such as the Traveling Salesman Problem.

The simulated annealing algorithm has also been used to optimally adjust weights during the training phase for feedforward networks [15]. It was compared to the gradient-based backpropogation algorithm in reducing the squared error between computed and desired outputs for a set of training pairs. It was proposed that the simulated annealing approach to training is superior particularly when neural networks are applied to generalization problems as opposed to recognition problems. This is because the ratio of number of nodes to number of training pairs is lower for generalization problems, and it is important to achieve global optimality during training rather than local optimality.

References

1. Metropolis, N., Rosenbluth, A.W., Rosenbluth, M.N., Teller, A.H.: Equation of State Calculation by Fast Computing Machines. Journal of Chemical Physics 21, 1087-1092 (1953)
2. Kirkpatrick, S., Gelatt, C.D., Vecchi, M.P.: Optimization by Simulated Annealing. Science 220, 671-680 (1983)
3. Geman, S., Geman, D.: Stochastic Relaxation, Gibbs Distributions, and the Bayesian Restoration of Images. IEEE Transactions on Pattern Analysis and Machine Intelligence PAMI6, 721-741 (1984)
4. Mitra, D., Romeo, F., Snagiovanni-Vincentelli, A.: Convergence and Finite-Time Behavior of Simulated Annealing. Proceeding IEEE Conference on Decision and Control (1985)
5. Lundy, M., Mees, A.: Convergence of an Annealing Algorithm. Mathematical Programming 34, 111-124 (1986)
6. Kincaid, R.K., Padula, S.L.: Minimizing Distortion and Internal Forces in Truss Structures by Simulated Annealing. Proceedings AIAA/ASME/ASCE/AHS Structures, Structural Dynamics, and Materials Conference, Long Beach, CA 1990
7. Chen, G-S, Bruno, R.J., Salama, M.: Optimal Placement of Active/Passive Members in Truss Structures Using Simulated Annealing. AIAA Journal 29, 1327-1334 (1991)
8. Balling, R.J.: Optimal Steel Frame Design by Simulated Annealing. ASCE Journal of Structural Engineering 117, 1780-1795 (1991)
9. Vanderplaats, G.N., Salajegheh, E.: New Approximation Method for Stress Constraints in Structural Synthesis. AIAA Journal 27, 352-358 (1989)
10. Hagar, K., Balling, R.J.: New Approach for Discrete Structural Optimization. ASCE Journal of Structural Engineering 114, 1120-1134 (1988)

11. May, S.A., Balling, R.J.: A Filtered Simulated Annealing Strategy for Discrete Optimization of 3D Steel Frameworks. Structural Optimization 4, 142-148 (1992)
12. Witte, E.E., Chamberlain, R.D., Franklin, M.A.: Parallel Simulated Annealing Using Speculative Computation. IEEE Transactions on Parallel and Distributed Systems 2, 483-494 (1991)
13. Bilbro, G.L., Hall, L.C.: A Provably Convergent Inhomogeneous Genetic Annealing Algorithm. SPIE Neural and Stochastic Methods in Image and Signal Processing 1766, 50-60 (1992)
14. Aarts, E.H.L., Korst, J.H.M.: Simulated Annealing and Boltzmann Machines. New York: Wiley 1989
15. Shekhar, S., Amin, M.B.: Generalization by Neural Networks. IEEE Transactions on Knowledge and Data Engineering 4 (2), 177-185 (1992)

Structured Multiresolution Neural Networks for Effective Processing of Signals

Stefanos D. Kollias

National Technical University of Athens
Department of Electrical and Computer Engineering
157 73 Zographou, Athens, Greece

Abstract. In this paper we present a class of structured neural networks that are able to effectively process, analyze, or classify one or two dimensional signals. Multiresolution signal decomposition is combined for this purpose with constructive network learning algorithms, resulting in hierarchical network structures which are efficiently trained and have good generalization capabilities. A simulation study is given which illustrates the ability of the networks to handle real-life problems.
Keywords. Multiresolution analysis, structured neural networks, signal processing

1 Introduction

Multilayer perceptrons have been widely examined in the neural network field as a tool for signal and image processing, especially for approximation, classification and interpretation purposes [14], [24]. A crucial aspect concerning the performance of multilayer network classifiers is generalization, i.e. the ability of the network to classify correctly input data which were not included in its training set. Results from various applications have shown that good generalization is a result of appropriate network design; a rather small network size can make the network learn incomplete solutions, while an unnecessarily large size may lead the network learn only the specific training samples and noise. A small number of interconnection weights (i.e., free parameters during training) should generally be used and any a-priori knowledge about the problem should be included in the network structure. It should be mentioned that some very good results have been obtained when structured networks are applied directly to signals and images [20],[7].

Backpropagation is a wide-spread learning technique for feedforward neural networks; it, however, suffers from various problems. A first has to do with network scaling, i.e., the impractical long learning times required for training networks of large sizes, as well trapping of network training in local minima, which also is more critical in networks of large sizes. There have been many approaches aiming at solving these problems. As far as speed of learning is concerned, a frequently used technique lets each synapse have its own learning rate parameter, increasing or decreasing its value according to the number of sign changes observed in the partial derivative of the error function with respect to the corresponding weight [17]. Other techniques adopted second order variants of backpropagation, using approximations of the Hessian matrix in the minimization procedure [20], [18]. These variants are not believed to bring a tremendous increase in learning speed, but converge reliably, without requiring extensive adjustments of the learning parameter.

A second crucial problem of backpropagation, which is closely related to network generalization is the need for a-priori specification of the network architecture, i.e., the number of hidden layers, as well as the number of nodes per layer. Since a-priori knowledge of the network architecture is very rarely available, the above unknown numbers are usually selected by a trial and error procedure. Two main categories of methods, namely pruning and constructive, have been proposed for automatic network structure specification during network training. All methods in these categories try to obtain a network which is as small-sized as possible, and at the same time, implements the required input-output mapping in a satisfactory way.

Pruning techniques [25] start training networks which include more neurons or connections than are actually necessary and use various criteria to reduce their number. Some of these techniques add a penalty measure of the network complexity to the conventional sum of squared output error criterion that is ususally minimized in backpropagation learning [4],[10], [16]. This penalty criterion helps to generate simple structured networks. Weight decay, which penalizes the complexity of the network by letting each weight decay towards zero at a rate that is proportional to its magnitude, is such a simple technique; weight elimination [30], or weight sharing and receptive fields architectures [20] are similar methods, which are also amenable to parallel implementations [19].

Constructive algorithms [27], on the other hand, first build an approximate model of low size and then add nodes to the network while learning more details, approaching, therefore, the optimum network size from below. Algorithms like node or network splitting [31], [13], as well as procedures such as the ones used by cascade-correlation [8] can be included in this framework.

Multiresolution network training is presented in this paper as a constructive approach, which starts by letting the network first process a low resolution representation of the signal and then move to higher resolution representations of it; an hierarchical procedure is constructed in this way and a weight transfer policy from the low to high resolution parts of the network is described which permits an effective structured decomposition of network training, dealing with a hierarchy of small-sized subnetworks.

Section 2 examines some of the most promising structured learning techniques that can be used to obtain appropriate small-sized multilayer networks for processing, classification, or interpretation of one and two dimensional signals. Section 3 refers to the multiresolution decomposition of signals in a sequence of approximation and detail low-resolution signals. Section 4 presents hierarchical structured networks for effective processing of signals based on multiresolution decomposition, while application of the above for classification of large sized 2-D signals is presented in section 5.

2 Structured Multilayer Networks

The problem of what constitutes a sufficiently large training sample and a suitable network architecture, for backpropagation networks, has been a subject of extensive research [2], [11]. Results indicate that the likelihood of correct generalization depends on the size of the hypothesis space, i.e. the total number of networks being considered, the size of the solution space, i.e. the set of networks which give good generalization, and the number of training examples.

Some theoretical bounds on the number of examples that are necessary to achieve generalization were given in [1]. Assuming that m random examples are chosen from some fixed, but arbitrary, probability distribution and to train a feedforward network with W weights resulting in a training error of, say $1-\frac{\varepsilon}{2}$, for m sufficiently larger than W, then there is large confidence that the network will correctly classify all but a fraction ε of future examples, belonging in the same distribution. A similar result can be shown, if a learning procedure is used, which starts with a larger number of weights, but during learning kills some 'useless' synapses. In general, using a net with W weights to fit fewer than $O(\frac{W}{\varepsilon})$ examples causes overfitting, so that good generalization cannot be quaranteed. Other bounds based on the Vapnik-Chervonenkis (VC) dimension have been proposed in the litterature [3],[15], which can offer some guide for designing networks, especially for classification purposes.

All this theoretical work still, however, has important shortcomings when applied in practice. Nevertheless, a conclusion that is easy to make from the above is that, since overfitting effects are due to oversized networks, an increase in the likelihood of correct generalization can be obtained, by minimizing the number of free parameters in the network. This, however, should be done, without reducing the size of the network to the point where it can no longer compute the desired function.

2.1 Pruning Techniques

The simplest form of pruning methods is exponential weight decay, which is implemented by adding a penalty criterion proportional to $\sum_i \sum_j w_{ij}^2$ to the conventional sum of squared output error criterion. Weight decay has some drawbacks, such as the fact that it takes infinite time for weak connections to completely fade away. Modification of the above penalty criterion, replacing the square of each weight by its absolute value, which is equivalent to a constant decay and introducing a forgetting factor in the weight updating [16] can overcome this problem, while preventing revival of completely faded connections. Similar results can be obtained by weight elimination techniques, which delete connections based on the contribution of the corresponding weights to the minimization of the quadratic error criterion. In case of the optimal brain damage technique [21], for example, this contribution can be easily expressed and measured by the product of w_{ij}^2 with the second derivative of the output error criterion with respect to w_{ij} . A comparison of various weight elimination techniques can be found in [12]. Other techniques include the addition of independent noise to the inputs of the network hidden units, resulting in skeletal network structures [23].

Another approach includes techniques that start by considering a large network and then impose constraints on the network topology. A technique that is frequently used in image recognition problems, restricts each neuron to have as input only a small portion of the input pattern, called a receptive field [20]. Thus, each hidden, or output, unit is connected to a subset of the units of the previous layer, resulting in a network with fewer interconnections between its units. Another technique in this category is weight sharing. This technique imposes equality constraints among the connection strengths, in the sense that several interconnections in the network are controlled by a single weight. In this way, the number of free parameters in the network is reduced, in an attempt to achieve better generalization, avoiding overfitting problems. Weight sharing can be used for shift-invariant feature extraction and is generally used in many pattern recognition problems. It should, however, be mentioned that the most frequently used technique in imposing weight equality constraints, when implementing weight-shared backpropagation, is the following: the increment of each weight, that is shared by different neurons, is computed independently by all corresponding neurons, while the weight is finally updated by the average of all computed increments. This technique may require a significant amount of communications among different processors, especially when implementing in parallel on-line forms of the weight-shared backpropagation algorithm [19].

Another structured network architecture, that is similar to weight sharing, are time-delay networks (TDNN). Such networks have the ability to represent relationships between events in time, and in classification problems, learn features or abstractions that are invariant under translation in time. To do this, a TDNN unit differs from a conventional NN unit, in that it receives not only the current input elements, but also delayed replicas of them. As a result, a TDNN unit has the ability to relate and compare its current input to the past history of events [29]. A time-delay network is composed of many such units, forming a multilayered structure that receives sequences of input patterns. Such networks have been used for speech recognition [29] and can also be applied in moving image monitoring or recognition. An equivalent way of investigating such networks is by using a spatially expanded input pattern plus some constraints on the weights. In particular, each collection of TDNN units is duplicated for each one frame shift in time. In this way, the whole history of activities is available at once. Moreover, the weights of the corresponding connections in the time shifted copies must be constrained to be the same. From the above it can be easily understood that time-delay networks would be equivalent to weight-shared networks, if spatial weight equality constraints were replaced by temporal ones.

2.2 Constructive Techniques

Constructive learning algorithms normally start with a small network; nodes and connections are added when the network is determined to have a low capacity that results in unnacceptable errors. After the network is expanded, only part of it is trained, while the rest is frozen. A well-known constructive algorithm is cascade-correlation [8]. This algorithm begins with a minimal network with no hidden units and then adds, during training, hidden units one at a time to create a multilayer network. Each new hidden unit is added in a way to maximize the correlation between the squared network output error and the activation function of the new unit. After the new hidden unit is added, all weights between the input and hidden units (as well as those among hidden units, if such connections exist) are frozen, allowing only hidden to output units to change.

In such constructive tecniques, freezing a specific part of the network implies that this part will not further participate in the minimization task, or that the information acquired at an early stage of the learning process will not be adjusted, or tuned, later. It is this " freezing" step which trains the network in a progressive way, with few parameters at a time, providing a fast and computationally efficient solution. In some cases, however, freezing a part of the network may limit the processing, e.g., classification, abilities of the whole network.

Node splitting in the hidden layers [31] is another constructive technique, which builds a large network from a smaller one. The method seeks for oscillations in hidden nodes of the small network, during training, which are then splitted in two parts using principal component analysis. Another constructive technique starts with a small subset of the training examples and a small network size and selectively expands the training set, as well as the network size and retrains the network until an acceptable generalization is acieved on the available data set [33].

3 The Multiresolution Decomposition

Representation of signals at many resolution levels has gained much popularity especially with the introduction of the discrete wavelet transform, implemented in a straightforward manner by filter banks using quadrature mirror filters (QMFs) [22],[28]. In image processing the above are equivalent to subband processing [32]. Signal decomposition is performed with an appropriate filter bank of decomposition (or decimating) QMF filters. An appropriate bank of reconstruction (interpolating) QMF filters guarantee perfect reconstruction of the original one-dimensional (1-D) or two-dimensional (2-D) signal from its subband components. Multiresolution decompositions result in approximation signals of low resolution that contain coarser information of the signal content and in detail signals which contain more information as resolution is gradually increasing. In the following we treat the more general case od 2-D signals. Multiresolution processing of 1-D signals can easily be derived by ignoring one of the two dimensions in the presented analysis.

Let x_0 denote an $N \times N$ 2-D signal. Using appropriate finite impulse response (FIR) perfect reconstruction filters $h_L(n)$ and $h_H(n)$, where $h_L(n)$ generally is a low-pass and $h_H(n)$ a high-pass filter, we can split the signal into four lower resolution $\frac{N}{2} \times \frac{N}{2}$ signals [5], as shown in Figure 1. Applying, for example, the low-pass filter $h_L(n)$ in the horizontal and then in the vertical direction of the original signal (let us consider the separable case, for simplicity), we get the *approximation* signal at the lower resolution level $j = -1$, denoted as x_{-1}^{LL}, where

$$x_{-1}^{LL}(m,n) = \sum_{k=1}^{N}\sum_{l=1}^{N} h_L(2m-k)h_L(2n-l)x_0(k,l) \tag{1}$$

By applying all other possible combinations of the above FIR filters shown in Figure 2, we get three lower resolution *detail* signals, denoted as x_{-1}^{LH}, x_{-1}^{HL}, x_{-1}^{HH}. As is described in [26], it is possible to use non-separable analysis (and synthesis) filters to perform the multiresolution decomposition. In this case, Eq. (6) takes the form

$$x_{-1}^{LL}(m,n) = \sum_{k=1}^{N}\sum_{l=1}^{N} h_{LL}(2m-k, 2n-l)x_0(k,l) \tag{2}$$

Perfect reconstruction of the original signal $x_0(k,l)$ can be achieved through synthesis of all four subband components, i.e., the approximation and the three detail signals. It is, however, possible to obtain an approximate reconstruction $x_0(k,l)$ of the original signal, by using only the approximation signal $x_{-1}^{LL}(m,n)$ and synthesis filter $f_{LL}(m,n)$ as follows

$$x_0(m,n) = \sum_{k=1}^{\frac{N}{2}}\sum_{l=1}^{\frac{N}{2}} f_{LL}(m-2k, n-2l)x_{-1}^{LL}(k,l) \tag{3}$$

If the signal decomposition procedure, described by Eqs.(1)-(2), is successively applied to the *approximation* signals we have a *multiresolution approximation* of the original signal, providing signals of continuously decreasing

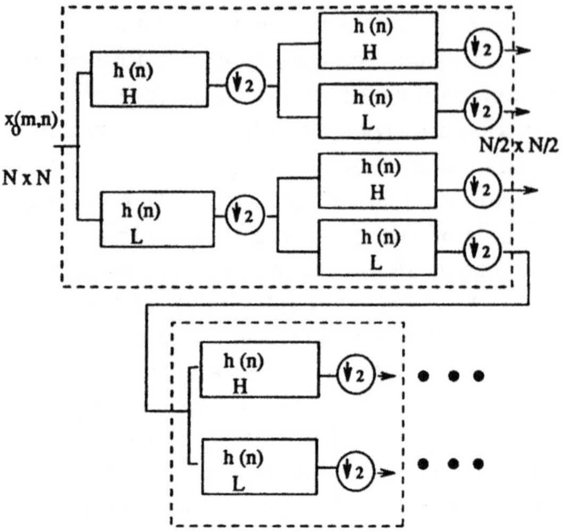

Figure 1: Multiresolution decomposition of a 2-D signal.

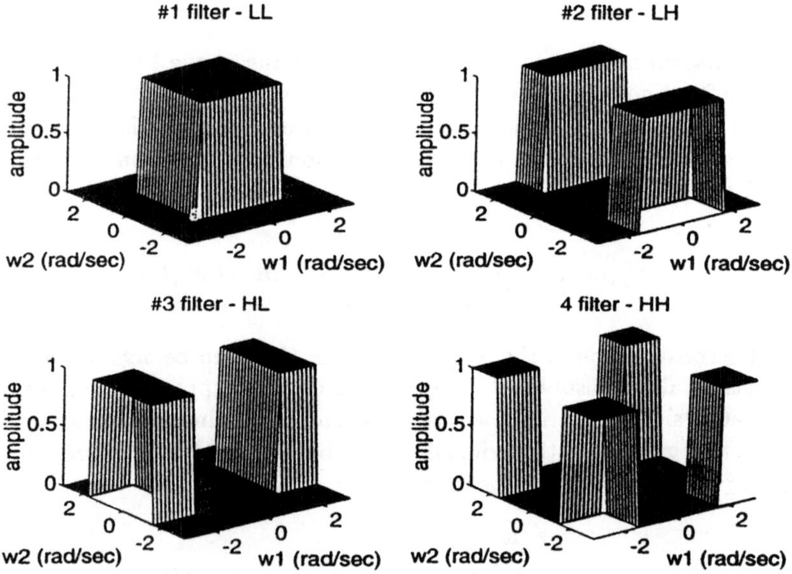

Figure 2: A set of (ideal) conventional subband analysis filters.

size. Such multiresolution representations can be very useful for signal processing at various resolutions; edge detection, segmentation and classification of 2-D images, as well as fast browsing through multimedia databases based on signal or image content, are examples where such representations can play a key role.

4 Hierarchical Networks for Signal Processing

Hierarchical neural network architectures are described next as an efficient scheme for processing the resulting multiresolution representations. Let us consider the image classification problem, while extensions to other signal processing tasks can be made in an analogous way. A feedforward multilayer network is used first to classify an *approximation* image of quite low resolution, trained by some backpropagation variant, including some pruning technique, such as weight decay. The hierarchical network is then recursively constructed to handle the image at higher resolution levels. More specifically, the proposed procedure starts by training a network at, say, resolution level j with $j \leq -1$, (network LL_j) to classify *approximation* images x_j^{LL} at that resolution level (see Figure 3). After training, the network performance is tested, using a validation set of *approximation* images at the same resolution level j. If the performance is not acceptable, training is repeated at the $j+1$ resolution level.

In this approach, it would be desired that the network at level $j+1$ (network LL_{j+1}) a-priori includes as much as possible from the "knowledge" of the problem acquired by the former network at level j. Some early results in this topic [9] suggested to use the computed weights of the low resolution network as initial conditions for the weights of the high resolution one. Using the fact that the information of the *approximation* image at level $j+1$ is equivalent to the information included at both the *approximation* and *detail* images at level j, we can train three more networks (LH_j,HL_j,HH_j), separately (or in parallel) from the former one, to classify the *detail* images at level j and let the network at level $j+1$ contain in its first hidden layer a number of units equal to the union of the first hidden layer units of all four lower resolution networks, as shown in Figure 4.

In the following we present forms that permit transfer of the generally large number of (already computed) weights between the input and the first hidden layer of the low resolution networks in corresponding positions of the high resolution network LL_{j+1} and keep these weights fixed during training of the high resolution network. A small number of nodes can be added to the first hidden layer of the LL_{j+1} network, while computation of the resulting new interconnection weights, as well as of the generally less complex upper hidden layers is performed then by training the corresponding parts of the high resolution network LL_{j+1}. The above-mentioned addition of hidden nodes can be performed sequentially during training, similarly to the cascade

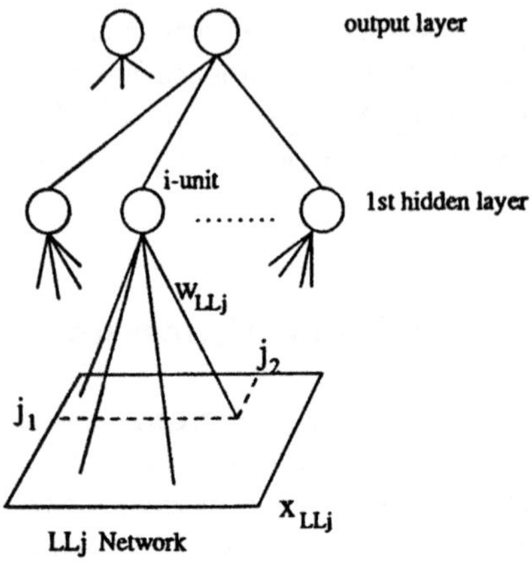

Figure 3: Classification at level j.

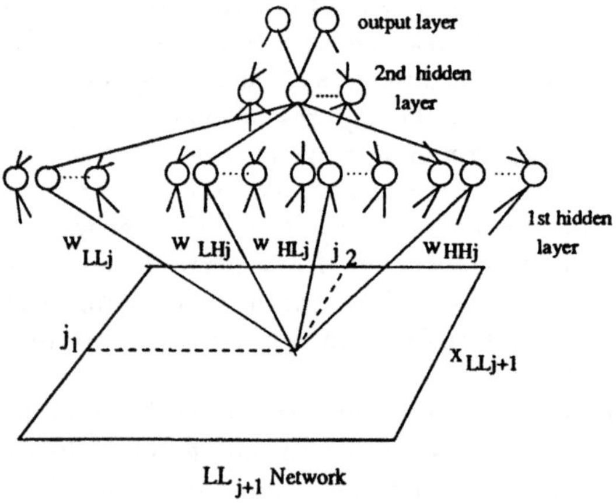

Figure 4: Classification at level $j + 1$, using weights from level j.

correlation methodology [8]. To implement the weight transfer, we impose the constraint that the inputs to the units of the first hidden layer of network LL_{j+1} be identical to the corresponding inputs of the units of networks at level j. In the case of network LL_j, for example, with $j = -1$, the input to each unit of the first hidden layer is

$$\sum_{j_1=1}^{L} \sum_{j_2=1}^{L} w_{-1}^{LL}(j_1, j_2) x_{-1}^{LL}(j_1, j_2) \tag{4}$$

where $w_{-1}^{LL}(j_1, j_2)$ is the weight connecting each hidden unit to the (j_1, j_2) pixel of the $L \times L$ image block at level -1. Then in the first hidden layer of the LL_0 network classifying each $N \times N$ image block, the input to the corresponding unit will be analogously

$$\sum_{k_1=1}^{N} \sum_{k_2=1}^{N} w_0^{LL}(k_1, k_2) x_0^{LL}(k_1, k_2) \tag{5}$$

If the computed values in the above equations are required to be equal to each other, then it can be easily shown using Eqs. (2), (4) and (5) that

$$w_0^{LL}(k_1, k_2) = \sum_{j_1=1}^{K} \sum_{j_2=1}^{K} h_{LL}(2j_1 - k_1, 2j_2 - k_2) w_{-1}^{LL}(j_1, j_2) \tag{6}$$

A similar form can be derived relating the weights in each network LH_j,HL_j,HH_j and the corresponding weights in network LL_{j+1}, as shown in Figure 4. The above forms permit computation of the generally large number of weights between network's input and first hidden layer be efficiently performed at lower resolution.

It should, however, be mentioned that training and use of all LL, LH, HL, HH networks at level j is not always meaningful; this is due to the fact that in most cases only some of the four subband images contain significant portion of the content of the original image. To overcome this problem, only one set of weights, corresponding to the low resolution image which contains the most significant part of the original image among the four subband low-resolution images, can be used for this purpose [26].

5 A Simulation Study

Let us consider a real-life application to examine the performance of the presented multiresolution neural network classifiers. Inspection of solder joints in printed circuit board manufacturing is such an application in which conventional pattern recognition techniques have not shown sufficient reliability. 2-D gray-scale images, showing either the height or the intensity as functions of the position across solder joints, were obtained by an optical laser scanner and used as signals to be classified in two categories; namely, good or poor solder joints, the latter containing insufficient amount of solder. An example of a good and a defective solder joint (height function) is shown in Figure 5. 100(23×23) input images, 50 of each category, were used as training data, while another set of 100 images, 50 of each category, was used to test the generalization ability of the networks.

In our first experiment we used a fully connected two hidden layer network with an input layer of 529 units, trained with the learning set of the original images. The large number of interconnection weights, in comparison with the small number of training patterns, resulted in poor learning and generalization performances. We then used a finite impulse response (FIR) 8-tap low-pass filter

$$h(n) = [0.0094, -0.0707, 0.0694, 0.49, 0.49, 0.0694, -0.0707, 0.0094] \quad (7)$$

and a corresponding high pass filter [5]

$$g(n) = (-1)^{1-n} h(1-n) \quad (8)$$

as well as subsampling by two, to get images at lower resolution levels. The *approximation* images of the solder of the first category shown in Figure 5, at three consecutive lower resolution levels, are shown in Figure 6. It can be verified that details of the original image are sufficiently preserved at resolution level $j = -1$; we, therefore, performed the classification task at this level. Figure 7 presents the *approximation*, as well as the three *detail* (11×11) images of the above image at level $j = -1$.

Instead of using four networks for transferring weights between levels (-1) and 0, we only used networks LL_{-1} and HH_{-1} to independently classify the approximation (LL) and one of the detail images (HH) at this resolution level. Each network had 121 units at the input layer, equal to the image size (11×11) at this level, and contained in total 1652 adjustable weights. Both networks required small training times to learn the training data set. We then tested the generalization performance of both networks. The LL_{-1} network classified the test data set with 80% accuracy, while the accuracy of the HH_{-1} network was 67%; this differnce is caused by the fact that the *approximation* image contains more information than the *detail* one at this level. Then we used the above-described procedure for transferring input-to-first-hidden-layer weights between levels (-1) and 0 and kept them frozen during training of the rest of the network, which only included 234 free parameters. The network, after training, classified the test set with 91% accuracy.

6 Conclusions

Structural learning algorithms and network architectures have been examined in this paper in an attempt to effectively process 1-D or 2-D signals

Figure 5: A good and a poor solder joint (23 × 23) image.

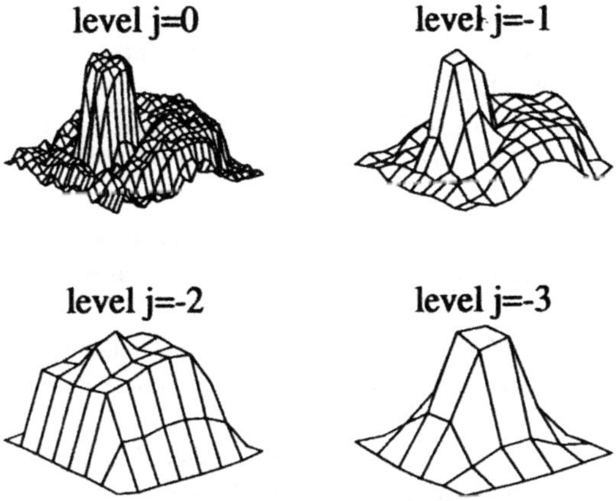

Figure 6: Original and approximation images at three lower resolution levels.

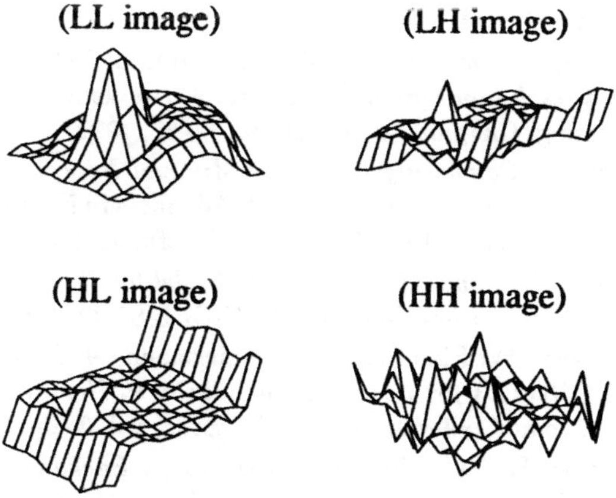

Figure 7: Approximation and detail images of the good solder joint.

of large sizes. A new constructive approach for generating hierarchical network structures and training representations of the signals at different resolutions has been presented; this approach, which is based on multiresolution signal decomposition can be quite useful for a variety of signal processing tasks, including analysis and signal recognition, as well as coding, storage and content-based retrieval of signals.

References

1. Baum, E. and Haussler, D., *What Size Net Gives Valid Generalization?*, Neural Computation vol. 1, pp. 151-160, 1989.
2. Baum E., *When Are n-Nearest Neighbor and Backpropagation Accurate for Feasible Sized Sets of Examples?*, Lecture Notes in Computer Science, Springer Verlang, vol. 412, pp. 2-27, 1990.
3. A. Blumer, A. Ehrenfeucht, D. Haussler and M. Warmuth, *Learnability and the Vapnik-Chervonkeis Dimension*, Journal of the ACM, vol. 36, pp. 929-965, 1989.
4. Chauvin Y., *A backpropagation Algorithm with Optimal Use of Hidden Units*, in D. Tourketzy (Ed.), Advances in Neural Information Processing Systems 1, Palo Alto, CA, Morgan, Kaufman, 1989.
5. M. Antonini, M. Barlaud, P. Mathiew and I. Daubechies, *Image Coding Using The Wavelet Transform* IEEE Trans. on Image Processing, vol. 1, pp. 205-220, 1992.
6. A. Delopoulos, A. Tirakis and S. Kollias, *Invariant Image Classification Using Triple-Correlation-Based Neural Networks*, IEEE Trans. Neural Networks, May 1994.
7. Esprit Project 2092 Handbook, *ANNIE: Neural Networks for Industrial Pattern Recognition Problems*, 1991.
8. S. Fahlman and C. Lebiere, *The Cascade-Correlation Learning Architecture*, in Advances in Neural Information Processing Systems 2, Morgan Kaufman, pp. 524-532, 1990.
9. C. Hand, M. Evans and S. Ellacott, *A Neural Network Feature Detector Using a Multiresolution Pyramid*, in Neural Networks for Vision, Speech and Natural Processing, Chapman and Hall, pp. 65-82, 1992.
10. Hanson S., *Comparing Biases for Minimal Network Construction with Backpropagation*, in D. Tourketzy (Ed.), Advances in Neural Information Processing Systems 1, Palo Alto, CA, Morgan, Kaufman, 1989.
11. Haussler D., *Generalizing the PAC Model for Neural Nets and Other Learning Applications*, Techn. Report, UCSC-CRL-89-30, 1989.
12. F. Hergert, W. Finoff and H. Zimmermann, *A Comparison of Weight Elimination Methods for Reducing Complexity in Neural Networks*, Proceedings of Intern. Joint Confer.on Neural Networks, vol. 3, pp. 980-985, 1992.
13. Y. Hirose, K. Yamashita and S. Hijiya, *A Backpropagation Algorithm Which Varies the Number of Hidden Units*, Neural Networks, vol. 4, pp. 61-66, 1991.
14. K. Hornik, *Approximation Capabilities of Multilayer Feedforward Networks*, Neural Networks, vol. 4, pp. 295-307, 1991.

15. D. Hush and B. Horne, *Progress in Supervised Neural Networks*, IEEE Signal Processing Magazine, vol. 10, pp. 8-39, 1993.
16. Ishikawa M., *A structural Learning Algorithm with forgetting of Weight Link Weights*, Proceedings of the IJCNN vol. 2, Washington D.C, June, 1989.
17. Jacobs, R., *Increased Rates of Convergence Through Learning Rate Adaptation*, Neural Networks, vol. 1, no. 4, 1988.
18. S. Kollias and D. Anastassiou, *An Adaptive Least Aquares Algorithm for Efficient Training of Artificial Neural Networks*, IEEE Trans. CAS vol. 36, pp. 1092-1101, 1989.
19. S. Kollias and A. Stafylopatis, *Parallel Implementations of Backpropagation Based on Network Topology*, in Parallel Algorithms for DSP, Computer Vision and Neural Networks, Wiley, pp. 233-258, 1993.
20. Y. LeCun, *Generalization and Network Design Strategies*, Connectionism in Perspective, North Holland, Switzerland, pp. 143-155, 1989.
21. Y. LeCun, S. Denker and S. Solla, *Optimal Brain Damage*, in Advances in Neural Information Processing Systems 2, Morgan Kaufman, pp. 598-605, 1990.
22. S. Mallat, *A theory for multiresolution signal decomposition: the wavelets representation*, IEEE Trans. PAMI vol. 11, pp. 674-693, 1989.
23. A. Murray and P. Edwards, *Synaptic Weight Noise During Multilayer Perceptron Training: Fault Tolerance and Training Improvements*, IEEE Transactions on Neural Networks, vol. 4, pp. 722-725, 1993.
24. Y. H. Pao, *Adaptive Pattern Recognition and Neural Networks*, Addison-Welsey, 1989. vol. 1, no. 3, July 1989.
25. R. Reed, *Pruning Algorithms - A Survey*, IEEE Transactions on Neural Networks, vol. 4, pp. 740-747, 1993.
26. A. Tirakis and S. Kollias, *Efficient Image Classification Using Multiresolution Analysis and Neural Networks*, Proceedings of ICASSP-93, Minneapolis, USA, 27-30 April 1993.
27. F. Smieja, *Neural Network Constructive Algorithms: Trading Generalization for Learning Efficiency*, Circuits, Systems and Signal Processing, vol. 12, pp. 331-374, 1993.
28. P.P. Vaidyanathan, *Multirate Systems and Filter Banks*, Prentice-Hall, Englewood Cliffs, NJ, 1993.
29. A. Waibel et al, *Phoneme Recognition Using Time Delay Neural Networks*, IEEE Transactions on ASSP, vol. 37, pp. 328-339, 1989.
30. A. Wigend, D. Rumelhart and B. Huberman, *Generalization by Weight Elimination with Applications to Forecasting*, in Advances in Neural Information Processing Systems 3, Morgan Kaufman, pp. 875-882, 1991.
31. M. Wynne-Jones, *Node Splitting : A Constructive Algorithm for Feedforward Neural Networks*, Neural Computing Applications, vol. 1, pp. 17-22, 1993.
32. J. Woods and S. O'Neil, *Subband Coding of Images*, IEEE Trans. on ASSP, vol. 34, pp. 1278-1288, Oct. 1986.
33. B. Zhang, *Accelerated Learning by Active Example Selection*, International Journal of Neural Systems, 1993.

The Graph Bi-Partitioning Problem: A Comparison of Search Methods

Bernard Manderick[1] and Hiroaki Inayoshi[2]

[1] Computer Science Dept., VUB, B-1050 Brussels, Belgium
[2] Computational Models Section, ETL, Tsukuba-shi, Ibaraki 305, Japan

Abstract. We assess the performance of the GA on the weighted graph bi-partitioning problem which is an NP-complete problem. The assessment is done in two ways. First, the GA is compared with other search techniques and second, the fitness landscapes to be optimized are quantified in different ways and these data are related to the GA-performance. The fitness landscapes studied can be tuned by changing the connectivity of the graph and the distribution of the weights. The effects of these changes are investigated.

1 Introduction

The genetic algorithm (GA) offers the promise of a widely applicable, robust global search strategy [4, 1]. Good GA-performance is a matter of finding a proper balance between exploitation and exploration. This in its turn is affected by GA-parameters like the selection strategy, the genetic operators and their corresponding rates.

Although a lot of theoretical and empirical knowledge has been accumulated, there is still little insight in the proper role of each of the components of the GA and how they interact to achieve effective search. Moreover, theory and practice seem to be at odds at certain points. For instance, what is the proper role of mutation? Is it merely a background operator whose main role is to recover lost alleles [4] or is it a much more important operator than previously thought as many experiments show, e.g. [13]. Or consider crossover disruption. The schema theorem seems to indicate that it should be minimized. But why then does uniform crossover perform so well on many problems while it is so highly disruptive [14, 9]. As a last example, consider the royal road functions which is a class of functions defined in such a way that the global optimum should be discovered easily by crossover [2]. Consequentially, it is hard to understand why random mutation hill-climbing outperforms the GA by one order of magnitude.

The GA can only be a competitive optimization technique for a whole range of problems if we have a clear understanding of the specific role of each of its components and how they interact synergetically.

We are convinced that such an understanding is impossible without a detailed analysis of the fitness landscape to be optimized [10]. We expect that the landscape will reveal when the GA effectively searches the whole space and when it gets stuck in local optima or faces other difficulties.

This is the first paper of a series where we take a bottom up approach and analyze empirically the contribution of the different components in the overall search, and this is done for qualitatively different fitness landscapes. Our goal is to end up with a number of guidelines telling us for which landscapes the GA is a competitive optimization technique and how to tune it to obtain maximal performance. This way, we hope to contribute to transform GAs from an art, which it very much still is, into a well understood technique.

In this paper, we assess the GA-performance on the weighted graph bi-partitioning or GBP-problem in two ways. First, it is compared with a number of other search techniques: three hill-climbing techniques [2], iterative improvement using the Kernighan-Lin heuristic [8], simulated annealing [11] and a hybrid technique combining the GA with the Kernighan-Lin heuristic. And second, fitness landscape corresponding with this problem is analyzed. In particular, we look at the distribution of local optima in the search space and investigate its impact on the efficiency of different search strategies especially the GA.

The GBP-problem was chosen for a number of reasons. First, it is an important combinatorial optimization problem known to be NP-complete [3]. Second, a powerful heuristic, called the Kernighan-Lin heuristic, exists which finds high quality local optima with a high frequency [8]. This heuristic serves as a bench mark for the other techniques. Third, the difficulties the GBP-problem poses can be tuned easily by changing parameters like the mean graph connectivity.

And finally, solutions can be encoded as bit strings in a straightforward way and the distance between these strings can be calculated easily. This makes the GBP-problem a perfect candidate for the genetic algorithm. This is not the case for other well-known combinatorial optimization problems like the traveling salesman problem and job-shop problems where permutations are a more natural genetic representation. Different distances can be defined on the set of all permutations and this gives raise to qualitatively different fitness landscapes for the same problem instance. Some of these distances are also difficult to calculate. Moreover, there is multitude of genetic operators which further complicates the picture.

The rest of the paper is organized as follows. First, we describe the optimization problem. Second, we describe how we quantify the fitness landscape to be optimized. Third, we compare the different search techniques considered. And finally, we conclude.

2 The Optimization Problem

The problem studied in this paper is a weighted graph bi-partitioning problem (or GBP-problem) with an even number $2N$ of nodes, $2N = 200$ in our case. The objective is to divide the nodes in two equal disjoint sets $A = \{a_1, a_2, \cdots a_N\}$ and $B = \{b_1, b_2, \cdots b_N\}$ ($\#A = \#B$) such that the total sum of weights $w_{i,j} > 0$, where $w_{i,j}$ connects the nodes $a_i \in A$ and $b_j \in B$, is minimized, i.e. the cost function $C(\{A, B\}) = \sum_{a_i \in A} \sum_{b_j \in B} w_{i,j}$ associated with the partition $\{A, B\}$ is minimized.

This minimization problem can be transformed in an equivalent maximization problem by defining the fitness function $f(\{A, B\}) = C_{max} - C(\{A, B\})$. Depending on the search technique, either the cost function C is minimized or the fitness function f is maximized.

The instances we have studied are characterized by the following tunable parameters:

1. The mean connectivity μ_p of the graph: for small μ_p the graph is sparsely connected while for large μ_p it is densely connected.

2. The standard deviation σ_p of the connectivity.

3. The distribution of the weights: In the experiments, the weights are distributed according to an exponential distribution. Other possibilities are normal and uniform distributions.

2.1 Search Space

In the following, each graph bi-partitioning (GBP) $\{A, B\}$ consisting of 2 equally sized node sets is called a bi-graph $\mathbf{b}$ and if the graph contains $2N$ nodes, then $\mathbf{b}$ is represented by a bit string of length $2N$ as follows: First, all $2N$ nodes are numbered from 1 to $2N$. If one of these nodes belongs to the subset A then the corresponding bit is set to 1, otherwise it is set to 0. Vice versa, each $2N$ bit string with an *equal* number of 0s and 1s represents a bi-graph $\mathbf{b}$. Therefore, we call this a bi-graph bit string or simply bi-graph and the set of all such strings is called the bi-graph space $\mathbf{B}$. From now on, depending on the context $\mathbf{b}$ denotes a bi-graph or its bit string representation.

This problem gives rise to the search space denoted by $\mathbf{B}_N$ (N is the dimension of the space) as follows: Two search points $\mathbf{b}$ and $\mathbf{b}'$ are said to be neighbors if and only if $\mathbf{b}$ can be transformed into $\mathbf{b}'$ by exchanging one 0 and one 1, i.e. one 0-1 exchange. It is clear that the mapping $d_B(\mathbf{b}, \mathbf{b}')$ defined for all $\mathbf{b}, \mathbf{b}' \in \mathbf{B}_N$ as the minimum number of 0-1 exchanges to transform $\mathbf{b}$ into $\mathbf{b}'$ is a distance in the usual sense. If $\mathbf{b}$ and $\mathbf{b}'$ differ in $2M$ of the $2N$ bits then $d_B(\mathbf{b}, \mathbf{b}') = M$.

In Table 1, we contrast the search space $\mathbf{B}_N$ of the bi-graphs with the space $\mathcal{Z}_2^{2N}$ of all bit strings of length $2N$ in which it is embedded: $\mathbf{b} \in \mathbf{B}_N$ when $\mathbf{b}$ contains an equal number of 0s and 1s. For $N = 200$, the size of the search space is $\#\mathbf{B}_N \approx 9.1 \times 10^{58}$ which is about 5.5% of the size of $\mathcal{Z}_2^{2N}$.

Finally, we get the fitness landscape by assigning each $\mathbf{b} = \{A, B\} \in \mathbf{B}$ the associated fitness $f(\mathbf{b}) = C_{max} - C(\mathbf{b})$.

Table 1: The search space B with the d_B-distance compared with the bit string space $\mathcal{Z}_2^{2N}$ in which it is embedded with the usual Hamming distance d_H.

	B	$\mathcal{Z}_2^{2N}$
Size	$\frac{(2N)!}{N!N!}$	2^{2N}
Neighbors	N^2	N
Diameter	N	$2N$
Density	$\frac{1}{\sqrt{\pi N}}$	1

3 Fitness Landscape Analysis

The methodology to analyze fitness landscapes is borrowed from [7]. Fitness landscapes can be quantified in a number of different ways.

1. The number of local optima and the size of their region of attraction is estimated.
2. The relation between the size of the region of attraction and the fitness of local optima is determined to see if there is a correlation.
3. The distribution of these optima over the search space is determined.
4. The autocorrelation function of the landscape and the correlation coefficients of the search operators are determined [10].

Together these measures give us a good picture of the difficulties a landscape might pose for a search technique and how the used search operators explore the landscape. This paper only concerns the distribution of optima.

In Figure 1, we show the distribution of local optima of a 200-node instance. The X-axis shows the cost of a local optimum and the Y-axis shows the distance from the optimum to the best local optimum that we have found. This figure clearly shows that the high quality local optima tend to be close to each other.

In Figure 2, we display the neighborhood of 2 bi-graph partitions (BGPs): one is a local optimum and the other is a random BGP. The X-axis corresponds with a node of the first subgraph and this node is then exchanged with each of the nodes of the second subgraph. The cost change is plotted in the Y-direction. This figure indicates that the landscape is quite smooth. The typical change in cost around a local optimum is about 7 % of cost of that GBP while it is about 2 % for a random GBP. A surprizing fact is that the mean connectivity μ_p of the graph to be partitioned has almost no effect on the fitness landscape: sparsely and densely connected graphs give rise to similar fitness landscapes. For instance, the autocorrelation function was the same for all landscapes.

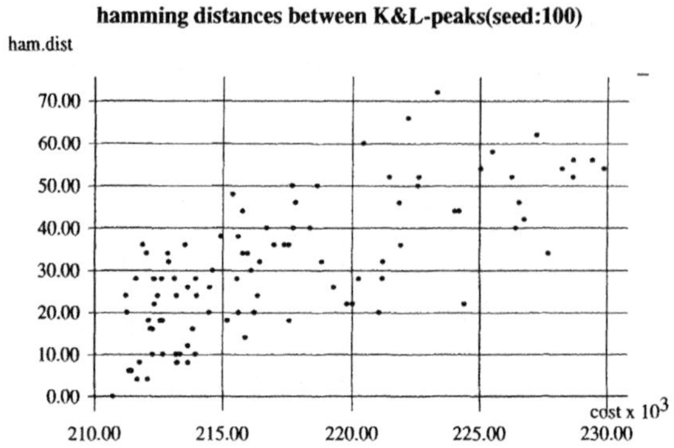

Fig. 1: The distribution of the local optima of a 200-node instance of the GBP-problem.

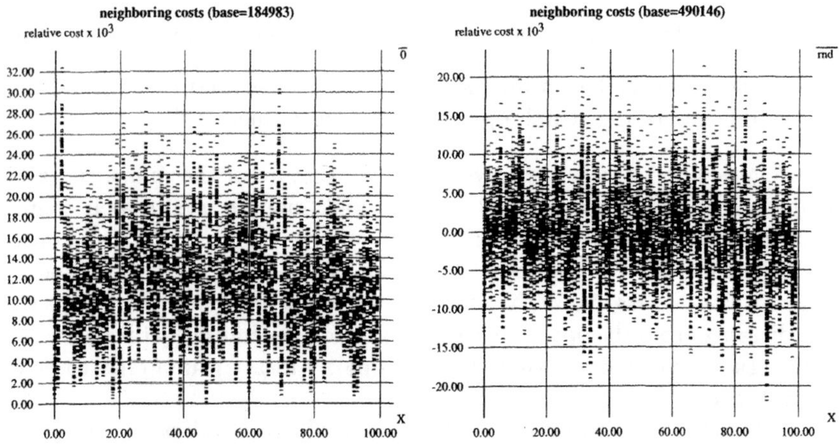

Fig. 2: The neighborhood around the global optimum (left) and a random GBP (right) of the GBP-instance in the previous figure. The cost of the global optimum and the random GBP are 184,983 and 490,146, respectively.

4 Comparison of Search Techniques

We have compared the following search techniques: three hill-climbing techniques [2], iterative improvement using the Kernighan-Lin heuristic [8], simulated annealing [11], Hopfield network [5] and a hybrid technique combining the GA with the Kernighan-Lin heuristic.

To apply each of these techniques one has to define: 1) a search space, 2) a move set that can be used to modify a search point and which defines a neighborhood relation and distance on that space, and 3) a fitness function. Together, they define a fitness landscape.

For all techniques, the search space and the fitness function were the space and the corresponding fitness function defined in Section 2.1. For all techniques but the Lin-Kernighan heuristic, the move set consists of the immediate neighbors of the current search point. As a result, all techniques are optimizing the same fitness landscape. The hillclimbers, the Hopfield net and simulated annealing are searching the immediate neighborhood while the Kernighan-Lin heuristic and the crossover operator can move further away from the current search point.

In this paper, we only consider iterative improvement using the Kernighan-Lin heuristic, the GA and a hybrid GA. For more details we refer to [6]. Suffice it to say that the hybrid GA outperformed the other techniques and that it was very consistent in the sense that almost always the global optimum was found.

4.1 Kernighan-Lin heuristic

The KL-heuristic for the graph GBP problem is as follows: Let $\{A, B\}$ be a partition of the graph where A consists of the nodes $\{a_1, a_2, \cdots a_N\}$ and B of the nodes $\{b_1, b_2, \cdots b_N\}$. Swap the nodes a_{i_1} and b_{j_1} for which gain g_1 in cost is maximal. Then, swap the nodes $a_{i_2} \in A \setminus \{b_{j_1}\}$ and $b_{j_2} \in B \setminus \{a_{i_1}\}$ for which the additional gain g_2 is maximized and this is continued until all nodes are swapped. Of course, the total gain $G = \sum_{i=1}^{N} g_i = 0$ since we end up with the same GBP as we started with. Now, we determine the k for which the partial gain $G_k = \sum_{i=1}^{k} g_i$ is maximal. Finally, the new GBP is obtained by swapping all a_{i_l} and b_{j_l} for $l = 1, \cdots, k$. This procedure is illustrated in Figure 3 for a 10-node graph.

The resulting GBP is then used as a new starting point for the above procedure and this is repeated until $g_1 < 0$. In this case, the GBP is a local optimum for the KL-heuristic. This heuristic is a very powerful one which finds high quality solutions with a high probability [8].

4.2 The Genetic Algorithm

The genetic algorithm (abbreviated as RGA) used has the following parameters: 1) the population size is ranging from $N = 40$ to $N = 100$, and 2) rank-based selection [15] is used since it combines two properties: First, the

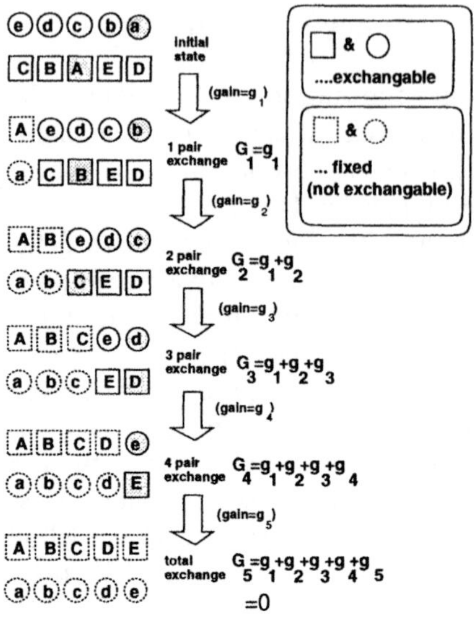

Fig. 3: The Kernighan-Lin heuristic for a 10 node GBP-problem. Given the partition $\{\{a, b, c, d, e\}, \{A, B, C, D, E\}\}$.

selection pressure remains constant during the search and second, the pressure can be tuned easily by changing the selection probabilities assigned to each rank. We used linear rank-based selection only. The usual genetic operators for bit strings are adapted in such a way that the offspring is again a bi-graph bit string.

Mutation of a bi-graph string $\mathbf{b}$ consists of one 0-1 exchange. The resulting string is one of its N^2 neighbors.

The crossover is a variant of uniform crossover. Take two parents $\mathbf{b}$ and $\mathbf{b}'$ and suppose that $d_B(\mathbf{b}, \mathbf{b}') = M$, i.e. $\mathbf{b}$ has at M locations a 0 where $\mathbf{b}'$ has a 1, and at M different locations $\mathbf{b}$ has a 1 where $\mathbf{b}'$ has a 0. At the remaining $2N - 2M$ locations both strings are equal.

Offspring is generated as follows: The $2M$ locations where the parents differ are filled at random with an equal number of zeros and ones, each other location inherits the common value shared by the two parents. The resulting offspring is again a bi-graph and its mean d_B-distance to each of its parents is $M/2$.

4.3 The Hybrid Genetic Algorithm

The hybrid GA (HGA) is the same as the GA above except that after each generation, the KL-heuristic is applied to all members of the population. This way, we get an improved generation consisting of local optima for the KL-heuristic which is then used by the GA to produce the next generation. This cycle continues for an number of generations.

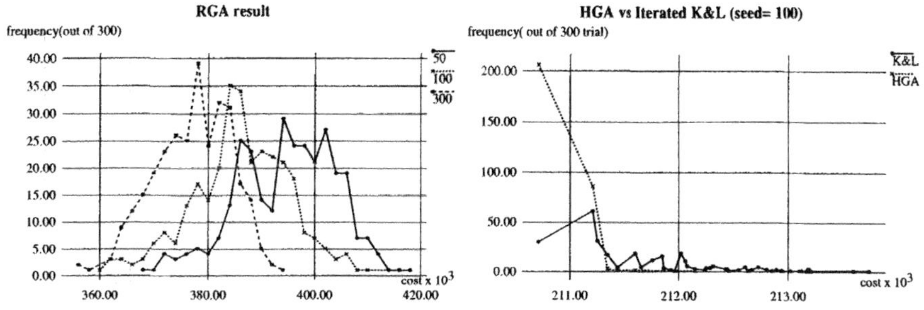

Fig. 4: Comparison of the GA-performance with that of the hybrid GA.

4.4 Comparison

We have done 300 runs for 8 different instances of the GBP-problem. The GA performance was quite poor on these instances. All other techniques that we have considered performed significantly better than the GA – cfr. the left part of Figure 4. The basic difference is that all the other techniques do some kind of hillclimbing and this seems to be important for smooth landscapes.

Iterative improvement using the KL-heuristic is a powerfull technique but it was consistently outperformed by the HGA. In most cases, HGA found the global optimum and in all other cases it found the second best optimum – confer the right part of Figure 4. In this figure, the X-axis stands for the best cost obtained in one trial, while Y-axis shows how many times the cost has obtained out of 300 trials. Here, one trial corresponds to a search in which less than 500 points are improved by the KL-heuristic and this for both HGA and iterated KL.

Our experiments indicate that crossover is a useful operator on the GBP-problem and Figure 1 explains why. The best local optima tend to be close to each other. As a result, genotypes located at high quality optima which are recombined might generate some offspring in the region of attraction of even better optima. And, that offspring will move to the better optimum due to hillclimbing. The distribution of local optima provides in a sense a "royal road" towards the global optimum.

5 Conclusions and Future Work

The results of our experiments presented here can be summarized as follows:

- Search combining the GA and local search is better than either technique alone. Moreover, this combination proved to be the best of the compared techniques.

- Crossover is a useful operator as a consequence of the distribution of the optima: the better optima tend to be in each neighorhood. Note that this usefulness has nothing to do with the lack of crossover disruption. Since it is randomly decided whether two nodes are connected or not, building blocks will have a large defining length on average and therefore they will be susceptible to crossover disruption.

Acknowledgments

While working on this project, Bernard Manderick stayed at ETL as a Science and Technology Agency Fellow.

References

[1] L. Davis, editor. *Handbook of Genetic Algorithms.* Van Nostrand Reinhold, New York, 1991.

[2] S. Forrest and M. Mitchell. Relative building-block fitness and the building-block hypothesis.

[3] M.R. Garey and D.S. Johnson. *Computers and Intractability: A Guide to the Theory of NP-completeness.* W.H. Freeman and Company, San Francisco, 1979.

[4] D. E. Goldberg. *Genetic Algorithms in Search, Optimization, and Machine Learning.* Addison-Wesley, Reading, 1989.

[5] J.J. Hopfield and D.W. Tank. Neural computations of decisions in optimization problems. *Biological Cybernetics*, 52:141–152, 1985.

[6] H. Inayoshi. *Simulations and Optimizations of Networks/Systems.* PhD thesis, Tsukuba University, 1992. In Japanese: Nettowa-ku sisutemu no simyure-syon to saitekika.

[7] S.A. Kauffman. *The Origins of Order: Self-Organization and Selection in Evolution.* Oxford University Press, Oxford, 1993.

[8] B.W. Kernighan and S. Lin. An efficient heuristic procedure for partitioning graphs. *Bell Systems Technical Journal*, 49(2):291–307, 1970.

[9] R. Caruana L. Eshelman and D. Schaffer. Biases in the crossover landscape. In Schaffer [12].

[10] Bernard Manderick, Mark de Weger, and Piet Spiessens. The genetic algorithm and the structure of the fitness landscape. In R. K. Belew and L. B. Booker, editors, *Proceedings of the Fourth International Conference on Genetic Algorithms.* Morgan Kaufmann, San Mateo, 1991.

[11] R.H.J.M. Otten and L.P.P.P van Ginneken. *The Annealing Algorithm.* Kluwer Academic Publishers, Boston, 1989.

[12] J. D. Schaffer, editor. *Proceedings of the Third International Conference on Genetic Algorithms.* Morgan Kaufmann, San Mateo, 1989.

[13] J. D. Schaffer, R. A. Caruana, L. J. Eshelman, and R. Das. A study of control parameters affecting online performance of genetic algorithms for function optimization. In Schaffer [12].

[14] G. Syswerda. Uniform crossover in genetic algorithms. In Schaffer [12].

[15] D. Whitley. The GENITOR algorithm and selective pressure: Why rank-based allocation of reproductive trials is best. In Schaffer [12].

A Unified Methodology for Model Synthesis and Recognition of 3-D Objects

A.K.C. Wong, W. Liu, P. Wong, and B. McArthur
Pattern Analysis and Machine Intelligence Laboratory, Systems Design Engineering, University of Waterloo, Waterloo, Ontario, Canada N2L 3G1

Keywords. 3-D model synthesis, object recognition, random graph, attributed hypergraph.

1 Introduction

A challenge in computer vision research is to automatically acquire 3-D models of objects through sensing, and, conversely, to recognize objects from their images based on previously synthesized 3-D models. Two typical types of sensory inputs are: 2-D intensity images obtained from CCD cameras, and range images obtained from range scanners [17]. Problems involving data of the first kind are generally more difficult and will be the focus of this paper.

Representative works on existing object modeling and/or recognition techniques include [25, 31], [12, 15], [2, 22, 23], [1, 11, 29, 34], [5, 16], and [9, 20, 21]. The fundamental questions are summarized below:

1. What representation, which can directly capture 3-D structures as well as handle uncertainties in sensor inputs and variations in object characteristics, should be used to model 3-D objects?

2. What processes are required to relate the features and feature relations extracted from images to the components of the 3-D model?

This paper presents an overview of a methodology which addresses both synthesis and recognition as a unified problem. As illustrated in Figure 1, key ingredients of the methodology are its graph representations and the associated synthesis, organization and search processes. Section 2 describes the fundamentals of graph representation of 3-D objects and the process for synthesizing wire frame models of objects from real world data. Section 3 describes our multi-view paradigm for object recognition. Section 4 describes how the various levels of information afforded by the paradigm can be captured by the Attributed Hypergraph Representation (AHR). Section 5 describes how object recognition is actually carried out based on the AHR framework. Finally, Section 6 summaries the key contributions and applications of the methodology.

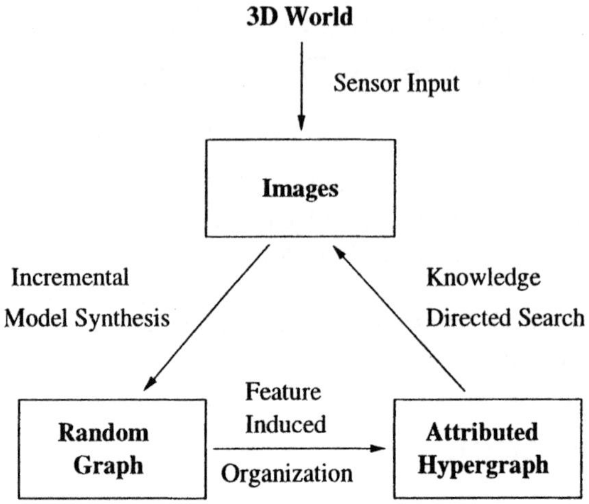

Figure 1: A unified approached for model synthesis and object recognition

2 Object Model Synthesis

Many object representation schemes are organized in the form of hierarchical tree structures [7]. More general relational models of objects are graph based with vertices representing geometric primitives such as points, edges and surfaces; and arcs denoting relations between primitives, such as connectivity. Semantic information can be added by attaching numeric, nominal, or logical attributes to the pattern primitives and relations. Such a representation is called an *attributed graph* [32], an *attributed relational graph* [29, 30], or a *structural description* [28].

Definition 1 *An* attributed graph (AG) *G is an ordered pair (V, E) of vertex set V and arc set E over $A_V \times A_E$ where A_V is a finite set of vertex attributes and A_E is a finite set of arc attributes.*

Figure 2 shows the AG representation (AGR) of an object by its surface, edge and corner features and their adjacency relations.

Definition 2 *The* graph morphism *of one graph G' onto another graph G is a one-to-one mapping of their vertex and arc sets which preserves the arcs to vertices incidences. The morphism is an* isomorphism *if $G' = G$ and a* monomorphism *if $G' \subset G$.*

Graph morphisms are used for matching and comparing AG's.

There are many effective algorithms for finding various types of graph morphisms [1, 27, 33, 36]. Their effectiveness is largely based on the use of constraints, such as geometrical and topological constraints, inherent in the vertex and arc attributes to assist in the pruning of inadmissible mappings.

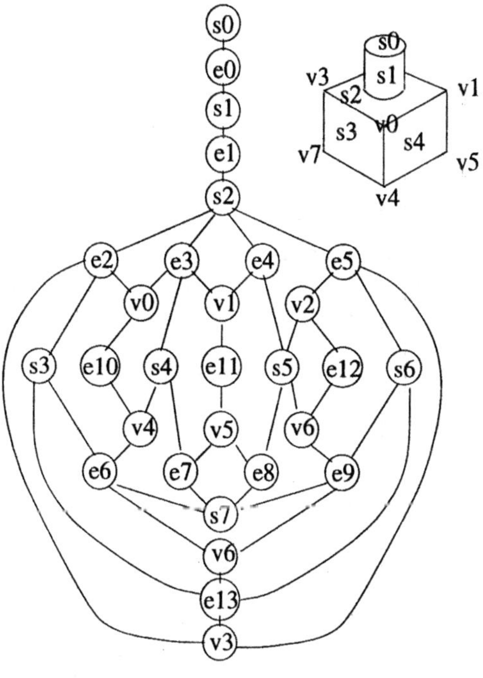

Figure 2: A surface-edge-vertex graph of an object.

Next, we introduce the concept of random graphs (RG) [32, 34], to allow for structural and measurement variations.

Definition 3 *A* random graph (RG) *is an ordered pair* (W, B) *of a random vertex set* $W = (\alpha_1, \alpha_2, \ldots, \alpha_n)$ *and a random arc set* $B = (\beta_1, \beta_2, \ldots, \beta_m)$ *in which the vertices* α_i *and arcs* β_i *are random variables.*

The random vertices and arcs may take on either continuous or discrete values as well as the null value. The purpose of the null value is to allow for missing vertices or arcs without having to modify the underlying graph structure.

Definition 4 *Let* $R = (W, B)$ *be a random graph. An* outcome *of* R *is an attributed graph* $G = (V, E)$ *such that the attributed vertices and arcs of* G *are given by* $v_i = \{\mu(\alpha_i) | i = 1, 2, \ldots, n\}$ *and* $e_j = \{\mu(\beta_j) | j = 1, 2, \ldots, m\}$, *for some graph monomorphism* μ.

Let

$$R = (\{\alpha_1, \ldots, \alpha_n\}, \{\beta_1, \ldots, \beta_m\})$$

be a RG and

$$G = (\{a_1, \ldots, a_n\}, \{b_1, \ldots, b_m\})$$

be an outcome graph with respect to a monomorphism μ. The probability of the occurrence of G is

$$P(G,\mu) = Pr\left(\bigwedge_{i=1}^{n}(\alpha_i = a_i) \bigwedge \bigwedge_{j=1}^{m}(\beta_j = b_j)\right) \tag{1}$$

Evaluation of (1) requires knowledge of the joint probability distribution of the random variables. This is impractical for real problems with high dimensionality. Wong *et al.* [32, 34] introduced the following first-order approximation which ensures that all RG outcomes will remain to be graphs:

1. The random vertices $\{\alpha_i\}$ are mutually independent.
2. Given values for $\{\alpha_i\}$, the random arcs $\{\beta_j\}$ are independent.
3. A random arc β is independent of any random vertex other than its endpoints which are denoted by $\sigma(\beta)$ and $\tau(\beta)$.

Based on these assumptions, (1) can be rewritten as

$$P(G,\mu) = \prod_{i=1}^{n} Pr(\alpha_i = a_i) \prod_{j=1}^{m} Pr(\beta_j = b_j | \sigma(\beta_j) = a_{j1}, \tau(\beta_j) = a_{j2}) \tag{2}$$

where $\sigma(\beta_j) = a_{j1}$ and $\tau(\beta_j) = a_{j2}$ are the values assigned to the head and tail of β_j. This formulation can be extended to continuous-valued attributes with the introduction of probability density functions [24].

2.1 Object Model Synthesis Using Random Graphs

A first-order RG can be used for model representation with its vertices corresponding to geometric features, such as points, lines, and surfaces; and its arcs corresponding to topological and spatial relationships between the geometric features. For wire frame models, corners and points formed by the intersection of extended edges are represented by the RG vertices and straight or curve line segments formed by the intersection of surfaces or object boundaries are represented by the RG arcs. An algorithm for incremental model synthesis using images captured by a single moving camera [24] has been developed.

Figures 3 and 4 provide an example of model synthesis based on corner and edge data extracted from 19 images of a sheet metal cover. The calculated first-order observation probabilities for vertices and conditional probabilities for arcs are given in Table 1. This example demonstrates how the RG model incorporates variations in observed structure caused by accidental feature alignments or errors in the feature extraction process. Model edge 19, which connects model vertices 2 and 11, is observed only in images 4 and 5 where

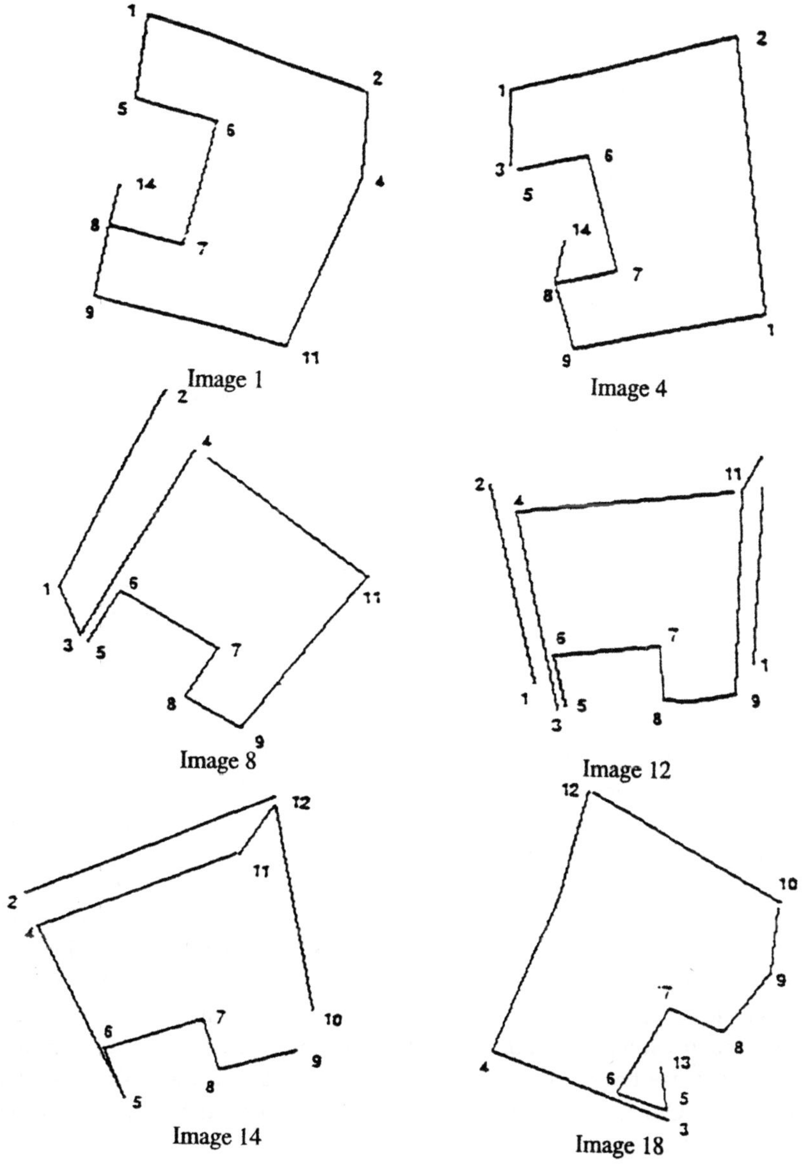

Figure 3: Sequence of views of an RTX robot cover

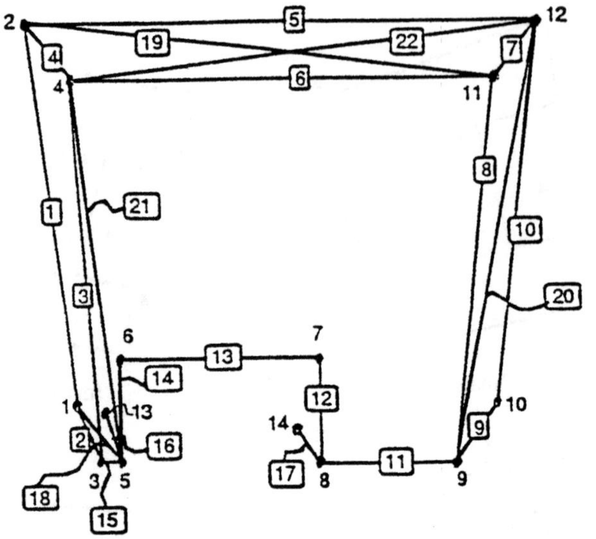

Figure 4: Random graph representation of an RTX robot cover

image vertices 2, 4, and 11 are collinear. Model edge 21, which is observed only in image frame 14, arises when image vertices 4 and 5 are incorrectly attached. None of the model graph edges 18 through 22, therefore, correspond to physical features on the cover. As shown in Table 1, the observation probabilities of the false edges are very low and thus presents only minor perturbations to the model structure.

In general, object models can be synthesized by integrating image data obtained by moving a single camera to a number of different viewpoints located about the object. The synthesis proceeds incrementally, i.e. data obtained at each new location is used to improve the estimate of the model. The synthesis process also makes use of knowledge on the transformation between each camera coordinate frame and the model coordinate system.

Thus far, we have discussed how object models can be synthesized from sensed data. Our method of recognizing objects in a scene based on previously synthesized models will be described in the following sections.

3 Multi-View Model for Object Recognition

As observed in [5], representing an object by its 2-D projections can lead to significant improvements in the performance of a model-based vision recognition system. However, existing multiple-view methods share the following weaknesses: 1) the viewpoint space is over partitioned (i.e. more views than

Vertex	$p(a)$	Arc	$p(b\|a_1, a_2)$	Arc	$p(b\|a_1, a_2)$
1	0.684	1,7,10	1.000	15	0.000
2	0.737	2	0.818	16,17	1.000
3	0.842	3	0.786	18	0.154
4	0.895	4	0.833	19	0.143
10	0.474	6	0.938	21	0.059
11	0.947	8	0.722	22	0.100
12	0.526	9	0.667		
13	0.105	11	0.947		
14	0.263	12 - 14	1.000		

Table 1: Vertex and arc probabilities for sheet metal cover model synthesis.

necessary for object recognition), and 2) there is no mechanism (i.e. transformation or indexing) to link the generated model views to 3-D geometric features so that model matching can be performed on the basis of visual and geometric constraints.

We have identified four levels of information which are useful for drawing inferences about 2-D images from 3-D models and vice versa. The four levels of model description are: 1) 3-D model features, 2) feature induced characteristic views (CV), 3) characteristic view classes (CVC), and 4) local image features. We therefore propose a procedure that utilizes: 1) the geometric measurements of 3-D object models; 2) the CV's, each of which contains a subset of model features visible from a common viewpoint; 3) a set of topological equivalence classes of the CV's; and 4) a set of local image features such as certain types of lines, curves and junctions.

We confine our attention to solids with planar, cylindrical or conic faces. For this class of objects, wire-frame models are adequate. The input raw models are generated by using the synthesis method described in Section 2.

Local image features can be used to initialize image-model matching. They form local topological groupings that can induce a set of possible CVC's for establishing mappings between features on the 2-D image and the 3-D model world. Information in each CVC could further direct the search for additional image features in establishing correspondence between 3-D model and 2-D image features. Figure 5 shows how model synthesis and the object recognition can be conducted by relating information at various levels. Descriptions of these four levels are given in the following sections.

3.1 3-D Model Features

The 3-D model features include points, lines or curves, and surfaces. These features are referred to as attributes and are expressed in the form of ordered pairs (a_n, a_v) where a_n is the attribute name and a_v the attribute value which

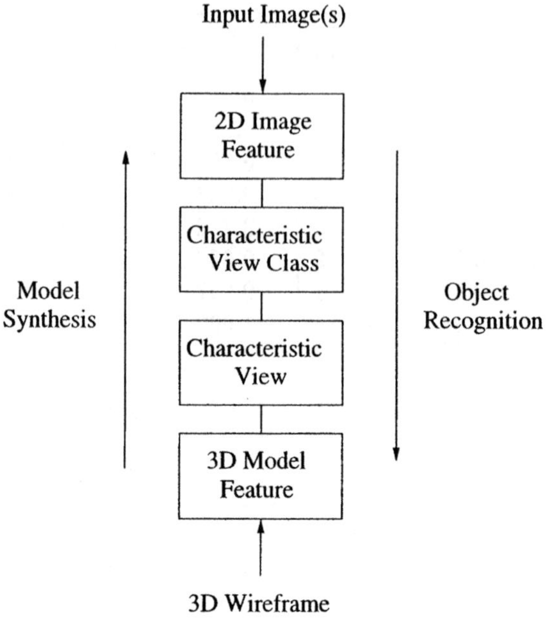

Figure 5: Object modeling at four descriptive levels.

can be a number, a symbol, a set of labels of other features, or a combination of these. According to their geometric and topological characteristics, they can be categorized on the basis of shape, relation, and 3-D position and orientation. An attribute may or may not be viewpoint dependent. For example, surface areas and edge lengths are viewpoint dependent whereas surface normal, radius, and adjacency relations are viewpoint independent.

Empirical observations show that the orientations of the feature in the model coordinate system are important for 3-D object recognition. Thus each model feature is assigned a direction vector known as its characteristic orientation (CO) which indicates the ideal camera direction for viewing the feature. Corresponding to each CO, a *feature-centered projection* (FCP) is defined. The FCP transformation determines how an object model is to be mapped into the image plane and is made up of a rotation matrix, a translation vector, and a perspective scaling.

3.2 The Characteristic View (CV)

The CV's of an object, originally defined in [5], describes the object using a set of its 2-D projections based on the CO of each model feature. Thus, the 3-D model features can be organized such that, given a candidate CO, all the features visible from the same viewing direction can be accessed efficiently. A natural way to do so is to register model features as a function of their CO's.

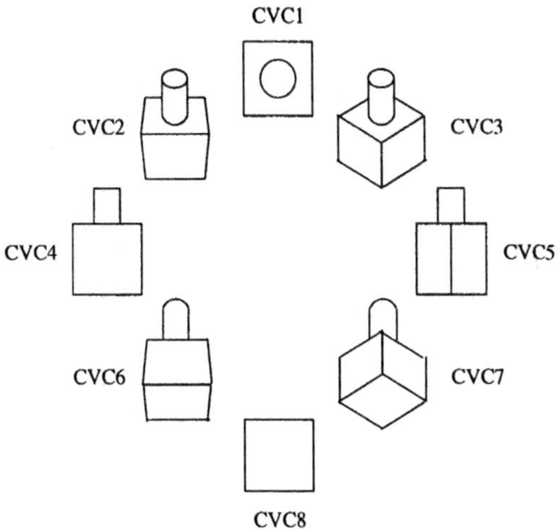

Figure 6: The characteristic view classes of the object in Figure 2.

Hence, the CV can be determined by three components: the viewing direction determined by three Euler angles (θ, ψ, ϕ), the FCP transformation, and the set of attributed model features visible (either fully or partially) under the projection. In general, since each CO of an object produces no more than one CV, the total number of CV's of a planar or quadric surfaced solid is equal to, or less than, the number of extracted 3-D model features.

3.3 Characteristics View Class (CVC)

Due to structural and topological similarities, CV's may be grouped into equivalent classes. It is shown in [10] that each CV can be associated to an image structure graph (ISG). Two CV's are said to belong to the same equivalent class if their ISG's are isomorphic. For example, the characteristic view classes of the object depicted in Figure 2 are given in Figure 6.

3.4 Image Features

It is desirable to extract the same types of features from the image that depicts the object's characteristic views and use them to guide the acquisition and extraction of image features for object recognition. For our purpose, three types of image features are considered: elliptical curves, junctions where two or more lines or curves meet, and local parallel, or symmetric lines, or curves, which together form a minimal closed boundary. In general, the set of 2-D image features of an object model contains all the distinctive 2-D image features extracted from the characteristic views, each of which is assigned attributes indicating its type, shape, topological relations with the neighboring features.

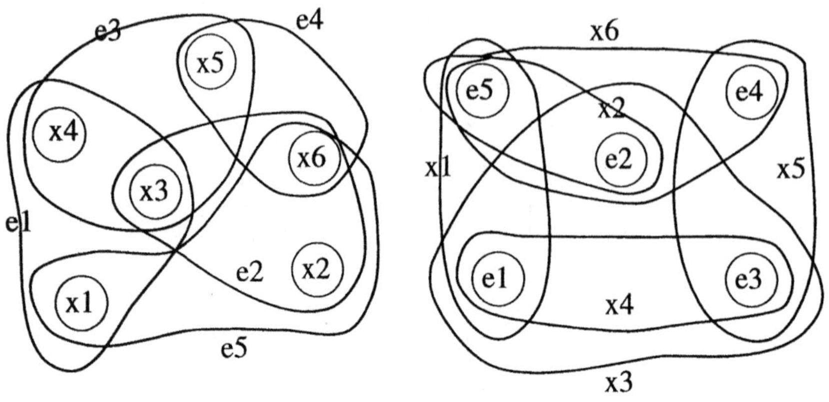

Figure 7: A hypergraph and its dual graph.

3.5 Summary of the Four Level Description Approach

The four level description approach provides direct structural information as well as multiple-view based feature groupings for interchangeable interpretation of the 3-D object and its 2-D projections from an input image. The local image features, which are easy to acquire from the input image but not reliable for object recognition and location, can be used to initialize the model matching at the image feature level. The identified image features will then form a cluster that can propose a set of characteristic view classes based on their combined co-occurrences. Since the triggering event for model matching in the 3-D domain involves a cluster of image features, the viewpoint sensitivity associated with the orientation of these features is minimized.

4 Attributed Hypergraph Representation

The four levels of model descriptions described above can be integrated into an Attributed Hypergraph Representation (AHR)[4, 17, 36]. Within this framework, a pair of dual AH's are defined. While the AH is used to represent the 3-D model features and the CV's, its dual is used to represent the CVC's and the 2-D image features. Since these two AH's share similar data structures, they can be readily converted from one to another.

Definition 5 *An* attributed hypergraph H *is an ordered pair* $(X, \mathcal{E})$ *over* $A_X \times A_E$ *where* A_X *is the finite set of vertex attributes and* A_E *is the finite set of hyperedge attributes. A set of attributes from* A_X *may be assigned to a vertex* x_i *in* X. *Likewise, a set of attributes from* A_E *may be assigned to an hyperedge* E_j *in* $\mathcal{E}$.

Definition 6 *Let* $H = (X, \mathcal{E})$ *be an AH.* $H^* = (X^*, \mathcal{E}^*)$ *is said to be the* attribute-induced dual hypergraph *of* H *if there exist*

$$\alpha : X^* \to \mathcal{E}$$

$$\beta : \mathcal{E}^* \to X$$

where α *and* β *are two abstract mappings which can be taken as a transformation, a compression, a mapping, or a feature indexing scheme.*

Note that H and H^* are dual to each other in the sense that the vertices and hyperedges of one may be respectively mapped into the hyperedges and vertices of the other.

Attributed hypergraphs can be used for object representation in the following way. First, an AH H is used to represent the 3-D object model with its vertices denoting the 3-D geometric features and its hyperedges denoting the characteristic views of the object. Then, the attribute-induced hypergraph H^* can be derived to represent the same object model but with its vertices denoting the CVC's and its hyperedges denoting the derived image features. This will enable certain implicit information in H, such as the feature visibility, to become explicit in H^*. The pair of attributed hypergraphs $H = (X, \mathcal{E})$ and $H^* = (X^*, \mathcal{E}^*)$ has the following properties :

1. $||X||$ = the total number of model features extracted from the object.
2. $||\mathcal{E}||$ = the total number of CV's of the object.
3. $||\mathcal{E}|| \leq ||\mathcal{V}||$ because two FCP's may produce the same CV's.
4. $||X^*||$ = the total number of CVC's of the object.
5. $||\mathcal{E}^*||$ = the total number of possible image features of the object.
6. $||X^*|| \leq ||\mathcal{E}||$ because each CVC contains at least one CV.
7. $||\mathcal{E}^*|| \leq ||X^*||$. This implies that the total number of image features for an object will be less than its number of 3-D geometric features.
8. In general, $(H^*)^* \neq H$ because the dual relationship is defined not only by the hypergraph topology but also on the basis of some mappings captured in the assigned attributes.

As shown in Figure 8, the relationship between 3-D model features and the CV's and that between the CVC's and the candidate image features are represented in the two AH's in the AHR. Meanwhile, the dual relationships between H and H^* in the AHR are defined by two mappings: one links the CV's to the CVC's based on the graph morphism (α) and the other links 3-D model features to its 2-D projection through feature indexing (β). These four mappings can be pre-calculated and used to achieve a fast object recognition through a knowledge-directed search process. Finally, the above observations lead to the following two conclusions.

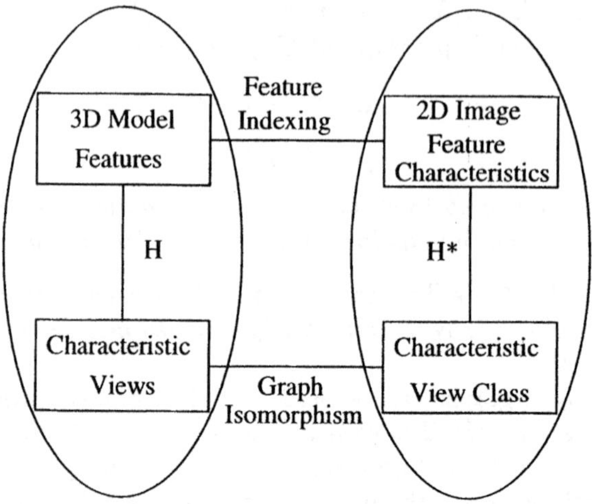

Figure 8: The abstract mappings involved in the AHR model.

1. For an AH $H = (X, \mathcal{E})$, there must exist an attribute-induced dual $H^* = (X^*, \mathcal{E}^*)$ over

$$\alpha : \mathcal{E} \to X^*$$

$$\beta : X \to \mathcal{E}^*$$

   if each x_i in X is covered by at least one hyperedge in $\mathcal{E}$ (i.e. there are no isolated features or CV's), and if each hyperedge intersects with at least one other hyperedge. Since the AH's are generated by conducting the feature-centered projection for every model face, edge and vertex, the above two conditions are met automatically.

   In view of this, given two abstract mappings α and β, the process of object modeling becomes a straight forward process. Different methods other than the feature-centered projection can also be used to generate the CV's as long as the two conditions are met.

2. Let $H_1 = (X_1, \mathcal{E}_1)$ and $H_2 = (X_2, \mathcal{E}_2)$ be two attributed hypergraphs, and $H_1^* = (X_1^*, \mathcal{E}_1^*)$ and $H_2^* = (X_2^*, \mathcal{E}_2^*)$ be their corresponding duals respectively defined on the same mappings (α, β). If H_1 and H_2 are isomorphic ($X_1 = X_2$ and $\mathcal{E}_1 = \mathcal{E}_2$), then their dual graphs are also isomorphic. If H_1^* and H_2^* are isomorphic, the hypergraphs H_1 and H_2 are isomorphic if and only if $\alpha = \alpha^{-1}$ and $\beta = \beta^{-1}$.

   This implies that the mapping of the 2-D projection of a model is the necessary but not sufficient condition for model matching. On the other hand, if there exists perspective invariance such that $\alpha = \alpha^{-1}$ and $\beta = \beta^{-1}$, a mapping of the 2-D projections of model features or the CV's confirms the recognition of the modeled object.

A procedure for constructing an AHR from the wire-frame model of an object is reported in [18]. Since the data source used to construct the AHR is the same as that sued during the object recognition task, as long as the AHR for an object can be constructed, the object should be recognizable under similar environments.

The computation complexity for the construction of an AHR is $O(n^4)$ which is similar to that for other multiple-view approaches. However, the memory requirement is $O(n)$ which is much less than other multiple-view based representations for which a few hundred views representing a simple object is not unusual. Since model construction can be done off-line to support the run-time object recognition tasks, memory requirement is often a more important concern than computation complexity.

5 Object Recognition

In model-based vision, object recognition is the process of matching previously acquired object models to a scene in order to identify the objects. The bottleneck lies in the establishment of correspondence between object model features and image features. Based on the AHR framework, we present a knowledge-directed search (KDS) strategy to assemble the model-to-image correspondence in an incremental manner using topological, visual and geometrical constraints.

5.1 Knowledge-Directed Search (KDS)

KDS combines the strategies of feature grouping, hypothesis refinement and graph morphism. It relies heavily on the context of search criteria used at the different stages of the search process. Since the AHR organizes the various constraints in a coarse-to-fine manner, the complexity of 3-D scene interpretation will be reduced.

KDS initializes the model matching hypotheses based on pre-compiled knowledge about image features and refines the hypotheses by applying, from loose ones to tight ones, the topological, relational, visual and geometrical constraints stored in the AHR framework. The search terminates whenever enough information is collected such that a pose hypothesis can be made and a back projection can be executed to verify the hypothesis.

With KDS, two information measures, namely reliability and plausibility, are used as criteria to guide the search. Reliability is used to select features/views for the hypothesis generation and refinement. Plausibility is used to select features/views for the hypothesis verification.

The steps of KDS involve: 1) selecting image features, 2) establishing a pose hypothesis, 3) estimating the CV's, and 4) verifying the hypothesis through back projection. The search criteria, plausibility and reliability, are instantiated by the information available at these steps from the object model and input image data.

An application of KDS for recognizing and location machine parts in a robot workcell is reported in [26].

5.2 Selecting Image Features

The selection of image features is the initial step of KDS. Two heuristics are used. The first is to choose the most unique features and the second is to choose the features which are most easily detected. They correspond to the criteria of reliability and plausibility in that the first heuristic selects the features with the lowest error rates in establishing hypotheses and thus maximizes reliability, while the second selects features with the highest detectability and thus maximizes the matching plausibility.

5.3 Establishing the Correspondence Hypotheses

For each selected image feature, KDS proposes a match between the image feature and an image feature type captured in the AHR. The match is then refined by selecting more image features using the relational constraints associated with the image features such as connectedness, collinearity, parallelism and relative positions to each other. This will establish some small groups of image features. Then, based on the concurrence values stored in the AHR, the image features in a group can vote for the most likely CVC.

Since a bigger group corresponds to a more reliable estimation, the reliability measure is set to be the size of the selected image feature group. However, the likelihood of back-tracking also increases as the size grows because more items can be outliers caused by noise or occlusions. In practise, an upper limit is therefore set for the size of the group depending on how many image features are normally required to advance the search into the 3-D space.

The AHR pf an object contains ranking of the CVC's according to their typicality. A CVC with a higher typicality will have a higher chance to appear in the image. This measure therefore reflects the plausibility of a view class hypothesis given that the view class has been assumed to be the one most likely to appear in the image.

Possible hypotheses are thus ranked according to a measure combining the feature group size and CVC typicality. The hypotheses are then examined in the order of in which they are ranked.

5.4 Estimating Characteristic Views

Each hypothesized CVC represents an equivalent class of CV's in the AHR. Each of these CV's is in turn associated with a pre-computed rotation and translation. The image features which suggest the hypothesis should form a consistent set after being transformed into 3-D space. This consistency requirement can be used to determine the best CV within the hypothesized equivalent class. A set of invariance rules based on properties such as colinearity, parallelism, convexity and cross-ratio is used to expedite the search.

5.5 Verifying through Back Projection

The final step in KDS is to verify the hypothesized match by a back projection of all object features supposedly visible in the image. Given a model-to-image feature correspondence hypothesis, it is conceptually straightforward to obtain the object's actual pose in the 3-D scene. The whole object is then projected onto the image plane to determine whether or not the image features included in the current hypothesis constitutes an acceptable recognition result. Therefore, a pose determination algorithm capable of quickly producing the pose transformation with the minimum number of feature correspondences is important for KDS.

Pose determination is a well studied problem. Algorithms employing analytic closed form solutions as well as numerical solutions can be found in [3, 6, 8, 13, 14, 19]. Based on these works, we have developed a collection of pose determination methods to address the issues of speed, accuracy and uncertainities about intrinsic camera parameters.

Our progress in this area include:

1. formulation of analytic algorithms to take advantage of three different image feature configurations, namely 1) a planar corner comprised of two fully visible lines plus one non-coplanar point, 2) a junction comprised of three lines which can be partially occluded; and 3) an elliptical curve (resultant from the projection of a circle) plus a fully visible line;

2. a hybrid strategy combining the advantages of the high speed offered by analytic algorithms and the improved accuracy (due to data redundancies) offered by numerical algorithms; and

3. a method to estimate the optimal values for the camera focal length and scaling factors.

5.6 Computational Complexity

The time complexity of KDS is $O(n^5)$ which is much better than the $O(n^m)$ for the general case. In practice, because all features and views are ordered, there is always a high possibility that the recognition will be achieved within the hypotheses generated by the first few selected image features. Thus, in typical cases, the actual recognition should take much less time than what theoretical analysis predicts.

6 Summary

This paper presents a new methodology for automatic synthesis of 3-D models of objects. Directed by the implicit features and feature relations of the synthesized model and the image, it automatically exploits the predicate, topologic and geometric relations/constraint for object recognition and pose determination. It demonstrates the feasibility of closing the loop of knowledge acquisition and the use of the acquired knowledge for inference and decision making in a fairly complex and clutter real world scenario without intermittent human interaction.

The uniqueness of the methodology are summarized below.

1. Unlike the neural network approach, it does not require significant hybrid effort to convert the structure representation into the neural network paradigms.

2. Unlike grammatic and/or rule based approaches, it does not require heavy human effort in designing and inducing the rule base for representing and recognizing complex structures.

3. It uses structural representations, namely attributed graphs and attributed hypergraphs, to furnish a much more flexible representation than trees and frames.

4. Its use of random graphs allows probabilistic variations in both structures and measurements of 3-D objects.

5. It is capable of constructing geometric models directly from sensing devices and transforming them into symbolic models for object recognition, pose determination and scene interpretation.

6. It provides robust and effective mappings to establish correspondences between structures using local and global constraints.

7. It uses various level of model descriptions to direct the mapping of image and model feature during the object recognition and pose determination processes.

8. It is relevant to disciplines including architectural, structural and mechanical design since the synthesized models can be directly related to virtual reality or CAD and inversely related back to the 3-D environment via 3-D vision and intelligent laser projection technology to aid construction in 3-D environments.

The proposed methodology has been applied to space robotics[35], mobile robots and flexible manufacturing.

References

[1] F.A. Akinniyi A.K.C. Wong, and D. Stacey, "A New Algorithm for Graph Monomorphism Based on the Projections of the Product Graph", *IEEE Trans. on SMC*, pp. 740-751, 1986.

[2] J. Aloimonos, "Unification and integration of the Marr paradigm," *Proc. of the Image Understanding Workshop*, DARPA, May 1989, pp. 507-551.

[3] S.T. Barnard, "Choosing a basis for perceptual space", *Computer Vision, Graphics, and Image Processing*, Vol.29, No.1, 1985, pp.87-99.

[4] C. Berge, *Graphs and Hypergraphs*, London, North-Holland, 1973.

[5] I. Chakravarty and H. Freeman, "Characteristic Views as a Basis of Three-Dimensional Object Recognition," *Proc. of SPIE 336 (Robot Vision)*, 1982, pp.37-45.

[6] H.H. Chen, "Pose determination from line-to-plane correspondences: existance condition and closed form solution", *IEEE Trans.on PAMI*, Vol.13, No.6, 1991, pp.530-541.

[7] G.J. Ettinger, "Large hierarchical object recognition using libraries of parameterized model sub-parts", *Pattern Recognition*, 1988, pp.32-41.

[8] M.A. Fischler and R.C. Bolles, "Random sample concensus: a paradigm for model fitting applications to image analysis and automated cartography", *ACM Communications*, Vol.24, 1981, pp. 381-395.

[9] Q.C. Gao and A.K.C. Wong, "A Curve Detection Approach Based on Perceptual Organization", *Pattern Recognition*, Vol. 26, No. 7, pp. 1039-1046, August 1993.

[10] Z. Gigus, J. Canny and R. Seidel, "Efficiently Computing and Representing Aspect Graphs of Polyhedral Objects", *IEEE Traans. on Pattern Analysis and Machine Intelligence*, Vol. 13, No. 6, pp. 542-551, June 1991.

[11] E. Gmur and H. Bunke, "3-D object recognition based on subgraph matching in polynomial time", *Structural Pattern Analysis*, R.Mohr, T.Pavlidis, and A.Sanfeliu (eds.), World Scentific Publi. Co., 1989, pp.131-147.

[12] B.K.P. Horn, "Obtaining shape from shading information", in *The Psychology of Computer Vision*, P.H.Winston (ed.), McGraw-Hill, New York, 1975, pp.115-155.

[13] R.M. Haralick et al., "Pose estimation from corresponding point data", *IEEE Trans. Systems, Man and Cybern.*, Vol.19, No.6, 1989, pp.1426-1446.

[14] T.S.Huang, A.N.Netravali, and H.H.Chen, "Motion and pose estimation using algebraic methods", in *Time-Varying Image Processing and Moving Object Recognition*, Cappellini, Ed., Amsterdam, The Netherlands: Elsevier, 1990, pp.243-249.

[15] T. Kanade and J.R. Kender, "Mapping image constraints: skewed symmetry, affine-transformable patterns, and the shape-from-texture paradigm," in *Human and Machine Vision*, J.Beck, et at.(eds.), Academic Press, 1983, pp.237-257.

[16] J.J. Koenderink and Van Doorm, "Internal representation of solid shape with respect to vision," *Biological Cybernetics*, 32(4), 1979, pp.211-216.

[17] S.W. Lu and A.K.C. Wong, "Recognition and shape synthesis of 3-D objects based on attributed hypergraphs," *IEEE Trans. on PAMI*, Vol.11, No.3, 1989, pp.279-290.

[18] W. Liu, "Object Recognition and Pose Determination Using Attributed Hypergraph Representation", Ph.D.Dissertation, Department of Systems Design Engineering, University of Waterloo, 1993.

[19] Y.Liu, T.S.Huang and O.D.Faugeras, "Determination of camera location from 2-D to 3-D line and point correspondences", *IEEE Trans.on Pattern Analysis and Machine Intelligence*, Vol.12, No.1, 1990, pp.28-37.

[20] D.G. Lowe, "Perceptual Organization and Visual Recognition, Ph.D.Dissertation, Stanford University", No.STAN-CS-84-1020, 1984.

[21] J.D. McCafferty, *Human and Machine Vision: Computing Perceptual Organisation*, Ellis Horwood Ltd., 1990.

[22] D. Marr and T. Poggio, "A computational theory of human stereo vision", *Proc. of the Royal Sociaty of London*, B204, 1979.

[23] D. Marr, *Computer Vision*, 1982.

[24] B.A. McArthur and A.K.C. Wong, "Random Graph Representation for 3-D Object Models", *SPIE Milestone Series Vol. MS72*, pp. 229-238, edited by H. Nasr in Model-Based Vision, 1991.

[25] L.G. Roberts, "Machine perception of three-dimensional solids," in *Optical and Electro-Optical Information Processing*, J.Tippett, et al. (eds.), MIT Press, 1965, pp.159-197.

[26] K.D. Rueb and A.K.C. Wong, "Knowledge-based visual part identification and location in a robot workcell", *International Journal of Mach.Tools Manufact.*, Vol.28, No.3, 1988, pp.235-249.

[27] R. Salay and A.K.C. Wong, " Genetic Algorithms for Solving the Largest Common Subgraph Problem" *Proc. 4th UNB AI Symposium,* Fredericton, NB., pp. 145-155, Sept. 1991.

[28] L.G. Shapiro and R.M. Haralick, "Structure descriptions and inexact matching," *IEEE Trans. on PAMI*, Vol.3, No.5, 1981, pp.504-519.

[29] W.H. Tsai and K.S. Fu, "Error-correcting isomorphisms of attributed relational graphs for pattern analysis," *IEEE Trans. on SMC*, Vol.9, 1979, pp.757-678.

[30] W.H. Tsai and K.S. Fu, "Subgraph Error-correcting isomorphism for syntatic pattern recognition", *IEEE Trans. SMC*, Vol.13, 1983, pp.48-62.

[31] D. Waltz, "Understanding line drawings of scenes with shadows," *Psychology of Computer Vision*, P.H. Winston (ed.), McGraw Hill, 1975, pp.19-91.

[32] A.K.C. Wong and M. You, "Entropy and distance of random graphs with application to structural pattern recognition", *IEEE Trans. on PAMI*, Vol.7, No.5, 1985, pp.599-609.

[33] A.K.C. Wong, M.L. You, and S.C. Chan, "An Algorithm for Graph Optimal Monomorphism", *IEEE Trans. on SMC*, Vol. 20:3, pp. 628-636, 1990.

[34] A.K.C. Wong, J. Constant, and M. You, "Random Graphs,'*Syntactic and Structural Pattern Recognition—Theory and Applications*, H. Bunke and A. Sanfeliu (eds.), World Scientific Publishing, Singapore, pp. 197-233, 1990.

[35] A.K.C. Wong, G.R. Heppler, D.N.C. Tse and K.D. Rueb, "Robotic Vision Technology for Space Station and Satellite Applications", *Acta Astronauticas Journal*, Vol. 29, No. 12, pp. 911-930, (1993).

[36] E.K. Wong, "Model matching in robot vision by subgraph isormorphism", *Pattern Recognition*, Vol.25, No.3, 1992, pp.287-303.

Summary of the Roundtable Discussion

Julia D. Biedermann
School of Engineering, University of Guelph
Guelph, Ontario, N1G 2W1, Canada
biedermn@net2.eos.uoguelph.ca

1 Preamble

The final session of this three day conference was a roundtable discussion that was organized in order to provide an opportunity for the participants to express what aspects of emergent computing they believe need to be discussed and further addressed. In advance, each participant was asked to suggest a possible topic for the discussion in the hopes of isolating a few prominent themes that the roundtable discussion could then focus on. Surprisingly, these inquiries resulted in many topics being suggested that the participants felt were important. This reflects the varying backgrounds of the participants and the fact that emergent computing applied to engineering problems is still in its infancy with much to be explored.

From the suggested topics three themes were selected for discussion. For each theme a panel of three or four participants was formed. Each panel participant was asked to speak for a few minutes on the particular theme being addressed by their panel. The following is a summary (paraphrased) of these panel discussions and the comments made by members of the audience.

2 Introducing Emergent Computing Methods into Engineering Practise and Software Development

Panel Speaker: Yuval Davidor

Yuval Davidor has a background in engineering but his work and research is more in the computer science side of genetic algorithms and evolutionary computing. He applies his work in industry through his company Schema-Evolutionary Algorithms, Israel.

"I was very surprised to see the extent to which the emergent computing methods of neural networks, genetic algorithms and simulated annealing has been caught by engineering in general. I was not aware of that and I was not aware that at the upcoming conference in Athens a good portion of the papers include genetic algorithms. It is a complete and wonderful surprise.

"Secondly, I did expect to see much more people from both domains present at this workshop. People from genetic algorithms through to people from neural networks. Because if the question is interaction and to see how do we bring those subjects together then we need representatives from all the disciplines.

"I had very strong feelings throughout the workshop that many wheels have been reinvented and reinvented and will be reinvented again in the future. I couldn't help the hopeless feeling that we have been spending a lot of time in the last decade in evolutionary computation to struggle with the questions that you are struggling with now. And I don't see why you should struggle with those questions again bacause it has already been digested in the literature or with the profession. You should venture out, contact the people who have done it, go to the literature, and isolate those domains that are still difficult to answer in your specific domain. I realize that when you come to a specific application you have speccific problems and these need to be answered. For instance, in Rafi's lecture he was discussing the need for variable length strings. Well, we [researchers in evolutionary computing] know about variable length strings and have had special models to handle variable length strings for the past five years.

"As another point, I wanted much more to have discussed at this workshop the type of domain within engineering that is justified or particularly of interest for the application of emergent computing techniques. It is a nice toy - genetic algorithms, neural networks and simulated annealing - but it is not always justified or necessary to use them. Therefore, we need to discuss together our experiences from engineering and emergent computing to isolate several subjects which in particular would benefit engineering. Doing that would help up focus. People developing evolutionary models could then keep in mind engineering problems rather than working on problems which we may just select to study and which may be completely detached from real world problems."

Panel Speaker: Rick Balling

Rick Balling is a civil engineer and a professor at Brigham Young University, United States. His research is in the area of structural optimization where he has had some success at applying the computing technique of simulated annealing.

"I saw a video once by a Professor from Stanford referring to artificial intelligence which I think is analagous to our situation. He likened artificial intelligence to a tiger in a cage. He chose wonderful applications throughout industry and particular instances where the technology had made great strides. I think here [NATO-ARW] we have heard wonderful reported occurrences of these [emergent computing] technologies, but still this powerful tiger seems to be in a cage in the sense that mainstream industry does not seem to be adopting a lot of these technologies. So, I guess the question that I have is how do we streamline this technology transfer? What types of tools, software and research are needed to make this more attractive to the mainstream of industry?"

Panel Speaker: Hauhau Lee

Hauhau Lee is an engineering researcher working for General Electric at their Corporate Research and Development Center, United States. He is concurrently working on his PhD in the area of neural networks.

"Perhaps I can provide an answer to Rick's question.

"I'd like to refer to my work experience where we used three types of search algorithms. Gradient-based. Expert systems - where the designer had the experience or knowledge about the design space. For this case, a large directed search was found to be very efficient. The last one, probably the most robust but also the most time-consuming, was a genetic algorithm or simulated annealing. For most designs, the designer's first choice is to do a knowledge directed search if they have the knowledge. And then, gradient-based. The last choice is to use a genetic algorithm to see if they can get a better solution. Apparently the genetic algorithm and simulated annealing is more robust in making provision for the whole design space but it just takes too long to get a good design. It is not that industry rejects this technology but rather that it is not as efficient as they expected. "

Panel Speaker: Jamshid Ghaboussi

Jamshid Ghaboussi is a civil engineer and professor at the University of Illinois at Urbana-Champaign, United States. His research is in the application of neural networks to modeling the nonlinear constitutive behaviour of materials.

"I have paltry comments regarding engineering applications of these emerging techniques. First, the engineering applications need to concentrate on real problems. The methods developed and tried on benchmark problems, small-scale problems and toy problems, really don't address the question of whether these methods are suitable or how they should be used in real engineering problems. Therefore, I feel it is important that right from the beginning we work with real engineering problems, real data and actual measurements.

"The second point is that most of the emergent-based methods are still at the very early exploratory stages. Major changes are occurring and will occur in this sphere. Therefore, it is very important that people involved in engineering applications of these methods keep this in mind and not take the current state as static. We must be prepared to accept new innovations which are coming down the line.

"Another point that I would like to make is that we take these methods that are developed by non-engineers and apply them to engineering problems but in the process we also do engineering-driven research. Specific requirements of engineering problems and applications dictate a certain amount of modification and development of these methods. This is very necessary but in doing this we must keep the origin of these models in mind."

Audience Response

Don Grierson

Professor of Civil Engineering. Research in structural optimization and genetic algorithms.

"I will relate Yuval's comments about the fact that a lot of the solutions to the problems that we seem to be struggling with here have been solved in the computer science community to an experience I had twenty years ago. At that time, I was told by people in the optimization community that they had all the software tools and solutions to our engineering problems. But when I came to them I found that either they didn't have the solutions or I couldn't understand what they were saying in their literature. It wasn't until I started associating with hybrid engineer-mathematicians that we were able to capture some of this information and use it effectively. The challenge then is to get more people, such as Yuval, involved in this area who have an appreciation for the problems we are dealing with and can relate solutions to these problems in a way we can understand and use."

Prabhat Hajela

Professor of Mechanical and Aerospace Engineering. Research in structural optimization and numerical algorithms.

"I think Jamshid's remark that engineers need to preserve the biological origins of these methods is a point of some argument. It is my belief that sometimes we don't understand nature like we think we do. We try to emulate some certain things that go on in nature and perhaps we are on the wrong track. This issue of trying to stay close to nature in applying these methods may be a good idea but can be dangerous. I give as a example that we didn't acutally have success with flight until we stopped flapping our wings."

Jerry Lin

Professor of Mechanical Engineering. Research in genetic algorithms.

"According to Hauhau, in order for these methods to become a practical engineering design optimization tool we have to cut down the computational cost to a level that the engineer in industry can afford. Cost has two components; one is the main power required to operate the optimization tool. According to my experience at GE, they like to use gradient-based optimization because they get a solution quickly but they have to repeat it a large number of times until the design stage limit is reached. So, in that concern, the gradient-based method is a local search method. If you want to use that approach you have to repeat the process a lot of times and that uses a lot of power. Although the genetic algorithm is not very popular to use in solving very practical engineering optimization problems there is a desire to develop it in order to reduce this use of power. Regardless of the methods used, I don't think a lot of research has been done on how to pick up the information that comes from previous generations or iterations to improve the search strategy."

Andrew Wong

Professor of Systems Design Engineering. Research in robotics.

"I feel that optimization search depends on how you use and represent knowledge. Today, many of the models ... are believed to have biological reality but there is still a big gap. We have to be careful of whether it is a belief or a rational assessment of the whole development "

Ian Parmee

Engineering design researcher. Research in adaptive search techniques.

"I agree entirely with Jerry Lin. We are saying to industry that we have this marvelous method that can solve your multi-parameter design problem.

Unfortunately the method that we have is a multi-parameter control problem that has to be solved in order to solve the multi-parameter design problem. We must get some form of intelligence within the search process that can use the information that has been generated through each generation to make that search more efficient and to remove the black magic element that the engineering designer does not want to see. Another point I'd like to make is that we have two optimization problems. We have the component optimization but we also have the optimization of the design process. It would be a significant contribution if we can reduce design lead times by using adaptive search techniques."

Bill Jenkins

Professor of Civil Engineering. Research in genetic algorithms.

"What has pleased me enormously at this conference is the diversity of the applications that have been presented to us. There are good engineering applications where the [genetic] algorithm is well employed in solving them and I don't think it matters if the algorithm is not applied in its purest form. If the engineering outlook is good, then OK. As the problems get more difficult then perhaps we have to dig deeper. We may have to go back and look at the biological principles again and see if there is anything there that can help us. We have to continue doing that. This is in itself an evolutionary process. Once you have lifted the principles out of the biological context and put them into engineering then it is the engineering that matters. And only when we are not making progress in that do we need to go back."

Panel Members Response

Jamshid Ghaboussi

"I completely agree with the comments that Bill just made. We need to keep in mind that these methods came from the biological field from non-engineers. While using them we need to keep in touch with that community to be aware of other new developments and new knowledge. In response to Prabhat's comments, even though we made progress after we stopped flapping our wings, we still kept our wings."

Hauhau Lee

"As an example, a turbine designer may be given one week to determine the optimum turbine design. He doesn't care whether the technique he uses is genetic algorithms or simulated annealing. The bottom line is that he needs to produce an optimum design.

"Regarding Prof. Wong's comments. I have knowledge in turbine design and know certain parameters are more sensitive to improve the efficiency [of the turbine]. That is the kind of knowledge that we can make use of in the search."

Yuval Davidor

"The really new thing about genetic algorithms is that you don't usually have to understand the problem in order to solve it. If you try to put your understanding and knowledge into the algorithm you prevent finding optimal or very good solutions. The inherent concept of genetic algorithms is that it can only work when you give it freedom. What we need then is a good shell and a good interface. This is where the engineering becomes so important; to know what to tie into it and what to allow to be played with. As proof, I have solved many problems in domains that I absolutely understand nothing about."

3 The Need for Interdisciplinary Interaction

Panel Speaker: Julia Biedermann

Julia Biedermann is a civil engineer and professor at the University of Guelph, Canada. Her research activities are in the application of object-oriented programming, artificial neural networks and genetic algorithms to structural design.

"Having recently come into this research domain I find it to be very much an interdisciplinary area and I have had a great deal of difficulty in finding information in a useable form. Through literature searches I found many references to papers in conference proceedings and journals but then was unable to actually get a hand on these. Especially international conference proceedings where only one of a few papers are of interest. It could take a very long time to get copies of these papers through the library system. That was frustrating.

"Recently I have joined a number of electronic newsletters and have found that this is a very good medium for exchanging information. There is not a lot of paper involved, it comes very quickly and you can scan it. You get email addresses of people that you can respond to for future information. I think that an electronic newsletter is one means of promoting interdisciplinary interaction.

This could provide abstracts of papers presented at conferences and in journals with the email address of the author. This way you could contact the author by email if you couldn't get a copy from your library. It would provide a means of advertising for positions. And also would be a good way to get your graduate students interacting where a lot of interaction actually does occur. Graduate students are quite open to communication by email. I know that I certainly was as a graduate student and used my email contacts for getting papers and information. The growth of electronic communication offers great potential as we are just moving into multi-media. The information exchange could be phenomenal. This workshop is a perfect starting point in order to execute such a scheme. We have approximately thirty people here and if each person got ten collegues then that could result in over three hundred subscribers. All that is needed is an editor and initial contributors."

Panel Speaker: Rafi Haftka

Rafi Haftka is an aerospace and mechanical engineer and a professor at Virginia Polytechnic Institute and Space Agency, United States. He is a long-standing researcher in the field of structural optimization and numerical algorithms.

"At a recent conference that I attended on genetic algorithms I had discussions about my work with a number of participants. I received lots of advice but all of it was different. That was somewhat frustrating.

"We need to find a mechanism for more of a balance of people at these conferences. The proximity of conferences is a very powerful tool for achieving this. For example, many of the people attending this workshop are also attending next week's Computational Structures Technology conference in Athens. If we could have organized this meeting right next to the conference [on genetic algorithms] that Yuval is organizing in Israel the chances are that there would have been more interaction with researchers from computer science. My suggestion is that in the future we try to create proximity in time and location between engineering conferences and computer science or applied math conferences."

Panel Speaker: Prabhat Hajela

Prabhat Hajela is a mechanical and aeronatical engineer and professor at Rensselaer Polytechnic Institute, United States. His research is in the area of optimization of mechanical systems and the use of genetic algorithms and neural networks in engineering design.

"Speaking as part of the organizing committee [for this workshop] I would like to say that the original intent was to bring in half the participants from computer science and the other fifty percent from engineering. While we were successful in getting people from neural nets we were only successful in getting Yuval [from evolutionary computing] to come to this meeting. Two others were supposed to come but in the last three or four months decided they couldn't make it.

"I have some suggestions on what we need to do to improve this interaction. I'd like to pick up on Julia's idea of using the electronic media. There is already a GA users group existing on the usenet, perhaps we could have a subset of that called GAENG. This genetic algorithms in engineering newsgroup could have interests from both communities.

"As part of the ICGA meetings perhaps we could have a special session at the ICGA for people from engineering doing interdisciplinary work. Then we would be right next to the people that are talking about genetic algorithms and they would get to see some of our applications.

"Finally, it may be of interest to us to put together a compendium, in the not too distant future, where we could have four or five specialized topics being written on by leading people in this business. These should not just be ideas that are accepted as given right now but new thinking as well. Perhaps this could serve as a breeding groung for new ideas."

Audience Response

Barry Topping

Professor of Structural Engineering. Research in neural networks and genetic algorithms in structural optimization.

"If you look back at the history of the development of the finite element method, which was originally developed by engineers, it is interesting to see what happened. Mathematicians came along and took it off in one direction and the engineers took it off in a practical direction. And now there are two paths in the literature. I think it is a futile task to try to correct this interaction. It happens at personal levels and on particular projects with true interactions occurring at random. I don't think it will work if we try to force it. People naturally diversify. People come together for awhile to work on a particular project and then eventually split up."

Vlasis Koumousis

Professor of civil engineering. Research in genetic algorithms in optimal structural design.

"I think that we need some solid basis in order to justify these methods in the context of interdisciplinary interaction. For example, the direction that was given by Rafi this morning for establishing reliability within these algorithms. Most of the papers on genetic algorithms that I have been looking at try to justify parameters, such as population size, on the basis of statistics. Statistics may not bring us to sound proof but will at least give us a good indication of whether we are moving in the right direction. Otherwise, as designers and engineers, we don't know our bounds in interfering with the main operators in genetic algorithms."

Ian Parmee

Engineering design researcher. Research in adaptive search techniques.

"At Plymouth we have an interdisciplinary group which includes computer scientists, mathematicians, psychologists and engineers, all working on applications of adaptive search in engineering design. It does work and I see this as essential in order to get it working right."

Andrew Wong

Professor of Systems Design Engineering. Research in robotics.

"I believe interaction is important and will not argue with that. Years ago I was in a similar situation. ... Therefore, eventually, we need to focus on the representation as well as the technique."

Panel Member's Response

Julia Biedermann

"With reference to Prabhat's comments, I don't think a newsgroup is the correct medium. Often, what you get in newsgroups is a lot of people arguing back and forth. I would like to see a newsletter where there is a lot more information disseminated on a specific topic. In this way, if you see something of interest you can contact that person directly to get more information. I think it would be a more fruitful exchange of information."

Rafi Haftka

"I take exception to what Barry said about not being able to force the issue [of interaction]. I believe we need to give it some encouragement. It is people like us that can organize sessions at other conferences or create proximity of dates and places. We cannot hope that these interactions will just happen by themselves."

4 Educating our Students in Emergent Computing Techniques

Panel Speaker: Don Grierson

Don Grierson is a civil engineer and professor at the University of Waterloo, Canada. He has a long history of research in structural optimization of structural systems with recent interests in the application of genetic algorithms and neural networks for conceptual design.

"If we are going to educate our students to use these techniques then we need to identify the engineering problems for which they are best used. Using standard toy problems and suggesting the use of genetic algorithms instead of a gradient-based search is not going to help in a transfer of these techniques to industry. We need to start looking at some of the tougher problems of engineering where these evolutive computing techniques can be effective and where other techniques haven't yet been found to be effective in order to properly educate our students. Then when they go out into industry they have a better understanding of how and when to apply these techniques to real world problems."

Panel Speaker: Mark Embrechts

Mark Embrechts is a professor of nuclear engineering and engineering physics at Rensselaer Polytechnic Institute, United States. He has long been doing research in artificial neural networks and has worked on applications of neural networks in the automotive industry.

"I think we all agree that it would be nice to teach these techniques to students and would be especially important to have it as a standard part of an undergraduate curriculum. Unfortuanately, it is not very easy to convince our colleagues that this really merits a course in undergraduate engineering. We all have to work towards this and when one of us succeeds it makes it easier for the rest of us. There are very good motivations for this besides the obvious ones.

Undergraduate students who took my graduate courses easily found good paying jobs. Items like neural nets and genetic algorithms spice up a resume.

"The question is how do we do this? We require a good text. Such a course may not deserve a three hour course [per week] but perhaps it could be done with a one hour course. My suggestion is to just do it."

Panel Speaker: George Thierauf

George Thierauf is a civil engineer and professor of civil engineering at the University of Essen, Germany. He has a research background in structural optimization, reinforced concrete structures, structural steel systems and applications of evolutionary computing methods.

"Even though I've been teaching for twenty-five years I have to think about this subject. Most of us teaching undergraduate courses are teaching optimization of engineering structures. I believe we should cover the whole field. We should continue to teach first and second order methods and methods which didn't use gradients at all, including these evolutionary techniques. We should have a look at the wideness of that field and accept that we are fiddling around searching for good solutions to engineering problems. If we restrict too much then we are losing. If there is one group only teaching genetic algorithms, the next one is concentrating on simulated annealing, and another group teaching evolutionary strategies, then they have a lot in common and it is very hard to say that one is better than another. That is a restriction and doesn't help us in solving our problem. We need to look at it in a very broad-minded way.

"We shouldn't worry so much about the cost of computing. We are always behind the development of better chips. Computers keep getting much faster and their price keeps dropping enormously. This is certainly connected to the progressive use of Darwin computing and the zero order techniques in optimization that have come up. It is also related to the rivival of these random search techniques. Therefore, the cost of computing is not so important but the cost of our engineering structures is.

"We need good contact with practising engineers. Everything we do costs a lot of money and we need a lot of resources. Our research will be more accepted if it is clear that it is of some use. We should try to teach practising engineers as well in order to gain acceptance of these techniques.

"As a closing remark, I think we need more meetings like these where we have small groups and limited discussions."

Audience Response

Grantham Pang

Professor of Electrical and Computer Engineering. Research in fuzzy neural networks for control systems.

"To provide comprehensive training for graduate students we need depth. Allowing students to explore more techniques and compare them is very important. They can then appreciate the strengths and weaknesses and develop a feel for each technique. In this way they will use the appropriate tool at the right time. Having benchmark problems in the different disciplines provides a means for comparison of the techniques.

"The development of software is also a very important aspect of a student's education."

Phillipe Trompette

Engineering researcher. Research in application of genetic algorithms to structural optimization problems.

"With respect to teaching our students, it is always easy to add material but very difficult, if not impossible, to remove material. What will be removed? A big question right now is what is the kernal to be teached?"

Bill Jenkins

Professor of Civil Engineering. Research in genetic algorithms.

"I hesitate to reveal the extent of this problem in the United Kingdom. We do not teach optimizing design in our undergraduate courses in general. Only in our post-graduate course is optimization taught. We have a long way to go and I'm not sure where we should start. Earlier, we dismissed the idea of optimizing apparently trivial engineering situations. I'm not against that. If that is the only way we can start then lets use it."

Prabhat Hajela

Professor of mechanical and aeronatical engineering. Research in genetic algorithms and neural networks in engineering design.

"Education is a two-way process. We [researchers] can learn something if we get interested parties to come back from industry and pose some problems to us. I have provided short courses to industry, including one on genetic algorithms. From that course we had someone come back with a problem for us to solve. And this problem revised my entire thinking about how I would go about applying the process."

Rafi Haftka

Professor of aerospace and mechanical engineering. Research in genetic algorithms for optimization

"We [Virginia Polytechnic] have been very successful in all three areas: getting rid of courses, introducing optimization and genetic algorithms at the undergraduate level, and doing more than trivial problems. We had a faculty retreat in our department about a month ago and almost painlessly we tossed out six credits from our engineering curriculum. I was surprised how painless this was. I think the retreat approach was very conducive to this type of discussion and decision making.

"Also, it's necessary to make sure that you are not teaching courses that are taught in other departments. The price to pay for this is that the other department may not teach the course exactly the way you would like it taught. But, it is good for the students to be exposed to the different ways of looking at problems.

"In terms of doing more complicated problems, I find that the tools that are available to the students are forcing me in this direction. I used to use very simple examples but now with students having access to tools such as Mathmatica, I need to give them much more complicated problems than I did five years ago."

Panel Response

Don Grierson

"In Canada, we too have a lot of inertia in our curriculum which I liken to a very slow evolutionary process with a very slow rate of convergence. I'm interested to see that Rafi has come up with a nice mutation operator called a retreat which has a very good probability of success."

George Thierauf

"The global situation is that we have to eliminate something in order to get something new in. In Germany, we were top in the world in having the longest duration of civil engineering courses. Unfortunately, our Ministry of Research and Education then set up a rule that limited the number of courses allowed "

NATO ASI Series F

Including Special Programmes on Sensory Systems for Robotic Control (ROB) and on Advanced Educational Technology (AET)

Vol. 114: Intelligent Systems: Safety, Reliability and Maintainability Issues. Edited by O. Kaynak, G. Honderd and E. Grant. XI, 340 pages. 1993.

Vol. 115: Learning Electricity and Electronics with Advanced Educational Technology. Edited by M. Caillot. VII, 329 pages. 1993. *(AET)*

Vol. 116: Control Technology in Elementary Education. Edited by B. Denis. IX, 311 pages. 1993. *(AET)*

Vol. 117: Intelligent Learning Environments: The Case of Geometry. Edited by J.-M. Laborde. VIII, 267 pages. 1996. *(AET)*

Vol. 118: Program Design Calculi. Edited by M. Broy. VIII, 409 pages. 1993.

Vol. 119: Automating Instructional Design, Development, and Delivery. Edited by. R. D. Tennyson. VIII, 266 pages. 1994. *(AET)*

Vol. 120: Reliability and Safety Assessment of Dynamic Process Systems. Edited by T. Aldemir, N. O. Siu, A. Mosleh, P. C. Cacciabue and B. G. Göktepe. X, 242 pages. 1994.

Vol. 121: Learning from Computers: Mathematics Education and Technology. Edited by C. Keitel and K. Ruthven. XIII, 332 pages. 1993. *(AET)*

Vol. 122: Simulation-Based Experiential Learning. Edited by D. M. Towne, T. de Jong and H. Spada. XIV, 274 pages. 1993. *(AET)*

Vol. 123: User-Centred Requirements for Software Engineering Environments. Edited by D. J. Gilmore, R. L. Winder and F. Détienne. VII, 377 pages. 1994.

Vol. 124: Fundamentals in Handwriting Recognition. Edited by S. Impedovo. IX, 496 pages. 1994.

Vol. 125: Student Modelling: The Key to Individualized Knowledge-Based Instruction. Edited by J. E. Greer and G. I. McCalla. X, 383 pages. 1994. *(AET)*

Vol. 126: Shape in Picture. Mathematical Description of Shape in Grey-level Images. Edited by Y.-L. O, A. Toet, D. Foster, H. J. A. M. Heijmans and P. Meer. XI, 676 pages. 1994.

Vol. 127: Real Time Computing. Edited by W. A. Halang and A. D. Stoyenko. XXII, 762 pages. 1994.

Vol. 128: Computer Supported Collaborative Learning. Edited by C. O'Malley. X, 303 pages. 1994. *(AET)*

Vol. 129: Human-Machine Communication for Educational Systems Design. Edited by M. D. Brouwer-Janse and T. L. Harrington. X, 342 pages. 1994. *(AET)*

Vol. 130: Advances in Object-Oriented Database Systems. Edited by A. Dogac, M. T. Özsu, A. Biliris and T. Sellis. XI, 515 pages. 1994.

Vol. 131: Constraint Programming. Edited by B. Mayoh, E. Tyugu and J. Penjam. VII, 452 pages. 1994.

Vol. 132: Mathematical Modelling Courses for Engineering Education. Edited by Y. Ersoy and A. O. Moscardini. X, 246 pages. 1994. *(AET)*

Vol. 133: Collaborative Dialogue Technologies in Distance Learning. Edited by M. F. Verdejo and S. A. Cerri. XIV, 296 pages. 1994. *(AET)*

Vol. 134: Computer Integrated Production Systems and Organizations. The Human-Centred Approach. Edited by F. Schmid, S. Evans, A. W. S. Ainger and R. J. Grieve. X, 347 pages. 1994.

Vol. 135: Technology Education in School and Industry. Emerging Didactics for Human Resource Development. Edited by D. Blandow and M. J. Dyrenfurth. XI, 367 pages. 1994. *(AET)*

Vol. 136: From Statistics to Neural Networks. Theory and Pattern Recognition Applications. Edited by V. Cherkassky, J. H. Friedman and H. Wechsler. XII, 394 pages. 1994.

NATO ASI Series F

Including Special Programmes on Sensory Systems for Robotic Control (ROB) and on Advanced Educational Technology (AET)

Vol. 137: Technology-Based Learning Environments. Psychological and Educational Foundations. Edited by S. Vosniadou, E. De Corte and H. Mandl. X, 302 pages. 1994. *(AET)*

Vol. 138: Exploiting Mental Imagery with Computers in Mathematics Education. Edited by R. Sutherland and J. Mason. VIII, 326 pages. 1995. *(AET)*

Vol. 139: Proof and Computation. Edited by H. Schwichtenberg. VII, 470 pages. 1995.

Vol. 140: Automating Instructional Design: Computer-Based Development and Delivery Tools. Edited by R. D. Tennyson and A. E. Barron. IX, 618 pages. 1995. *(AET)*

Vol. 141: Organizational Learning and Technological Change. Edited by C. Zucchermaglio, S. Bagnara and S. U. Stucky. X, 368 pages. 1995. *(AET)*

Vol. 142: Dialogue and Instruction. Modeling Interaction in Intelligent Tutoring Systems. Edited by R.-J. Beun, M. Baker and M. Reiner. IX, 368 pages. 1995. *(AET)*

Vol. 144: The Biology and Technology of Intelligent Autonomous Agents. Edited by Luc Steels. VIII, 517 pages. 1995.

Vol. 145: Advanced Educational Technology: Research Issues and Future Potential. Edited by T. T. Liao. VIII, 219 pages. 1996. *(AET)*

Vol. 146: Computers and Exploratory Learning. Edited by A. A. diSessa, C. Hoyles and R. Noss. VIII, 482 pages. 1995. *(AET)*

Vol. 147: Speech Recognition and Coding. New Advances and Trends. Edited by A. J. Rubio Ayuso and J. M. López Soler. XI, 505 pages. 1995.

Vol. 148: Knowledge Acquisition, Organization, and Use in Biology. Edited by K. M. Fisher and M. R. Kibby. X, 246 pages. 1996. *(AET)*

Vol. 149: Emergent Computing Methods in Engineering Design. Applications of Genetic Algorithms and Neural Networks. Edited by D.E. Grierson and P. Hajela. VIII, 350 pages. 1996.

Vol. 152: Deductive Program Design. Edited by M. Broy. IX, 467 pages. 1996.

Vol. 153: Identification, Adaptation, Learning. Edited by S. Bittanti and G. Picci. XIV, 553 pages. 1996.

Vol. 154: Reliability and Maintenance of Complex Systems. Edited by S. Özekici. XI, 589 pages. 1996.

GPSR Compliance

The European Union's (EU) General Product Safety Regulation (GPSR) is a set of rules that requires consumer products to be safe and our obligations to ensure this.

If you have any concerns about our products, you can contact us on ProductSafety@springernature.com

In case Publisher is established outside the EU, the EU authorized representative is:

Springer Nature Customer Service Center GmbH
Europaplatz 3
69115 Heidelberg, Germany

Batch number: 10755053

Printed by Printforce, the Netherlands